AF781704

Environmental Degradation in Industrial Composites

This book is dedicated to Sarah, Guitty, Francis and Jon

Environmental Degradation in Industrial Composites

Céline A. Mahieux

ELSEVIER

AMSTERDAM • BOSTON • HEIDELBERG • LONDON • NEW YORK • OXFORD
PARIS • SAN DIEGO • SAN FRANCISCO • SINGAPORE • SYDNEY • TOKYO

ELSEVIER B.V.
Radarweg 29
P.O. Box 211
1000 AE Amsterdam
The Netherlands

ELSEVIER Inc.
525 B Street
Suite 1900, San Diego
CA 92101-4495
USA

ELSEVIER Ltd
The Boulevard
Langford Lane
Kidlington,
Oxford OX5 1GB, UK

ELSEVIER Ltd
84 Theobalds Road
London WC1X 8RR
UK

© 2006 Elsevier Ltd. All rights reserved.

This work is protected under copyright by Elsevier Ltd., and the following terms and conditions apply to its use:

Photocopying
Single photocopies of single chapters may be made for personal use as allowed by national copyright laws. Permission of the Publisher and payment of a fee is required for all other photocopying, including multiple or systematic copying, copying for advertising or promotional purposes, resale, and all forms of document delivery. Special rates are available for educational institutions that wish to make photocopies for non-profit educational classroom use.

Permissions may be sought directly from Elsevier's Rights Department in Oxford, UK: phone (+44) 1865 843830, fax (+44) 1865 853333, e-mail: permissions@elsevier.com. Requests may also be completed on-line via the Elsevier homepage (http://www.elsevier.com/locate/permissions).

In the USA, users may clear permissions and make payments through the Copyright Clearance Center, Inc., 222 Rosewood Drive, Danvers, MA 01923, USA; phone: (+1) (978) 7508400, fax: (+1) (978) 7504744, and in the UK through the Copyright Licensing Agency Rapid Clearance Service (CLARCS), 90 Tottenham Court Road, London W1P 0LP, UK; phone: (+44) 20 7631 5555; fax: (+44) 20 7631 5500. Other countries may have a local reprographic rights agency for payments.

Derivative Works
Tables of contents may be reproduced for internal circulation, but permission of the Publisher is required for external resale or distribution of such material. Permission of the Publisher is required for all other derivative works, including compilations and translations.

Electronic Storage or Usage
Permission of the Publisher is required to store or use electronically any material contained in this work, including any chapter or part of a chapter.

Except as outlined above, no part of this work may be reproduced, stored in a retrieval system or transmitted in any form or by any means, electronic, mechanical, photocopying, recording or otherwise, without prior written permission of the Publisher. Address permissions requests to: Elsevier's Rights Department, at the fax and e-mail addresses noted above.

Notice
No responsibility is assumed by the Publisher for any injury and/or damage to persons or property as a matter of products liability, negligence or otherwise, or from any use or operation of any methods, products, instructions or ideas contained in the material herein. Because of rapid advances in the medical sciences, in particular, independent verification of diagnoses and drug dosages should be made.

First edition 2006

ISBN-13: 978-1-85617-447-3
ISBN-10: 1-856-17447-6

Typeset by Integra Software Services Pvt. Ltd, Pondicherry, India. www.integra-india.com
Printed in Italy

Cover illustration courtesy of L/M Glasfiber, Denmark

Working together to grow
libraries in developing countries

www.elsevier.com | www.bookaid.org | www.sabre.org

ELSEVIER BOOK AID International Sabre Foundation

CONTENTS

LIST OF FIGURES

LIST OF TABLES

LIST OF CASE STUDIES

ACKNOWLEDGEMENTS

A very special thank you to Professor Reifsnider for his scientific insights and for his kindness. Thank you to those who have invested time to help me with this work: Prof. Y. Jack Weitsman, Prof. John C. Fothergill, Prof. Scott W. Case, Prof. Ever Barbero, Prof. David Allen, Dr. Nikhil E. Verghese, Dr. Geoff Smaldon, Jonathan Medding at Esec, Alain Champier at Alstom, T. Kunz at Alstom, C. Rückert at Airbus, R. Schmidt at Aerodyn, R. Heierli at IBM, S. Broust Nielsen at LM Glasfiber and all of you who have sent me contributions.

1

INTRODUCTION

1.1 INTRODUCTORY CASE STUDY: WINDMILL BLADES

The wind market has been experiencing a drastic growth since 1995. Increasing needs for renewable energy output, but conflicting space limitations, have driven power producers to significantly increase the size of single wind turbines. For example, the 5 MW Brunsbüttel wind turbine (Germany) equipped with the world's largest blades was recently commissioned and now provides energy for 5000 households.

This increase in blade size is not without introducing new technical challenges. Indeed, the power increases proportionally to the square of the blade length, when blade mass and bending load increase in proportion to the cube of the blade. Reinforced polymer matrix composites naturally offer light-weight solutions to the blade industry. The use of polymer composites for large wind turbines also offers further advantages such as tailorable properties in main loading directions and good fatigue resistance.

The design of turbine blades is complex and the multilayered composite structure generally requires a labor-intensive manufacturing process (Figure 1.1). Three aspects are usually considered at the design stage: aerodynamics, structure and lay-up. The LM glasfiber blade concept is shown in Figure 1.2. The different aspects can be combined such as aerodynamic and load carrying functions, which are coupled in one element.

Windmill blades are exposed to a large number of sustained and occasional environmental loads such as bending, vibrations, UV, cold and heat, moisture, impact, lightning etc. Such loads combined with large dimensions perfectly illustrate the challenge of durability assessment in composite materials.

Blade specifications generally mandate a 20-year lifetime. To guaranty the blade lifetime, the manufacturers generally revert to large series of costly material testing. Small-scale tests are performed on coupons in the laboratory (Figure 1.3) mainly at the preliminary stage. UV, heat, cold, water or salt exposure parameters are varied independently and then combined. To account for scaling effects, full-scale prototype testing is also required for design validation and final certification.

Figure 1.1. Composite blade manufacturing. (Courtesy of LM Glasfiber.)

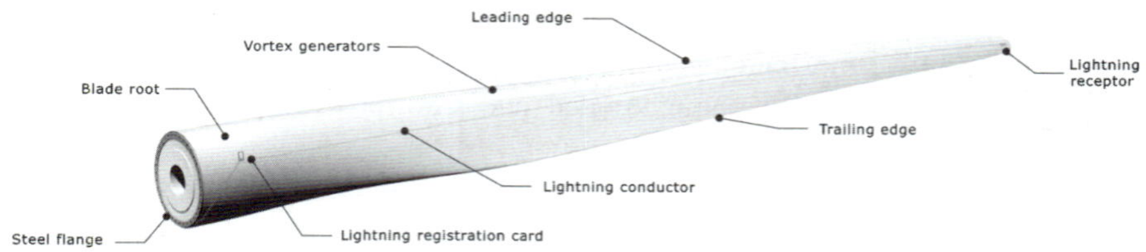

Figure 1.2. Blade concept. (Courtesy of LM Glasfiber.)

Such tests can be spectacular and require significant investments. For example, blades are usually equipped with lightning protection systems. Real size lightning tests performed in high voltage laboratory to validate simulation responses are shown in Figure 1.4. Certification testing is mandatory and includes static and dynamic tests on a full-scale prototype. The static experiment is performed at 110% of the design load (wind gust/Seventy Year Wind). The deflections generally involved in such testing are very large due to the length of the blades: Figure 1.5 shows a 15 m tip deflection resulting from the testing of a 51.5 m blade. The turbine blades are also subjected to turbulence-induced fatigue loads. The number of cycles over a 20-year lifetime is expected between 10^8 and 10^9 load cycles. Certification therefore requires the dynamic testing of a full-scale blade, in which the blade undergoes 5 million cycles (over a 6–8 month period). The blade is excited (flat and edge wise) by a counterweight using the blade's eigenfrequency. The resulting deflection is typically 4–5 m in each direction. The strain gage results are documented and remitted to the customer.

Like wind turbine blades, most industrial composite products generally specify a guarantied lifetime. Therefore a careful durability assessment is a necessity in many (if not all) development projects. To date, there is, unfortunately, no lifetime

Figure 1.3. Fatigue testing of coupon composite for windmill blade application. (Courtesy of LM Glasfiber.)

Figure 1.4. Lightning experiment on composite windmill blade. (Courtesy of LM Glasfiber.)

Figure 1.5. Blade bending test. (Courtesy of LM Glasfiber.)

prediction recipe applicable to all polymer composite materials. The durability study of composite materials is probably one of the most challenging as well as fascinating fields offered by current industrial technologies.

In order to reduce development costs and risks, the use of carefully selected analytical tools and appropriate accelerated testing procedures is required in order to evaluate the effects of prolonged environmental exposure. The purpose of the present text is to provide the reader with basic elements enabling the proper analysis of the effects of environmental loads on polymer composites. This cannot be done without a fair amount of theoretical content. However, case studies are developed to illustrate the different chapter topics. In addition, special emphasis is placed on common pitfalls and useful tips are provided to reduce development time while maintaining a safe design approach.

1.2 INTRODUCTION TO ENVIRONMENTAL DEGRADATION IN COMPOSITE MATERIALS

The term *composite* designates a very broad class of materials, made of several (at least two) components. By opposition to alloys, the composite constituents are generally distinct at the macroscopic level (with the exception of nano-composites). Composites such as wood can be found in the nature. However, composite materials are often engineered to answer specific requirements.

Three large composite classes can be distinguished, namely metal matrix composites [1,2], ceramic matrix composites [3] and polymer matrix composites. The rules governing the materials degradation are very specific to these composite classes and the current book focuses strictly on polymer matrix composites. Composites in the present text will therefore systematically refer to polymer matrix composite materials (PMCs).

Despite such a restriction, the field of polymer composites still remains broad. Indeed, the nature and geometry of the materials constituents define to a large extent the response of the composite to external mechanical and environmental loads. A short overview of the different types of composite materials is provided at the beginning of this chapter and common manufacturing processes briefly introduced (Section 1.3).

Composite materials are by definition inhomogeneous at the microscopic scale, i.e. their properties vary from one sample location to the other. Composite materials can also be designed to provide different and hopefully optimal responses in various load directions: such materials are referred to as anisotropic. Section 1.3.3 of the introductive chapter concentrates on summarizing some specificities of Polymer Matrix Composites.

Considering the wide variety of composite materials, it is not surprising to find a very large but fragmented application market. The market, reviewed in Section 1.4, illustrates the diversity of environmental load cases seen by the different composite products.

Degradation processes and durability assessment methods will be developed within this book. Case studies along the text illustrate complex environmental load situations on real composite products, covering a broad range of applications for composite materials including windmill blades (Chapters 1, 5 and 6), bearings, cryogenic tanks for aerospace applications, mass transportation (Chapter 2), house ducting, fuel cells, sewer pipes, marine applications (Chapter 3), high voltage equipment (Chapter 4), bridges and oil and gas applications (Chapter 6).

The tools to tackle such situations are presented in the various chapters. Indeed, Chapter 2 introduces temperature effects on composite materials. Simultaneously, time-dependent behavior is presented and viscoelasticity introduced. Chapter 3 focuses on the effects of liquid and gas exposure on polymer composites. High electrical fields and radiation effects are presented in Chapter 4. A method for the introduction of these different effects in micromechanical and macromechanical models is proposed in Chapter 5. Cycling, loads combination and durability assessment schemes are finally developed in the last chapter.

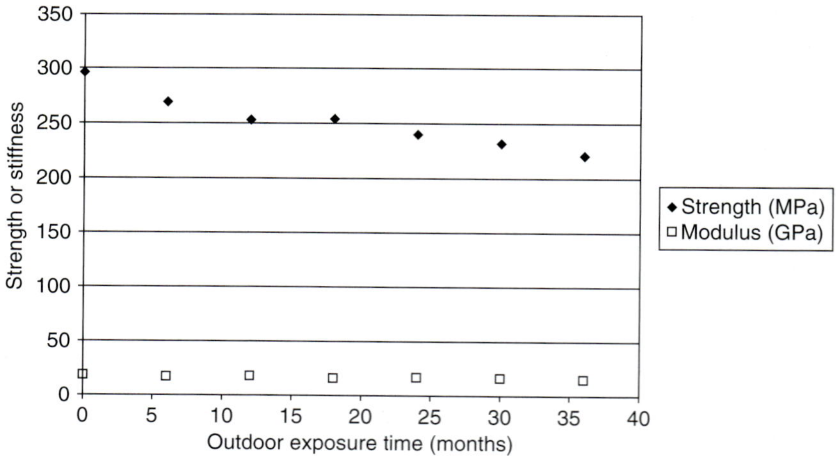

Figure 1.6. Effect of outdoor exposure on the mechanical properties of light gray 128 glass cloth with high temperature resistant polyester. Data from Rugger [4].

With the exception of this introductory section, all chapters contain a paragraph on testing methods and a final tool kit, in which major equations, related assumptions and importance are summarized.

For clarity purposes, the scope of the present text was concentrated on loads mainly associated with weathering. Indeed, exposure to an outdoor environment (high and low temperatures, UV radiations, humidity, mechanical loads etc.) can greatly affect the composites properties, such as strength (see Figure 1.6).

The effects of wear were voluntarily excluded from the discussion. Specialized texts on this area are available in the literature [5]. The more exotic field of biological degradation was also excluded from the book's scope. D.V. Rosato and R.T. Schwartz review the effects of biological products (feces, urine, flatus, sebum, sweat, vomit, algae, fungi and bacteria) in [6]. Enzyme attacks are similar to many degradation processes explored in this book and generally result in the scission of the matrix into smaller molecules. It is interesting to note that polymers are new to nature. This novelty characteristic confers some additional resistance to traditional fungus and bacteria aggressions. However, nature keeps constantly adapting and evolves to produce new enzymes able to induce molecular scission. An excellent description of the main degradation processes can be found in [7].

1.3 COMPOSITE MATERIALS: GENERAL DEFINITIONS

Composites generally comprise a reinforcing constituent such as fibers and a binding material, also called *matrix*. We have already mentioned the infinite possibilities of materials variations leading to a large number of very diverse composites even for the restricted area of polymer-based materials. It is then natural for common classifications to be based upon matrix type (thermoset or thermoplastic), reinforcement type and composite structure.

1.3.1 Classification

1.3.1.1 Classification by polymer type

With more than two-thirds of the composite market, thermoset materials represent the main polymer class for composite matrices. The broad thermosetting family includes polyesters, alkyds, epoxies and phenolics. Thermosets are polymers that can undergo substantial crosslinking reactions under the action of heat, catalysts or UV radiations. The three-dimensional structure thus obtained is irreversible and thermosets generally cannot be recycled.

Thermoplastic materials in turn can be repeatedly melted and reshaped without significant losses in original properties. The properties of the thermoplastic materials are however strongly influenced by the materials degree of crystallinity. Thermoplastic families include styrene polymers, acrylics, cellulosics, polyethylenes, vinyls, nylons and the various fluorocarbon materials [8].

1.3.1.2 Classification by reinforcement type and geometry

A large choice of fibers and fillers is today available to engineers. Glass and carbon fibers represent the two major classes of reinforcement for advanced composites. However, examples of more exotic reinforcement such as mica will be explored in the following chapters.

Glass is an amorphous material (i.e. non-crystalline). Glass has been used for different purposes since thousand of years and various glass compositions have been developed to better suit the different applications: E-glass (E for electrical) is based on CaO-Al_2O_3-SiO_2. Its excellent processability enables the drawing of long fibers for relatively low costs. S-glass on the other hand is based on SiO_2-Al_2O_3-Mg and exhibits a higher stiffness and strength. Unfortunately, the higher temperature resistance of S-glass also translates in more difficult fiber manufacturing processes, yielding higher fabrication costs. Developments at the time of writing in the field of glass fibers, especially with respect to corrosion resistance enhancement, are further discussed in Chapter 3.

Carbon fibers are generally used for more advanced applications. Carbon fibers can be developed from different precursors such as rayon, cellulose and polyacrylonitrile (PAN). Manufacturing processes are generally complex and details kept proprietary. The price of carbon fibers still remains high and their use is generally confined to key components where the high stiffness to weight ration enables significant cost reductions.

For calculation purposes especially, it might also be meaningful to classify composites with respect to their reinforcement geometries. Multiple reinforcement geometries exist including continuous fiber reinforcement, fabrics, short (mat) reinforcement, particulate reinforcement and occasionally three-dimensional reinforcement. Depending on the application, such reinforcements can be included in various composite structures such as single layers, laminated structures, or sandwich composites.

1.3.2 Manufacturing

Not only materials nature but also manufacturing processes influence the final response of the composite to the environment. Manufacturing processes for composites are very diverse and include hand lay-up, filament winding, pultrusion, resin transfer molding and forming.

Hand lay-up is broadly used for all kind of laminated structures ranging from skis to boat hulls. It allows for a large flexibility and the production of very complex three-dimensional shapes. Typical production rates by manual labor are in the range of 0.5 kg/h [9] excluding the curing step. This subsequent step is generally achieved by autoclaving.

Autoclave manufacturing is key to the composite industry. Its development historically started in the early twentieth century with the use of steam-pressure-based autoclaves for the building, food and rubber industries. This later drove the development of hot-air autoclaves, which are now used in various types and sizes by the aerospace or the leisure industry. The increased demand on temperature led to the development of different types of air circulation (longitudinal flow, transverse flow, turbulent flow). The current state of the art enables a temperature control at each point in the range of ±2°C for large autoclaves. The largest hot-air autoclaves are generally requested by airplane manufacturers. Indeed, a 6.10 m diameter, 24 m length autoclave was recently delivered to Airbus for part processing up to 70 bar and 650°C [10] and a 9.1 m diameter, 18.2 m length, 363 ton autoclave is currently designed for manufacturing selected parts of the future Boeing 7E7 [11].

Hand lay-up however generates incompressible base costs and further business optimizations often require the use of automated processes. Pultrusion is one of the cheapest manufacturing processes for large series production of parts with constant cross-sections such as beams. The Creative Pultrusion fiber-glass roving-based composite decks studied in Chapter 6 are examples of pultruded products (Figure 1.7). In this fully automated process, the fibers are pulled through a wetting tank and then into a heated die, where the resin is cured [9].

Filament winding is another automated manufacturing process, more adapted to the fabrication of composite tanks and revolution structures. Typically, in a filament winding process, the mandrel is rotating while the head aligns the pre-impregnated continuous fiber strands at the required angle (Figure 1.8). Helical winding, in which the head travels along the length of the mandrel, is most probably the most common alternative for filament winding. The thermoset matrix is generally simultaneously heated and cured during the application.

For thermoplastic systems, a similar process called tape fiber placement can be used where the reinforced tape is wound around the mandrel: pressure and high

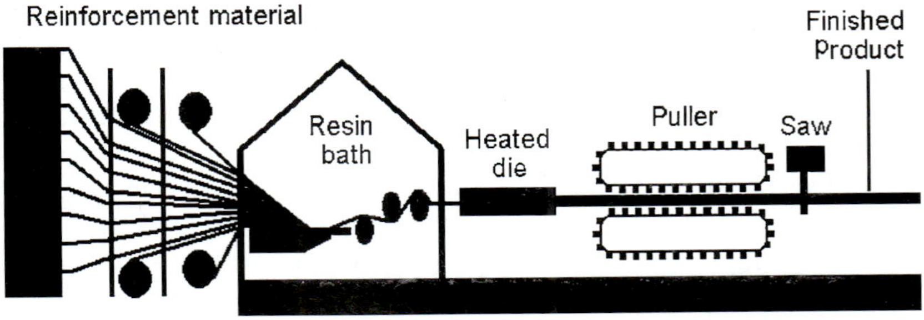

Figure 1.7. Pultrusion process. (Courtesy of Creative Pultrusions, Inc.)

Figure 1.8. Filament winding: Fiber delivery system. (Courtesy of Venus Magnum Products.)

temperature (infrared or laser heating) is applied at the head nip and provides an on-line consolidation. This process was thought of for large thermoplastic structures but was in an experimental stage at the time of writing.

Smaller thermoplastic parts also offer molding alternatives. This process can be used for medium series. The problem is generally the prohibitive cost of the mold and recent developments tend toward the use of cheaper multiple-use disposable mold elements (such as silicon-based counterparts).

The major development efforts in the composite manufacturing area were so far very much focused on closed mold methods. These developments were mainly driven by boat manufacturers in search of cost-efficient automated manufacturing processes for small series, such as resin infusion. In a resin infusion process, dry reinforcement and core materials are placed into a mold and covered by a vacuum bag. Vacuum is then applied, drawing the resin through the part [12]. Belonging to this type of processes are the patented SCRIMP (Seeman Composites Resin Infusion Manufacturing Process) and SPRINT. In this later method, dry fabrics and solid polymer films are alternated during the lay-up process. Vacuum and then temperature are applied to the composite allowing flow and curing of the composite. This process can result in high quality (low void content) thick laminates [13].

Other closed mold applications include vacuum molding, in which the resin is injected in the mold in shots, to a maximum pressure of 1 bar (0.1 MPa), while vacuum is simultaneously applied at room temperature. However, flow and cure rates are limited. Those can be improved in processes such as resin transfer molding (RTM, Figure 1.9). The RTM also involves the injection of the resin but this time

Figure 1.9. RTM injection equipment. (Megaject RTM-Pro RTM Injection machine, courtesy of Plastech.)

at a higher pressure (typically 0.2–0.4 MPa) with the simultaneous application of temperature through the mold tools. Different variations on the RTM process were developed in the past years and include vacuum-assisted resin injection (VARI), vacuum-assisted resin transfer molding (VARTM) and resin infusion under flexible tooling (RIFT). The RTM processes were strongly automated over the past years and can offer high volume fraction composites as well as Class A surface finish required for the automotive or boating industry. However, due to the thermal and pressure loads applied to the tool, the tools are significantly more expensive than in a standard resin infusion process.

1.3.3 Technical Specificities of Composite Materials

1.3.3.1 Inhomogeneity and anisotropy

We have mentioned in Section 1.2 that composite materials are made of at least two distinct components. The presence of two distinct phases create local differences in the materials properties. The material is intrinsically heterogeneous. Damage and degradation in composite materials will therefore often be strongly influenced by local processes.

Additionally, composite materials are often characterized by a significant amount of anisotropy. For example, a unidirectional carbon fiber epoxy material will be much stiffer under an axial load (in the fiber direction) than in the transverse direction (perpendicular to the fiber direction). Even random reinforcement is usually distributed in-plane and lead to different properties in plane and through the thickness. On the one hand, this anisotropy enables the optimization of the composite part to directional loads. On the other hand, this anisotropy contributes to more complexity in the assessment of the damage mechanisms and in their impact on the composite responses.

1.3.3.2 Non-linearity

An additional complexity in dealing with polymer composites is the preponderance of non-linearities in the materials behavior. We will encounter non-linearities in many chapters of this book. Indeed, stress–strain behavior under quasi-static loading can lead to non-linear curves due to the contribution of the non-linear matrix (Chapter 2) or due to a progressive (Chapter 5) damage, such as ply failure. The response of polymer composites is also often dependent on time but not always according to a linear relationship (see Section 5.3.2). Environmental factors such as high temperatures or moisture often contribute to reinforce the non-linear characteristics of the material. We will constantly see throughout this text that non-linearities cannot be neglected *a priori*. The magnitude of the deviations from linearity should be carefully assessed if the use of simplified linear models is considered. Indeed, the use of even large safety factors might not be sufficient to cover for the effects of power-laws.

1.3.3.3 Environmental dependence

Most polymer matrices are characterized by the presence of amorphous phases. Due to this non-equilibrium state, polymer composites are particularly sensitive to environmental factors such as temperature, time, exposure to liquids, gases, electrical fields and radiation. Static and dynamic mechanical loads can interact with the environmental parameters and accelerate the degradation process. Defects along the matrix and reinforcement interface further amplify the action of environmental factors. In this book, we therefore propose to investigate the effects of the main environmental factors first individually then in combination with other parameters.

1.4 ADVANCED COMPOSITE MARKET

Considering the infinite choice of materials combinations, it is no surprise that composite materials can be used for many and diverse applications resulting in a very large but fragmented market. The world composite market including raw materials, intermediate, equipment, distribution and processing was estimated in 2004 to be in excess of €40 billion [14]. In a report at the time of writing on the composite world market [14], thermosets were confirmed as the leading choice for matrix material with 70% of the total composite volume. Glass fibers also dominate the reinforcement market with around 89% of the total volume (82% of value) against 0.6% for carbon fibers (13% of value) while natural fibers have a non-negligible 10% volume share. The contribution of Aramid fibers is around 0.4% in volume (5% of value). Raw materials producers and equipment manufacturers (additives, mold, machinery and software) create 29% of the total value added. Intermediate processors (prepreg producers, pellet producers and fabric manufacturers) represent 9% of the total value added and independent distributors 5%. Final processors have the largest value contribution at 57%, see Table 1.1.

The spectrum of composite end-users is very broad and encompasses nearly all industrial fields. The volumes in Table 1.2 evidence a surprising result: the aerospace industry traditionally thought of as the main user of composite materials represents only 3% of the total composite volume (17% in value). For this reason, this book strongly emphasizes examples of other applications such as construction and civil engineering, which represent a volume share of 30% and a value share of 21% [14]. The automotive industry is also identified as a main end-user of polymer composites (Figure 1.10).

However, the performance and cost requirements set on the material for mass production have limited, so far, the use of composite materials for common car applications. For this reason, the car industry not only focuses on polymers but also considers other alternatives such as aluminum or magnesium to reduce the car weights.

For more than 40 years, racing (such as Formula One) cars have been all-composite making extensive use of carbon fibers. Carbon fibers are also used in

Table 1.1. 2002 Worldwide composite market – Volume and value per manufacturing process [14].

Process type	Process	% of volume	Value added (Eurobn)
Manual	Manual molding	10	2.2
Manual	Spray molding	10	2.2
Manual	Tape laying	6	1.4
Compression	SMC	10	1.9
Compression	GMT	3	0.6
Injection	BMC	9	1.7
Injection	Thermoplastic injection molding	25	3.7
Injection	RTM	3	0.6
Injection	RIM	1	0.2
Continuous	Pultrusion	10	1.8
Continuous	Laminating	8	1.5
Continuous	Filament winding	5	1
Other		0.1	

Table 1.2. 2002 Worldwide composite market – Volume and value per application [14].

Application	% of volume	% Value
Construction and civil engineering	30	21
Automotive	25	23
Industrial equipment	10	8
Electronics	9	6
Sport	8	11
Shipbuilding	6	6
Electrical	5	3
Aerospace	3	17
Consumer goods	1	0.5
Medical	1	2
Railroad	1	1
Windmills	1	2

high-end cars such as the Porsche Carrera GT, the Peugeot 607, F40 and F50 Ferraris. On the other end of the spectrum, the average European car contains around 30 kg of polymer matrix composites, 70% of them being short-fiber reinforced thermoplastics [15]. The use is mainly focused on the interior body: dashboards, door, roof panels and rear window shelves.

The requirements for the exterior are more difficult to meet in a cost-effective way for composite materials. While good energy absorption, light weight and design flexibility (function integration) are positive points in favor of polymer composites,

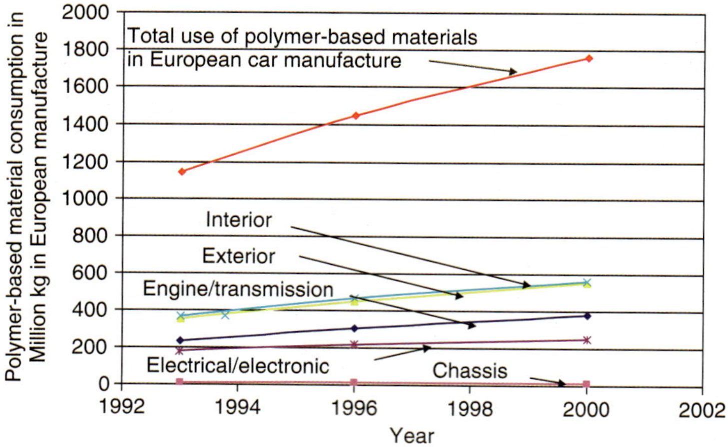

Figure 1.10. Use of polymer-based materials in European car manufacture [16].

problems with mass manufacturing and surface finish still hinder the widespread use of composite materials as structural elements in car manufacturing. Nevertheless, the use of sheet molding compound in 2002 in the automotive industry could be estimated in the order of a non-negligible 175×10^6 Kg [17]. Indeed, polymer-based materials can be tailored for optimum acoustic, mechanical vibrations and shock absorption – properties that match the driver's main requests such as performance, safety and comfort. Today, SMCs are used for body panels (hood), quarter panels, wings and deck lids [17]. For example, the Lincoln aviator uses a sheet molded composite for front wings, the Dodge Viper VGX sports car uses carbon/glass hybrid composites for the windshield surround panels, the inner door panels, wing support systems and headlamp supports, the Cadillac XLR has SMCs hood, door surrounds, doors, tonneau cover, roll bar and roofs, the Mercedes Maybach uses a SMC trunk lid and the Hummer H2 contains a SMC front end, hood and wings [18].

The directives on recyclability constantly put the use of thermoset materials for composite applications in jeopardy. The EU directive 200/53/EC on end-of-life vehicles (ELVs) forces in 2006 a minimum level of reuse and recovery at 80% to increase up to 95 and 85% respectively in 2015 [19]. Solutions for the recycling of thermoset composites is limited. The most practical solution is probably the burning of the material for heat production. This has pushed car manufacturers to consider the use of thermoplastic materials. Pioneering in this field is the company Saint-Gobain Vetrotex featuring the TWINTEX® glass fiber/thermoplastic pellets based on commingled glass fiber and thermoplastic filaments aiming principally at the automotive market for bodywork, spoilers, door panes and roof panels (Figure 1.11) [20].

The assessment of the effects of environmental loads on the materials is a challenge common to all industries, from automotive to aerospace applications. The

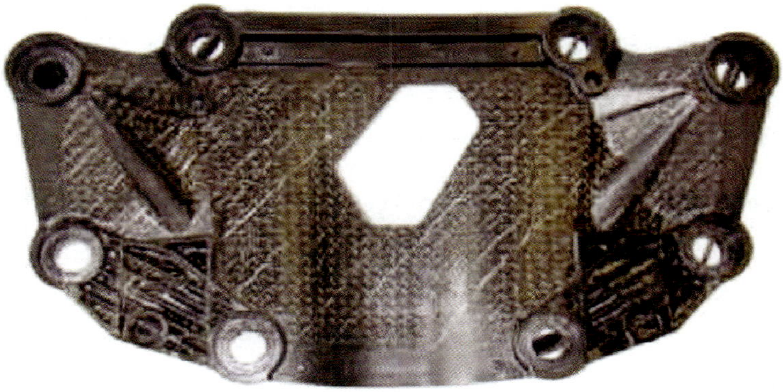

Figure 1.11. Hybrid glass mat/±45° Twintex® Fabric rear box sub-frame 4 × 4 Volvo 70. (Courtesy of Saint-Gobain.)

following chapters are intended to provide the engineer with basic answers and methodologies in order to approach the study of environmental degradation in the most systematic and efficient way possible.

REFERENCES

1. Clyne, T.W., P.J. Withers, D.R. Clarke, S. Suresh and I.M. Ward, *An Introduction to Metal Matrix Composites*. Cambridge Solid State Science Series, Cambridge University Press, 1993.
2. Fridlyander, J.N. (Ed.), *Metal Matrix Composites*. Chapman & Hall, 1994.
3. Warren, R., *Ceramic-Matrix Composites*. Blackie Academic and Professional, 1992.
4. Rugger, G.R., Weathering, in D.V. Rosato and R.T. Schwartz (Eds), *Environmental Effects on Polymeric Materials*. Interscience Publishers, New York, 1968.
5. Friedrich, K., *Friction and Wear of Polymer Composites*. Elsevier Science Ltd, 1986.
6. Rosato, D.V., Other properties and characteristics, radiation, in D.V. Rosato, R.T. Schwartz (Eds), *Environmental Effects on Polymeric Materials*. Interscience Publishers, New York, 1968.
7. Schnabel, W., *Polymer Degradation: Principle and Practical Applications*. Hanser International, 1981.
8. unistates.com/rmt/explained/glossary/rmtglossaryt.html.
9. Gutowsky, T.G., *Advanced Composites Manufacturing*. John Wiley & Sons, 1997.
10. The history of autoclaves. *JEC-Composites*, January 2004, n6, 64–65.
11. Thermal Equipment Corp. to build Boeing autoclave for 7E7. *High Performance Composites*, July 2004, 13.
12. Cost effective infusion of sandwich composites for marine applications. *Reinforced Plastics*, December 2002, 31.
13. Shut it! Move into closed moulding. *Reinforced Plastics*, May 2002, 18–24.
14. Composite materials: An extremely fragmented global market. *JEC-Composites*, April 2004, n8, 26–43.
15. Toursel, S., Will the ideal automobile be made of composite. *JEC-Composites*, October 2003, n4, 39.

16. Report P-023U Advanced Polymer Composites. BCC. 1999.
17. New technologies against paint defects, *JEC-Composites*, October 2003, n4, 50–53.
18. ACA reports new composite components on 2003 vehicles. *JEC-Composites*, October 2003, n4, 54.
19. Underscoring commitment to the automotive industry. *JEC-Composites*, October 2003, n4, 34.
20. Car makers increase their use of composites. *Reinforced Plastics*, February 2004, 26–32.

2

EFFECT OF TEMPERATURE ON POLYMER MATRIX COMPOSITES

2.1 INTRODUCTION

Dealing with the short- and long-term effects of temperature on polymer-based materials is probably one of the most difficult challenges for neophytes and experienced composite part designers alike. The effect of temperature can dramatically change the instantaneous response of the composite: mechanical, electrical and optical properties can undergo order of magnitude changes over a 100°C temperature change and moisture diffusion (Chapter 3) can vary by as much as three orders of magnitude. Polymer composites can exhibit rubber-like behavior (think of your car tires which are steel reinforced composite) as well as an elastic, metal-like behavior at very low temperatures. In the usual composite operating temperature range however, the material exhibits an intermediate behavior between the response of a solid and a liquid. This has consequences on the instantaneous as well as long-term response of the composite. For example, a polymer composite undergoing a tensile test at elevated temperature is very likely to have a response significantly different than metals. This difference in the instantaneous response of the material will be assessed in the description of viscoelasticity in Section 2.2.3. Furthermore, polymer-based materials are characterized by a constant molecular motion in the amorphous state even under static loads. The re-arrangement rate is a function of many parameters including stress level, length of the macromolecules and temperature. Generally speaking, the re-arrangement rate increases with temperature. This means that a polymer composite tested today and after 10 years will have different properties and therefore mechanical, thermal, optical and electrical responses (this statement is valid for all polymer matrices operated below the glass transition with the exception of fully crystalline polymers). Therefore, designers should always expect and anticipate changes in the materials behavior after a few years of operation. The question that immediately comes to mind is of the significance of those changes. The magnitude of the changes is tightly linked to the operation and storage temperatures of the composite parts. Quantifying the long-term response of polymers is a difficult task addressed later in this chapter.

Designing with metals or polymer composites therefore calls for different rules and tools. Safety coefficients are also not proper safeguards as polymeric behavior is often characterized by non-linear laws. The present chapter gives all necessary basic elements to understand and model polymer-based material behavior. It introduces viscoelasticity, models for creep and relaxation, the time–temperature equivalence concept, instantaneous temperature effects, degradation and physical aging concept. Special emphasis is put on the limits and pitfalls of accelerated testing. Finally, common test methods are introduced.

Industrial case studies are included throughout the chapter. Cryogenic tanks for re-launchable space vehicles illustrate the difficulties associated with operating conditions at very low temperature. Fire requirements for mass transportation such as train or plane illustrate the other end of the temperature spectrum.

If there is need to convince you further of the importance of the present chapter, let us consider an industrial example.

Case study: Carbon-fiber polyetheretherketone (PEEK) coating for hydrogenerator bearings

Hydrogenerators are large electrical machines that help convert mechanical energy from the water turbine rotation into electrical energy which is sent to the grid and finally to your home and businesses (Figure 2.1).

Vertical machines can be very large with a typical thrust between 50 and 3000 ton. Such machines are often supported by hydrodynamic thrust bearings.

Figure 2.1. Mühleberg Hydro Power Plant (Switzerland). (Courtesy of BKW FMB Energy AG.)

The largest thrust bearing to date is the one designed and installed on the Yangzi river in China's gigantic 3-Gorges project (6000 ton load capacity, 75 rpm rotation speed). The outer diameter of the bearing is in excess of 5 m.

Hydrodynamic thrust bearings support the rotor via tilting pads immersed in oil. Under the action of the rotating forces, the stationary pads tilt to an equilibrium position as pressure builds in the oil film between the pads and the rotor. Typical pad surfaces can reach 1 m^2 while the oil-film thickness separating the pad from the rotor is only of a few tens of microns! The design of such bearings requires extremely fine tolerances for planarity and surface state. This is usually achieved using steel pads coated with a tin–lead alloy called *Babbitt*. The Babbitt can easily be machined to the desired finish. In case of accidental contact with the rotor, the coating also insures that no damage occurs to the (expensive) rotor.

The losses in the bearings are directly proportional to the pad surface. Therefore, utility companies can achieve major savings if the bearing admissible load is increased. For large hydrogenerator bearings, the admissible load with conventional Babbitt material is generally limited to 4 MPa (5 MPa in the case of complex supporting systems). In order to achieve higher values, Russian and Chinese industrials turned their attention in the 1970s to polytetrafluoroethylene (PTFE, Teflon®) as an alternative coating material. A one-to-one replacement was performed on the pads in which the 4–8 mm thick metal coating was replaced by PTFE bonded on the steel base. Bonding was achieved by mechanically pressing the polymer into a copper mesh. The admissible mean load could immediately be increased to 7 MPa. However, under the effect of high compressive loads and at an operating oil temperature of 80°C, the pad coatings started showing extensive signs of creep. The reductions in the coating thickness combined with wear induced risks of contact between the steel base and the rotor.

To solve such long-term durability issues, the industry turned to composites. Fillers and fibers of all kinds were added to the Teflon® in order to increase the creep resistance of the pads and therefore the lifetime of the bearings. For example, the dimensional integrity of the pad coating could be significantly improved by adding carbon fibers. However, some bearings showed unexpected hydrodynamic behaviors and occasional failure of the bearing especially at higher temperatures: unexpectedly, the short-term behavior of the filled polymers was significantly different than in the past.

Recent investigations have shown that the hydrodynamic behavior of the bearing is strongly influenced by the coating material [1]. This is mainly explained by the local deformations of the polymer-based coating. While the metal coating exhibits very limited deformation, the composite coating can experience large strains under load. This deformation is time and temperature dependent. At room temperature, the deformation is negligible and the hydrodynamic response of the bearing is similar to the metal-coated one. At normal operation temperature however (80°C), the deformations are in excess of a few microns and become significant for the bearing. Furthermore, the presence of random reinforcement creates variations in the coating thickness and disturbs the planarity of the pad surfaces. Short-term behavior therefore dictates a need for smaller reinforcement contents and sizes while long-term dimensional stability at elevated temperature requires creep resistance mainly achievable by increasing fiber content and length. A compromise could be found using a PEEK matrix and a 15% 3 mm carbon-fiber reinforcement. The optimized reinforcement mixture does not induce short-term instabilities and the reinforced PEEK provides outstanding creep resistance. Such a bearing has been in operation since 2003 in the Mühleberg PowerPlant in Switzerland (Figure 2.2).

This example illustrates several challenges raised by the use of composite materials:

- The composite acted very differently from the traditional metal part, even in unexpected areas such as the hydrodynamic response.
- The long-term behavior of the original polymer coating at elevated temperature was unacceptable due to extensive creep. The suppliers could not reliably provide durability guaranties.
- The short-term behavior was also strongly influenced by local phenomenon that changed with small variations in operating temperatures. Material selection had to be re-considered fully taking into account the specific (viscoelastic) nature of the composite.

Figure 2.2. Carbon-fiber/PEEK bearing. (Courtesy of Alstom.)

2.2 POLYMER MATRIX COMPOSITES VERSUS METALS

2.2.1 Stress–Strain Curves

The behavior of most metals and other crystalline materials is usually well understood. This scientific understanding allows for precise and reliable lifetime calculations. The mechanical behavior is predictable to a large extent. A typical response to a tensile stress for such materials is shown in Figure 2.3.

Region 1 is the elastic region. Removal of the load in this region leads to a recovery of the initial strain. Region 2 characterizes the plastic deformation. The decrease in stresses beyond a certain deformation level is explained by a decrease in the sample cross-section (necking). Finally, rupture occurs at the end of region 2. For most industrial applications, the materials are used in region 1 and calculations rely on the elastic properties of the material. In this region, Hooke's law relating the stress (σ) to the strain (ε) and Young's modulus (E) is precisely applicable.

$$\sigma = E \times \varepsilon \tag{2.1}$$

However, not all materials exhibit the presence of a linear stress–strain relationship. This is, for example, the case of gray cast iron, polymers and many composites (including concrete). The reasons for non-linearity in composites are multiple and range from the non-linear response of one of the constituents to the development of an irreversible damage such as fiber debonding.

Long-fiber carbon- or glass-reinforced composites are often considered as *fiber* dominated composites, implying that mechanical behavior and failure are mainly driven by the fiber response to loading. But if we take a closer look at micromechanical calculations and durability schemes (Chapters 5 and 6), we realize that the matrix properties are in fact required in those calculations. The present text

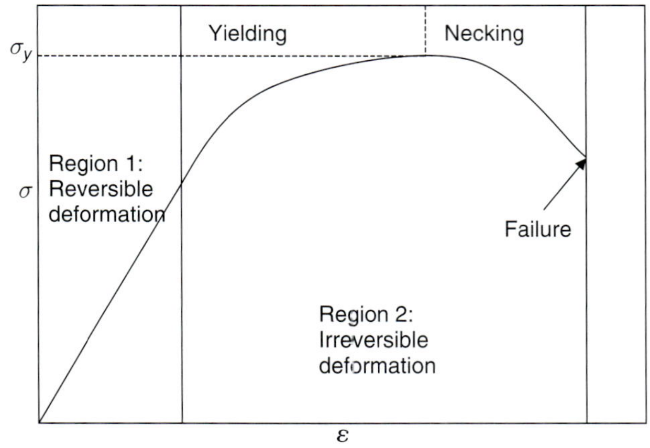

Figure 2.3. Typical stress–strain curve for an elastic material.

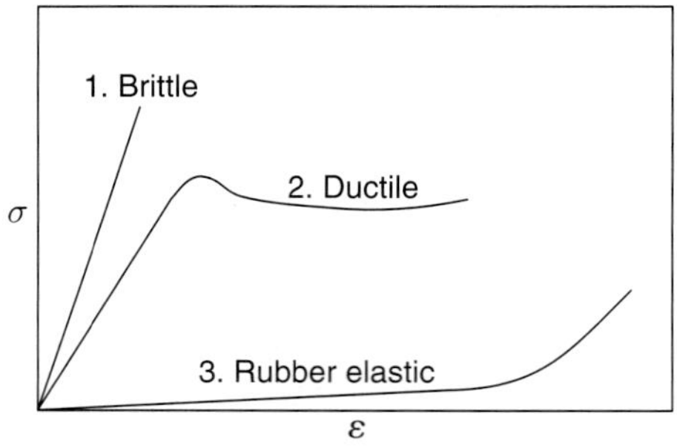

Figure 2.4. Typical stress–strain curves for polymers and polymer matrix composites.

focuses on polymer matrix composites, and polymers can exhibit a multitude of stress and strain curves (Figure 2.4). Unlike strictly elastic materials illustrated in Figure 2.3, polymer-based materials can show limited or even no linearity in the stress–strain behavior.

For industrial applications, a wide range of polymer matrices can be used. Curves 1 and 2 are representative of a brittle and ductile polymer response respectively. Curve 3 is the typical response of an elastomeric material. Elastomers and rubber elasticity are detailed in ref. [2–4] and extend well beyond the scope of this book.

2.2.2 Ductility versus Brittleness

The ductile/brittle aspect of polymer matrix composites depends on many factors and most composites can exhibit the two behaviors depending upon environmental conditions. The failure of a brittle or ductile material is fundamentally different: brittle fracture occurs *fast* and is often characterized by instability. Materials in the brittle state have a low fracture toughness (generally below $1\,\text{MPa}\,.\,\text{m}^{1/2}$ [5]), a low impact strength [6] and show little signs of damage before failure. On the other hand, ductile failure is more progressive. For most industrial applications, materials with a certain ductility and that give early damage warnings are preferred (see Chapter 6). The conditions influencing the ductile/brittle aspect of composites can be grouped into two categories. The first category includes environmental parameters such as temperature (in general ductility (resp. brittleness) is increased by raising (resp. decreasing) the temperature), strain rate (ductility is increased by decreasing the strain rate) and solvents (Chapter 3 evidences the increased ductility of composites with high water content etc.).

The second category influencing the brittleness of the composite includes parameters intrinsic to the material: the nature of the polymer matrix, fillers and fibers.

The nature of the polymer matrix is strongly dependent upon its degree of cross-linking. Highly cross-linked materials are usually very brittle. Therefore, thermoplastic materials are generally more ductile than epoxies. The molecular structure and stiffness of the polymer chain also contribute to the brittleness of the polymer phase. In the polymer itself, brittle failure typically occurs via crazing after a 1–2% elongation. However, a polymer in a ductile state undergoes a molecular slip that is characteristic of yielding and experiences a strain to failure usually higher than 2%.

Ductility can also come from the presence of multiple phases in the material. Epoxies, for example, can be toughened using liquid rubber modifiers, thermoplastic modifiers, reactive diluents modifiers and inorganic modifiers such as glass beads and silica. A good review of toughening methods and effects is available in the literature [7]. Significant increases in the facture toughness can be obtained, but those modifications are often detrimental to other properties such as glass transition temperature or flexural stiffness.

The presence of larger fibers such as glass or carbon can also modify the toughness of the material. The volume fractions, spatial distribution and quality of the interface influence the response of the material. For example, a crack initiated in the matrix might find it easier to propagate around a fiber rather than through a fiber. This longer crack path contributes to an increase in the toughness of the material. Composite toughening mechanisms such as crack bowing, crack deflection, debonding, pull-out, wake toughening, and microcrack toughening are discussed in the literature [8]. Such phenomena can also lead to significant non-linearity in the materials stress–strain curves. It is possible that linear-elastic polymers combined with linear-elastic fibers may result in non-linear stress–strain curves under load as well. The ductility of the composite can therefore be tailored not only through the choice of the composite constituent nature, but also through proper selection of the reinforcement size and fraction.

2.2.3 Viscoelasticity – Definition

Depending on the temperature, the same polymer can be brittle and behave elastically (low temperature) or be ductile and behave viscoelastically (high temperature). The composite properties and response therefore strongly depend upon temperature and other environmental parameters. This is probably one of the main differences between metals and polymer matrix composites. In order to perform a safe polymer matrix composite industrial design, it is necessary to understand and thoroughly consider the viscoelastic response of the composite. Time and environment are key factors in the long-term behavior of composite products. As a first step, we propose to understand the basic mechanisms that affect property changes in the composite (Chapters 2–5). We will then integrate those concepts into a global life prediction scheme (Chapter 6).

The time component should always be considered when dealing with composite materials. Indeed, we have mentioned in Section 2.1 that the state of composite materials changes with time. The goal is now to quantify this change to evaluate its significance. Time and temperature are equivalent to some extent for polymer-based materials and it is difficult to isolate their effects. A similar equivalence also exists for time and loading rates. The response of the polymer composites at long times, low loading rates or high temperatures have similarities. For example, polymer-based materials can exhibit rather different behaviors depending upon the temperature and the strain rates:

- A glass-like elastic behavior at low temperatures and/or high strain rates (instantaneous response).
- A liquid-like behavior (viscous) at high temperatures and/or under low strain rates.
- An intermediate behavior at moderate temperatures and strain rates.

This intermediate behavioral option is the reason for classifying such materials as *viscoelastic*.

We therefore need a third axis to the traditional stress–strain curve in order to consider time. Figure 2.5 shows the strain response of three materials subjected to tensile loading. The elastic material deforms instantaneously under stress. After removal of the stimulation, the original geometry is fully recovered. Such purely elastic deformation is fully reversible. For a viscous material, the strain linearly increases with time, as long as the load is applied. Once the stress is released, the material does not *shrink* back. The deformation of the purely viscous material is totally irreversible. Viscoelastic materials such as polymer-based materials have a behavior comprised between those two extremes: the material experiences first a limited instantaneous deformation upon stress application. The strain will then keep on increasing with time in a non-linear manner. Once the stress is released, the material experiences an instantaneous dimensional recovery limited to the elastic contribution of the material.

We will now spend a significant part of this chapter on understanding viscoelasticity. Not accounting for viscoelasticity when working with polymer

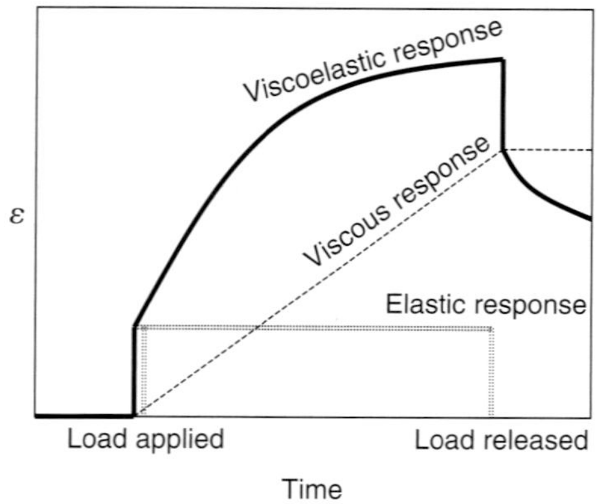

Figure 2.5. Elastic, viscous and viscoelastic strain with time.

matrix composites can be identified as the predominant factor for the failure of development projects involving the use of composites in traditional industries. In the traditional industries (heavy machinery, energy etc.), most designers are used to working with linear elastic metals. Non-linearity and time dependency are often disregarded or only considered for design and lifetime assessment by adding a safety coefficient. It is to be noted that this approach is extremely dangerous: the safety margin expected from the use of a simple safety coefficient applied to linear relationships can easily be exceeded in real operation due to the exponential relationships driving stresses and strength evolution in the composite (see chapter 6).

Viscoelasticity is a complex field, well documented in the literature [9–15]. The elements provided here are very basic and should give starting points for proper testing and modeling. More elaborated models [12–15] are available and should be used whenever in-depth theoretical modeling is necessary.

2.3 MODELING CREEP, RELAXATION AND TIME-DEPENDENT RESPONSE TO CYCLIC LOADS IN POLYMERS AND COMPOSITES

2.3.1 Creep versus Stress Relaxation

A necessary and early step in analytical and finite element calculations is to explicitly relate the strains experienced by the material to the state of stresses. For most polymer matrix composites, this relationship varies with time. The molecular re-arrangement leading to time-dependent behaviors is referred to as relaxation. In the industrial context, the word creep is used more frequently. However, at least two experimental cases and therefore two modeling methods should be distinguished: in a *creep* experiment, a *stress* is *applied* to a sample and the *deformation* is *measured*. In the second case, a *deformation* is *applied* to the sample and the *stress* is *measured*. These later conditions define a stress *relaxation* experiment. It is not uncommon that a composite experiences a situation between creep and stress relaxation. However, most of the time one condition will prevail. For example, if the geometry of the part is constrained (little deformation allowed), stress relaxation will be pre-dominant.

2.3.2 Models for Creep and Stress Relaxation: Introduction to Viscoelasticity

In a pure elastic case, Hooke's law holds (Equation (2.1)) and the material can be thought of as a spring of modulus E (instantaneous deformation, Figure 2.6).

In the case of a viscous liquid, the strain response is delayed when compared with the application time point of the load (stress). The mechanical analog is a dashpot (Figure 2.7) of viscosity η where:

$$\sigma = \eta \frac{\mathrm{d}\varepsilon}{\mathrm{d}t} \tag{2.2}$$

Figure 2.6. Spring element.

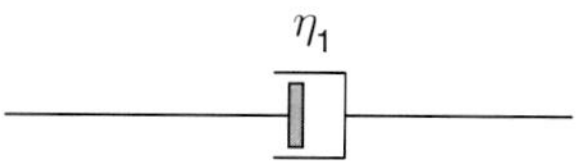

Figure 2.7. Dashpot element.

Figure 2.8. Maxwell model.

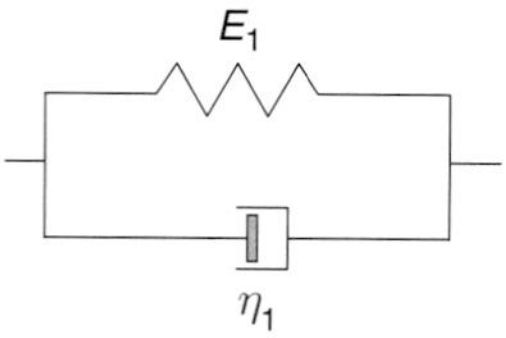

Figure 2.9. Voigt model.

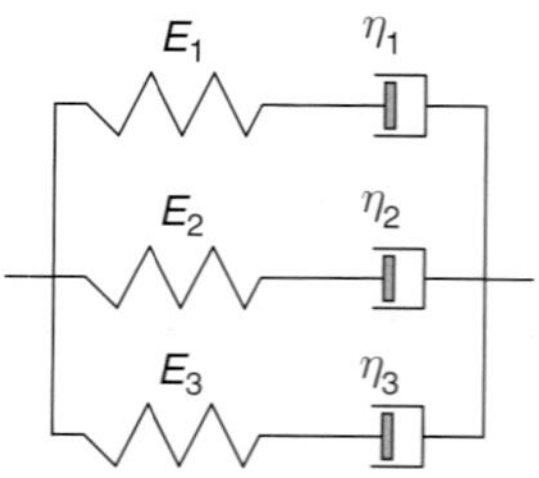

Figure 2.10. Maxwell–Wiechert model.

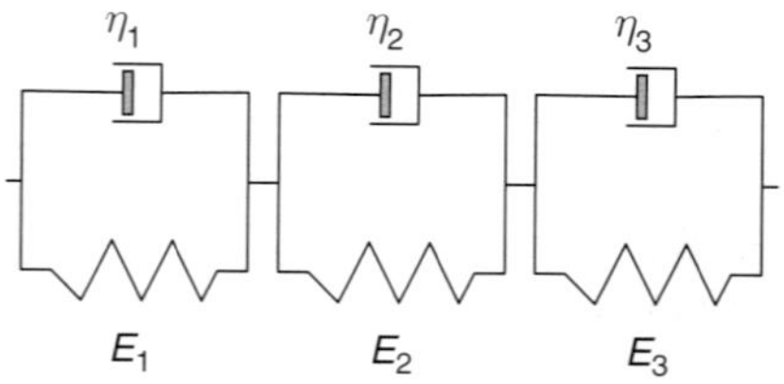

Figure 2.11. Voigt–Kelvin model.

Intuitively, the response of the polymer-based material will be a combination of these two elements. The combination possibilities are infinite. We propose to focus here on the most basic and commonly used models (Figures 2.6–2.11).

The solutions for the Maxwell (Figure 2.8) and Voigt (Figure 2.9) models are given above. In most cases, these models suffice to describe creep or relaxation to the level or precision needed in an industrial context. If this was not the case and if the models would not fit the data to an acceptable level, more complex combinations of springs and dashpots can be used such as the Maxwell–Wiechert (Figure 2.10) or Voigt–Kelvin (Figure 2.11) models. Those models shown for three elements only can be extended to as many elements as needed.

The derivation of the following equations, solutions to the various models for different boundary conditions can be found in Aklonis and MacKnight [11]. All models are not applicable to all boundary conditions: for example, the Voigt model should not be used for stress relaxation situations as it would lead to a constant modulus (independent of time), which contradicts observations of viscoelastic behaviors. Therefore, only relevant equations are reported here. The Equations 2.3 and 2.4 should enable engineers to fit an analytical formulation to a wide range of experimental data. When this is not possible, engineers can revert to the more complex models [11].

Let us fix the following notations:

- η the viscosity of the dashpot
- E_o the modulus of the spring
- D_o the compliance of the spring. For the spring, compliance and modulus are inversely proportional:

$$D_o = 1/E_o \tag{2.3}$$

- τ the relaxation time of the element. τ is related to the dashpot viscosity and the spring modulus by:

$$\tau = \eta/E_o \tag{2.4}$$

- z the number of elements in series or parallel (for Maxwell–Wiechert and Voigt–Kelvin models)
- $D(t)$ the creep compliance of the element
- and $E(t)$ the tensile stress relaxation modulus of the element.

With the help of these quantities, we can now specify the models for creep and relaxation in polymer-based materials.

2.3.2.1 Creep

During a creep experiment, the stress applied on the sample is a constant and the strain varies with time. The creep compliance D can be expressed as a function of time according to Equation (2.5) for a Maxwell model, according to Equation (2.6) for a Voigt model, according to Equation (2.7) for a Voigt–Kelvin model. The criterion for selecting the model is mainly based on the quality of the fit to the data. Equations (2.5), (2.6) and (2.7) should cover the large majority of materials and creep loading conditions.

The creep compliance for a Maxwell element can be written as:

$$D(t) = D_o + (t/\eta) \tag{2.5}$$

The creep compliance for a Voigt element is:

$$D(t) = D_o(1 - \exp(-t/\tau)) \tag{2.6}$$

Finally, the creep compliance for a Voigt–Kelvin element becomes:

$$D(t) = \sum_{i=1}^{z} D_{o_i}(1 - \exp(-t/\tau_i)) \tag{2.7}$$

2.3.2.2 Relaxation

In a stress relaxation experiment, the deformation is kept constant and the changes in stresses are measured (fixed geometry experiments). The tensile stress relaxation modulus E then varies as a function of time. The Maxwell and Maxwell–Wiechert models are appropriate to describe the time-dependent response of the material subjected to relaxation loading conditions. The use of the Voigt model on the other hand would lead to a modulus erroneously constant over time:

The relaxation modulus for a Maxwell element is:

$$E(t) = E_o \exp(-t/\tau) \tag{2.8}$$

The relaxation modulus for a Maxwell–Wiechert element is:

$$E(t) = \sum_{i=1}^{z} E_i \exp(-t/\tau_i) \tag{2.9}$$

As for creep experiments, the model (Equation (2.8) or (2.9)) will be selected by fitting the results to the experimental data.

2.3.2.3 Dynamical loading

Cycling is extensively covered from a comprehensive damage perspective in Chapter 6. In the present chapter, we will focus only on the viscoelastic contribution to the response of the polymer matrix composite under cycling mechanical loads.

To account for cyclic loading, sinusoidal loads can be introduced in the simple models presented previously:

$$\sigma(t) = \sigma_o \exp(i\omega w) \tag{2.10}$$

where σ_o is the amplitude of the stress and ω the frequency. The resulting strain is also sinusoidal of frequency ω. We have already mentioned the dashpot effect in which the response of a polymer to an applied stress is delayed. Therefore, under cyclic load, the strain lays behind the stress. By convention, the phase difference between the stress and the strain is noted as δ. The stress can thereby be split into two – in-phase and out-of-phase components σ' and σ'':

$$\sigma' = \sigma_o \cos\delta \tag{2.11}$$

and

$$\sigma'' = \sigma_o \sin\delta \tag{2.12}$$

In the case of dynamic experiments, it is necessary to introduce new quantities to measure creep and relaxation. The resulting *dynamic* modulus is called the *complex modulus* and noted as E^*. Respectively, the *dynamic* compliance is called the *complex compliance* and noted as D^*. By convention we write:

$$E^* = E' + iE'' = \sqrt{E'^2 + E''^2} \tag{2.13}$$

where E' is the storage modulus and E'' is the loss modulus (out-of-phase component).

$$E' = E^* \cos \delta \tag{2.14}$$

$$E'' = E^* \sin \delta \tag{2.15}$$

The loss factor (commonly called *tangent* δ) is the ratio of the out-of-phase and in-phase components of the complex modulus:

$$\tan \delta = \frac{E''}{E'} \tag{2.16}$$

E' represents the amount of energy stored in the material during the deformation and E'' represents the energy dissipated during the deformation. The loss tangent defined above quantifies the angle between the in- and out- of phase components during cycling. In an industrial context, E' and tangent δ are commonly used. Analysis based on E'' are somehow marginal.

Aklonis and Macknight [11] derive the solutions for the various elements submitted to sinusoidal dynamic experiments: the storage and loss moduli for a Maxwell element resulting from a cyclic sinusoidal loading are given by Equations (2.17) and (2.18).

$$E' = \frac{E_o \omega^2 \tau^2}{1 + \omega^2 \tau^2} \tag{2.17}$$

$$E'' = \frac{E_o \omega^2 \tau^2}{1 + \omega^2 \tau^2} \tag{2.18}$$

Equations (2.19) and (2.20) express the storage and loss compliances for a Voigt element resulting from a cyclic sinusoidal loading:

$$D' = \frac{D_o}{1 + \omega^2 \tau^2} \tag{2.19}$$

$$D'' = \frac{D_o \omega \tau}{1 + \omega^2 \tau^2} \tag{2.20}$$

If necessary, the storage and loss moduli resulting from a cyclic sinusoidal loading can be calculated using more elements and following the Maxwell–Wiechert model:

$$E' = \sum_{i=1}^{z} \frac{D_{o_i}}{1 + \omega^2 \tau_i^2} \tag{2.21}$$

$$E'' = \sum_{i=1}^{z} \frac{D_{o_i} \omega \tau_i}{1 + \omega^2 \tau_i^2} \tag{2.22}$$

Finally, the storage and loss compliances resulting from a cyclic sinusoidal loading can be derived for a Voigt–Kelvin element:

$$D' = \sum_{i=1}^{z} \frac{D_{0_i}}{1+\omega^2\tau_i^2} \tag{2.23}$$

$$D'' = \sum_{i=1}^{z} \frac{D_{0_i}\omega\tau_i}{1+\omega^2\tau_i^2} \tag{2.24}$$

During cyclic loading, heat might be generated in the material (viscoelastic heating). Indeed, the cyclic loading of a viscoelastic materials was found to yield hysteresis loops in the stress–strain relationship. A large part of the dissipated energy was also found to convert into heat. In adiabatic conditions, the resulting temperature rise might be significant and strongly depend upon the loading rates [16].

"In cross-linked and crystalline polymers, the effect of frequency on loss properties is minimal. In other cases, frequency effects are associated with the activation of motions of the 'back-bones' of macromolecules and maxima correspond to resonance-like behavior" [17]. If we consider a viscoelastic sample under a cyclic mechanical load, the dissipated energy will generate heat in the material. If heat cannot be evacuated out of the sample at a sufficient rate, the core temperature of the sample will increase. If the input is strain, then the complex modulus (E^*) will decrease as the temperature increases, so the stress will drop and it will be possible for the sample to reach equilibrium with the surroundings. If input is stress, then the complex compliance (D^*) will increase as the temperature increases. The sample will then experience thermal runaway, which might lead to melting of the material [17].

2.3.2.4 Dynamic versus static moduli

It can be useful for practical or understanding purposes to mathematically relate the dynamic and static properties. Aklonis and MacKnight [11] derive the following relationships between the dynamic moduli and the static modulus. Note that these solutions are not unique and other relationships are also available [18].

$$E^*(\omega) = \int_0^\infty \omega(\sin\omega s)E(s)\mathrm{d}s + i\int_0^\infty \omega(\cos n\omega s)E(s)\mathrm{d}s \tag{2.25}$$

$$E'(\omega) = \omega\int_0^\infty \sin\omega s E(s)\mathrm{d}s \tag{2.26}$$

$$E''(\omega) = \omega\int_0^\infty \cos\omega s E(s)\mathrm{d}s \tag{2.27}$$

Practically, these equations can be inverted using Fourier transforms [11,19]. Generally speaking, storage modulus versus time and tensile modulus versus time

Table 2.1. Tensile versus storage modulus for selected polymers and composite

Material	Tensile modulus at 20°C (E)	Storage modulus at 20°C (E')
Polymethylmethacrilate (PMMA)	3.4 GPa [20]	1.3 MPa (10^6 g/mol) [21]
Polyetheretherketone (PEEK)	3.1–8.3 GPa [20]	1.0 GPa (38×10^3 g/mol) [21]
Carbon-fiber polyphenylenesulfide (AS4/PPS)	16–55.2 GPa [20]	10.7 GPa ($V_f = 50\%$) [21]

curves are qualitatively similar. Table 2.1 compares some values of tensile and storage moduli (measured by DMA at 20 Hz).

2.3.2.5 *Important consequences on composites*

As a general rule, creep should always be investigated when dealing with polymer matrix composites. Kerr and Haskins [22] display in the literature interesting creep data on various composites. Figure 2.12 shows creep data for a six-unidirectional-layer-laminate $[0°]_6$ graphite/epoxy composite. This is an example of the least time sensitive case where the load is parallel to the fiber direction (±45° lay-ups would exhibit much more creep). Such composites (Figure 2.12) are typically considered as fiber dominated and creep is often neglected. Indeed, creep at 121°C is negligible for 100 hours. However, a sudden change (followed most likely by failure) occurs after 100 hours. Neglecting creep would lead to an infinite lifetime prediction!

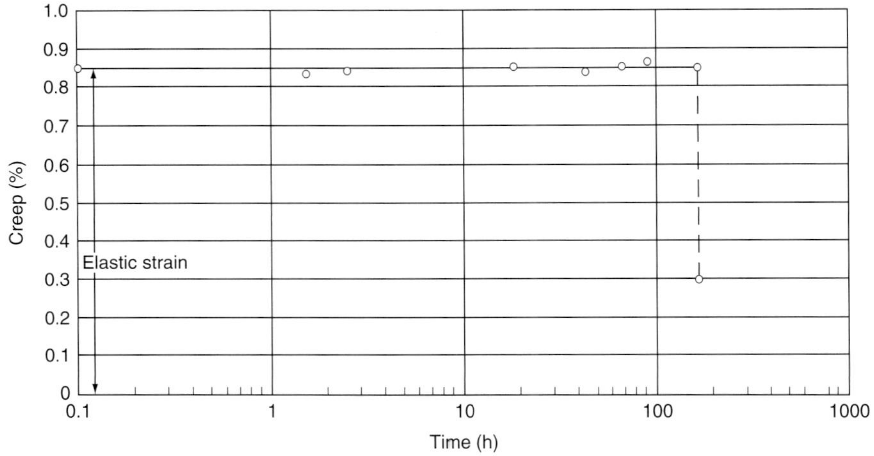

Figure 2.12. Creep Strain for $[0°]_6$ AS/30501-5 Graphite Epoxy at 121°C. (Data and graph from Kerr and Haskins [22].)

Failure after constant exposure to temperature and stress is referred to in the present text as stress rupture and will be further discussed in Chapter 6. The data presented in Figure 2.12 also means that accelerated testing with a duration of 100 h at 121°C would not have predicted such an abrupt change of behavior. We will come back on the very important matter of accelerated testing in Section 2.5.4.

2.4 TRANSITIONS AND KEY TEMPERATURES

Most fibers traditionally used as reinforcement in polymer matrix composites do not show dependence upon temperature, in temperature ranges seen by the operating composite. Carbon fibers, for example, can be used at temperatures in excess of 2500°C if protected from oxygen [8] and epoxy-based composites are usually operated below 250°C. Therefore, the changes in the materials mechanical properties with temperature are mainly driven by the changes in the polymer, which properties are very sensitive to temperature variations.

At the molecular level, temperature and time show some type of equivalence. Therefore, polymer and composite properties can vary significantly with time. The present part of this chapter is dedicated to the understanding of the polymer characteristic temperatures, keys in defining transitions between regions in which the materials response to stimulation will be significantly different.

A polymer or its composite subjected to an elevated temperature exhibits changes in its instantaneous mechanical properties. This change is very large for a composite, the response of which is driven by the matrix, for example, in the case of short and random reinforced polymers or in the transverse testing of unidirectional polymer composites. Figure 2.18 illustrates this concept and shows drastic drops in the composite (here AS4/PPS) storage modulus during a dynamic mechanical analysis (DMA) experiments (see Section 2.8.3 for definition).

Contradicting common prejudice however, the influence of temperature is not only limited to matrix-dominated testing and operation. In the fiber direction (i.e. tensile tests on unidirectional composites), the composite also experiences significant modulus changes with temperature as illustrated in Figure 2.13. Therefore, the effects of temperature on a composite part should always be accounted for.

2.4.1 The Four Regions of the Master Curve

Most polymers and polymer matrix composites experience a marked drop in their modulus as the temperature rises. This drop is mainly due to the changes in the matrix properties with temperature. Therefore, the following parts mainly focus on the matrix.

The stiffness versus temperature curve can have different shapes depending upon the nature of the polymer. The modulus versus temperature curve for a typical polymer exhibiting a secondary relaxation is illustrated by Figure 2.14. This curve is

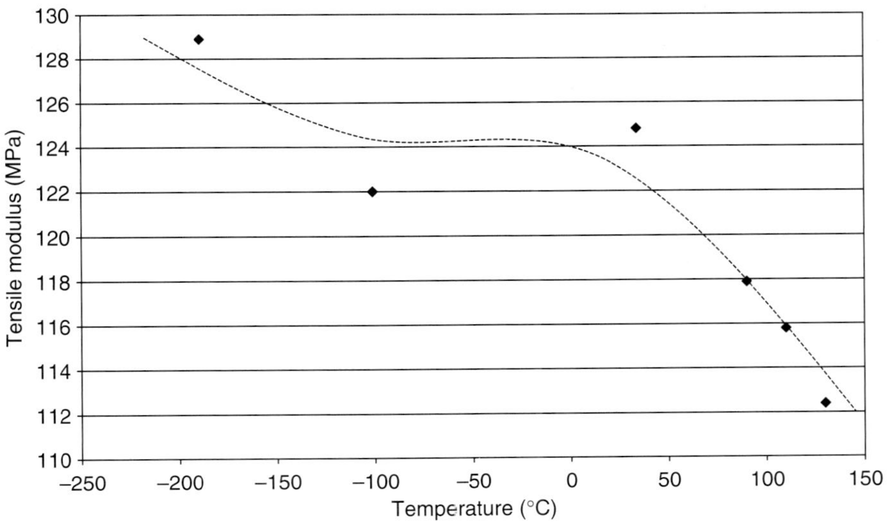

Figure 2.13. Unidirectional carbon-fiber reinforced vinyl ester composite with polyurethane interface tested parallel to the fiber direction – Stiffness versus temperature. (Data from Walther [23].)

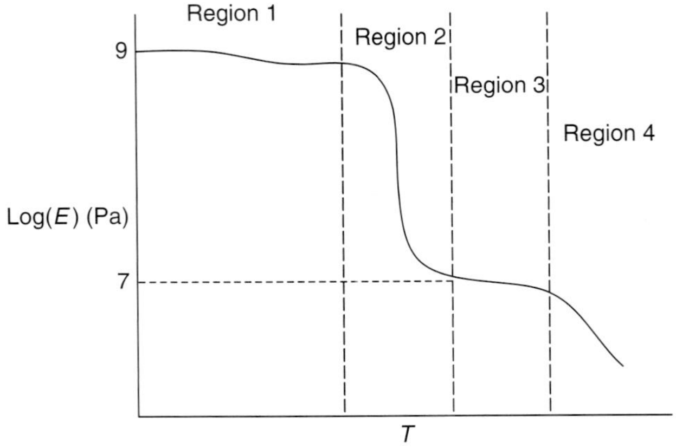

Figure 2.14. Modulus versus temperature for a typical polymer.

traditionally divided into four distinct regions: the glassy stage, the glass transition region, the rubbery stage and the rubbery flow.

2.4.1.1 Glassy stage

The first region of the modulus (E) versus temperature (T) curve is referred to as the glassy region. This region is characterized by fairly stable values of E. For many polymers, this value is around 3 GPa. In this region, molecular motion is rather restricted. The molecules act as if they were *frozen* and mainly display vibrational motions.

However, the presence of peaks during thermomechanical tests, for example, evidences the presence of molecular motion, even for low temperatures. As the temperature increases, the polymer can undergo several transitions. Typically, the first transition is called γ *relaxation*, the second is termed β *relaxation* and the third is referred to as the glass transition (T_g) or the α *transition* [24]. This convention is modified for some semi-crystalline polymer matrices such as PE, PP, PEO, POM that exhibit the presence of a specific transition, α_c, before the glass transition (called β for these exceptions), attributed to re-orientational motions within the crystals.

The γ and β transitions (secondary transitions) reflect molecular motions occurring in the glassy state (below T_g). In the glassy region, the thermal energy is much smaller than the potential energy barriers to large-scale segmental motion and translation, and large segments are not free to jump from one-lattice site to another [25]. Secondary relaxations result from localized motions. The secondary relaxations can be of two types [11]: side group motion or the motion of few main chains. It is difficult to establish a general model for these relaxations due to the molecular specificity (nature of the side groups). However, the crankshaft mechanism attributed to different authors [26,27] illustrates most of these mechanisms (Figure 2.15).

The rotation of a part of the main chain can be activated by temperatures lower than the glass transition temperature. This is the case in many polymers containing $(CH_2)_n$ sequences where the number of monomers equals four or greater [11] that exhibit a transition at −120°C. Even larger units can relax [28]. Group motion can be illustrated by the β relaxation of poly(alkyl methacrylate)s. If the radical is a cyclohexyl ring, a relaxation at 180 K can be observed at 1 Hz.

The study of β relaxations can be complex and depends on the chemistry of the material. We will cite, for example, the detailed descriptions of β relaxations given

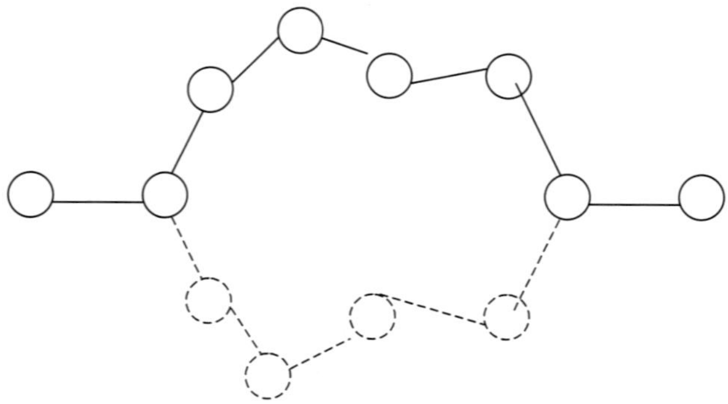

Figure 2.15. Crankshaft mechanism.

by Bartolotta [29] et al. for polyethylene (PE) oxide–iron thiocyanate polymeric complexes (where the influence of the salt content is also discussed) and for thermosetting systems by Wang [30] (influence of cure extent on β relaxation, explained by the difference in the micromechanisms before and after gelation due to cross-linking). Study of the secondary relaxations also appear in the study of PEEK by Krishnaswamy [31] and poly(methyl methacrylate) by Muzeau [32].

According to Aklonis and Macknight [11], a time–temperature or frequency–temperature superposition scheme can be applied to these relaxations. However, the following Arrhenius equation:

$$\log a_{\mathrm{T}} \propto -\frac{H_{\mathrm{a}}}{2.303RT} \tag{2.28}$$

where a_{T} is the time–temperature shift factor, H_{a} is the activation energy, R is the gas constant and T the temperature, must be applied instead of the traditional WLF equation [33] (detailed in Section 2.5.2). The change in viscosity due to temperature in the glassy state depends on the presence of a *hole* for the polymer segment to move into. Therefore, the presence of an energy barrier justifies the use of an activation energy. The log a_{T} versus $1/T$ plots for a secondary relaxation will be a straight line (not a curve as in the WLF case).

Below the glass transition temperature, the molecules are in a *non-equilibrium* state. Densification, also referred to as physical aging, can be observed over very long periods of time. The molecules tend to re-organize (decrease the free volume) and try to reach the equilibrium state. Several theories have been established concerning this complex process where "relaxation time depends on entropy and free volume, while the rate of change for both is controlled by the changing relaxation time. The relaxation process is coupled with thermodynamic change" [34]. Different theories have been established to describe this process (see Matsuoka [34] for details) and are beyond the scope of this book. However, this topic will be developed in a practical perspective in Section 2.4.1.5.

Most polymers and composites are being used in the glassy state, as the modulus is rather high and the mechanical values constant over a defined temperature range. In this region, most polymers (with the exception of chopped fiber polymeric composites which tend to exhibit creep over their lifetime) will generally exhibit an elastic behavior simplifying design and calculation tasks.

2.4.1.2 Glass transition region

The second region is called the *glass transition region* (also called *α transition*): the modulus of the material drops significantly (is generally divided by 10^3). The glass transition region is characterized by a steep drop in the polymer instantaneous or storage modulus. "Qualitatively, the glass transition region can be interpreted as the onset of long-range, coordinated molecular motion. While only 1–4 chain atoms are involved in motions below the glass transition temperature, some 10–50 chain atoms attain sufficient thermal energy to move in a coordinate manner in the glass transition" [24]. In mechanical analysis tests (see Section 2.8.3), the glass transition temperature is given by the peak of the loss tangent or the inflexion point in the modulus versus temperature resulting from quasi-static experiments. The glass transition temperature can also be determined by the point of discontinuity in the C_p (heat capacity), α (volume coefficient of expansion), or G'' (loss shear modulus) versus temperature curves [24].

The laws describing the material in this region are complex [35–41] and the materials properties change drastically. The material also has a very high creep rate. Most industrial applications will not use materials undergoing the glass transition during operation or off-duty. However, some specific applications will make use of this complex region: car and truck tires can be designed to operate within the glass transition temperature. In this region, the tangent delta (loss factor) reaches a maximum providing natural damping.

The glass transition region extends from the beginning of the modulus drop until a plateau. The breadth of the region can vary from one polymer to the other but is often around 20°C (rule of thumb for the glass transition region breadth). The temperature at which this transition occurs depends on the material. The factors influencing the glass transition temperature are further discussed in Section 2.4.2.1. Finally, it is to be noted that several theories are available to model and explain this transition region. However, the various theories (free volume theory, thermodynamic theory, kinetic theory [35–41]) are beyond the scope of a book focused on industrial applications and the explanations provided here are essentially phenomenological.

2.4.1.3 Rubbery stage

The third region is the rubbery stage. This region shows most often a stable value of E (around 3 MPa). This plateau corresponds to the long-range rubber elasticity. The length of the plateau increases with increasing molecular weight. The end of the plateau is characterized by the presence of a mixed region: the modulus drop becomes more pronounced but not as steep as in the liquid flow region. Short times are characterized by the inability of the entanglements to relax (rubbery behavior) while long times allow coordinate movements of the molecular chains (liquid flow behavior). The concept of reptation describing molecular motion in this region was initially introduced by De Gennes [42]. The polymer chain relaxation in the rubbery stage can be thought of as a wormlike (reptation) movement around obstacles or by a chain movement restricted inside a tube. The diffusion coefficient of the chain in the gel (D) is proportional to the molecular weight (M):

$$D \propto M^{-2} \tag{2.29}$$

Therefore, the relaxation time is found to be proportional to the third power of the molecular weight. Finally, this model leads to a proportionality of the steady-state viscosity proportional to the third power (3.4 empirically [24]) of the molecular mass while the modulus and the compliance are independent of the molecular weight (the number of entanglements are large for each chain and can occur at constant intervals). This concept has been particularly successful and will be given a special attention in our analysis (Section 2.4.1.5). For semi-crystalline polymers, the height of the plateau depends upon the percent of crystallinity of the material [43]. An increasing crystalline phase content leads to a higher modulus because crystallites act like physical cross-links by tying the chains together. For cross-linked materials, the plateau remains flat at a height corresponding to [44]:

$$E = \frac{\rho RT}{M_c} \tag{2.30}$$

where M_c is the molecular weight between cross-links.

Fillers and fibers are also impediment to reptation and help elevate and lengthen the rubbery plateau. If a material is completely cross-linked, this plateau extends up to the degradation temperature. Figure 2.16 shows the effect of cross-linking on the modulus versus temperature curve of polyisoprene [45].

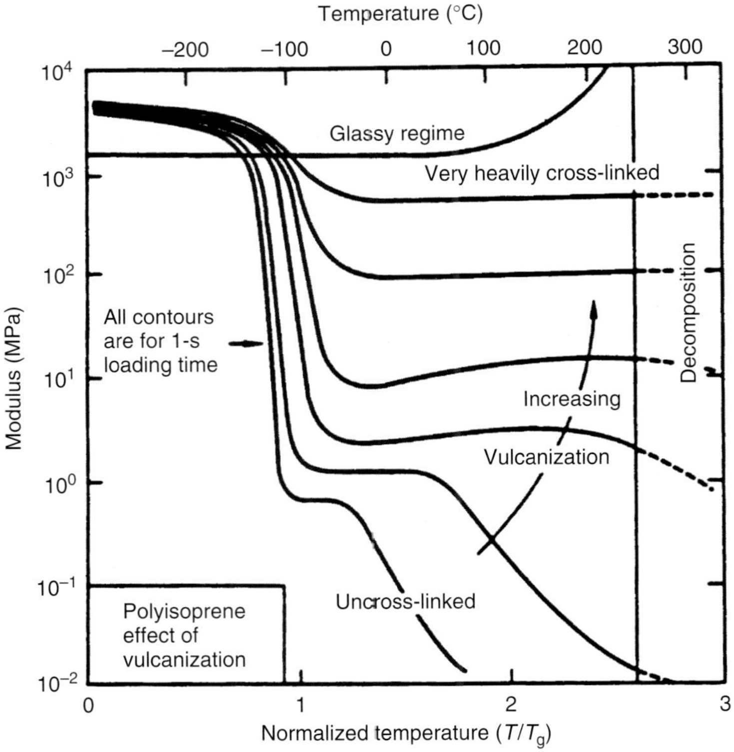

Figure 2.16. Influence of cross-linking on the modulus versus temperature curve. (Copyright 1986, *Engineering Materials* 2, by M.F. Ashby and D.R.H. Jones, Pergamon Press [45].)

By opposition, a linear polymer shows a progressive drop in the modulus. The length of the plateau in this case strongly depends upon the molecular weight of the polymer: the longer the molecules, the more the difficulties are encountered for coordinated motion and therefore the longer the rubbery plateau. Finally, for semi-crystalline materials, this plateau extends until the melting temperature of the polymer.

2.4.1.4 Rubbery flow

The last region is characterized by another sharp drop in the modulus. In the transition region, rubbery flow can be observed in which the materials response is very dependent upon strain rates. When the temperature is further increased, the material exhibits the properties of a liquid. If the melting temperature for semi-crystalline materials was not reached, the crystalline clusters still impart some rigidity to the material and impede to some extend the molecular flow of the amorphous phase. Fibers also contribute to a slowing down of the molecular flow.

Ultimately, the polymer becomes a viscous liquid and the modulus of the material drops dramatically. According to Sperling [24], the modulus of semi-crystalline polymers decreases quickly until it reaches the modulus of the corresponding amorphous material. Cross-linked polymers do not exhibit such a behavior, due to the presence of chemical primary bonds. The modulus of the polymer remains constant until degradation.

The use of polymers and composites at such temperatures is rather unusual for composite applications and relate to the science of rheology. We will therefore not detail those regions further.

2.4.1.5 Instantaneous versus time-dependent stiffness

Stiffness in composites can change due to a modification of the composite constituents (such as physical aging, Section 2.5.3) or microscopic and macroscopic damage (such as cracks). We will leave the later for discussion in Chapters 5 and 6 and we will for now focus on the stiffness changes in the matrix due to the environment or more specifically due to temperature. Indeed, we will find useful in the next chapters to express the instantaneous modulus as an explicit function of temperature over the entire operation range (from glassy to flow states). By instantaneous, we restrict the applicability of the model to *fast* loading: loading rates well in excess of the polymers relaxation rates.

The derivation of the modulus–temperature model is straight forward [46–48] and leads to the following general equation:

$$E = \sum_{i=1}^{N} H_i \exp\left(-\left(\frac{T}{T_{\mathrm{ref}_i}}\right)^{m_i}\right) \tag{2.31}$$

The number of transitions in a polymer depends on the nature of the material. In the present case, we will consider one to three transitions ($1 < N < 3$). The H_i

coefficients (magnitude of the transition steps) can be obtained by different means. We can subtract the value of the material stiffness before and after the transition, leading to Equation (2.32) (case of a material that does not undergo any transition prior to flowing), Equation (2.33) (material with two transitions, e.g. glass and flow) and Equation (2.34) (material undergoing three transitions, e.g. β, glass and flow):

$$E = E_1 \exp\left(-\left(\frac{T}{T_3}\right)^{m_3}\right) \tag{2.32}$$

$$E = (E_1 - E_3)\exp\left(-\left(\frac{T}{T_2}\right)^{m_2}\right) + E_3 \exp\left(-\left(\frac{T}{T_3}\right)^{m_3}\right) \tag{2.33}$$

$$E = (E_1 - E_2)\exp\left(-\left(\frac{T}{T_1}\right)^{m_1}\right) + (E_2 - E_3)\exp\left(-\left(\frac{T}{T_2}\right)^{m_2}\right) + E_3 \exp\left(-\left(\frac{T}{T_3}\right)^{m_3}\right) \tag{2.34}$$

The T_i (Figure 2.17) correspond to the transition temperatures in Kelvin (measured by DMA or DSC, see Section 2.8). The E_i (Figure 2.17) represent the intantaneous stiffness of the material at the beginning of each plateau or region. E_1 is the instantaneous modulus at very low temperature, E_2 is the instantaneous modulus immediately after the β transition, E_3 is the instantaneous stiffness at the beginning of the rubbery plateau. A reliable experiment leading to consistent values for stiffness measurements is the ultrasound method. The modulus obtained in this fashion corresponds to a uniaxial experiment performed at very high strain rate or cyclic frequency.

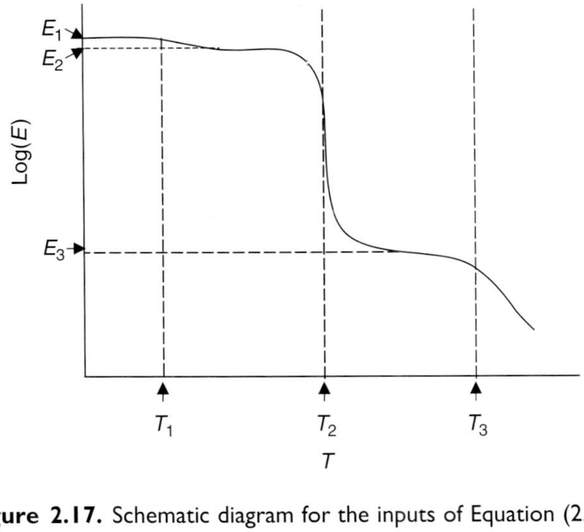

Figure 2.17. Schematic diagram for the inputs of Equation (2.34).

The drops in modulus in the different regions (H_i) represent the *importance* of the relaxation processes. These values depend on the chemistry of the polymer (stiffness of the backbone), molecular weight, crystallinity and degree of cross-linking. An increased crystallinity produces only a slight increase in the glassy state stiffness but can result in a large rise in the value of the modulus of the rubbery plateau. Therefore, the magnitude of the glass transition step will decrease significantly as the crystallinity is increased. Increased molecular weight and cross-linking will stiffen the material to a lesser extent.

The last parameters (m_i) are Weibull moduli and are assumed to correspond to the statistics of the bond breakage. To allow rotations of side groups for the secondary relaxations, the strengths of the bonds that need to be broken depend on the relative position of the side groups with respect to the other molecular chains. Therefore, there is a wide distribution of bond strengths and we expect m to be small. Reptation involves translation of the main chains. If the material is very homogenous (narrow distribution of bond strengths) as in the case of amorphous materials, we expect m to be very large. However, this parameter depends on the degree of impediment of the molecular motion (cross-linking, molecular weight and crystallinity, presence of fibers etc.). If the movement of the molecular chains is severely restricted at precise locations (by cross-linking, etc.), we expect m to be low (approaching a Boltzman distribution). For cross-linked materials, the slope of the drop in the viscous flow region decreases with increasing degree of cross-linking. For heavily cross-linked materials, the flow region can even disappear. The m_i coefficients were found to be more or less constant for the materials ($m_1 = 5$, $m_2 = 20$, $m_3 = 20$ seems to fit a very large number of materials). It is however recommended to fit the best coefficients for the chosen materials.

This model was successfully applied to a wide range of polymers and composites as illustrated by the case of carbon-fiber reinforced polyphenylenesulfide (AS4/PPS) in Figure 2.18. Level of crystallinity or reinforcement can be accounted for in the model [46–50]. The influence of the microstructure on the input parameters is summarized in Table 2.2.

Equation (2.31) is an *engineering* model, very useful for systematically considering temperature in micromechanical and macromechanical calculations (Chapters 5 and 6). However, it considers only very fast loading rates. Explicit relationships between stresses, strains and strain rates are given by the viscoelastic models of Section 2.3. Figure 2.19 shows the importance of combined time and temperature on the mechanical response of Polymethyl Methacrylate (PMMA). The data clearly illustrates a displacement toward higher temperatures and a broadening of the glass transition for short times (or high loading rates).

2.4.2 Transition Temperatures

Polymers and composites have very different behaviors and therefore mechanical or electrical responses under temperature. The responses such as modulus can vary by several order of magnitudes within a few degrees. Moving from one region to

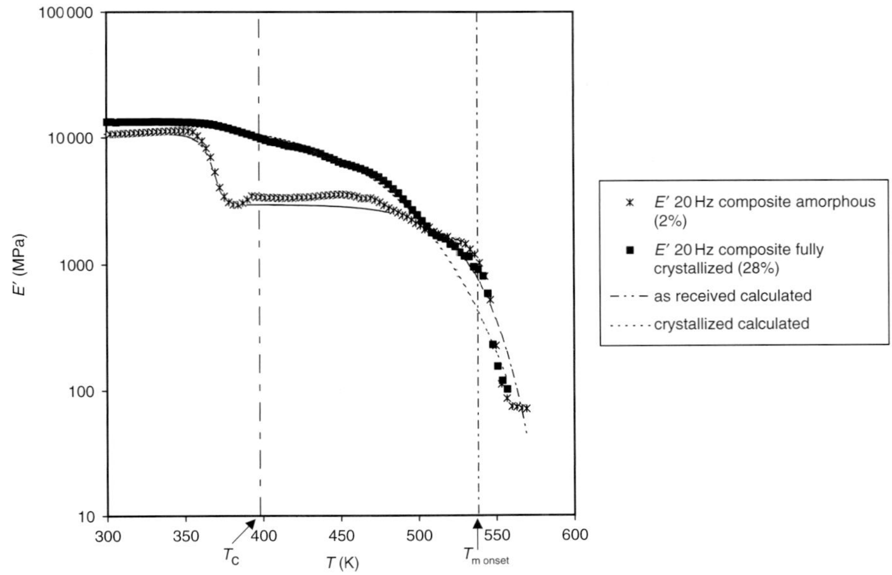

Figure 2.18. Experimental and theoretical results for various crystallinities of carbon-fiber polyphenylenesulfide (AS4/PPS) composite.

Table 2.2. Dependence of the input parameters on the microstructure

Depend on ⮫	Molecular weight	Crystallinity	Filler content
E_i	Yes	Yes	Yes
T_i	Yes	Yes	Yes
m_i	No	Yes	Yes

the other (e.g. from glassy state to rubbery state) is only possible by crossing a transition temperature. It is therefore necessary to understand the basic physical phenomenon causing those transitions in order to select or design industrial parts.

2.4.2.1 Glass transition temperature

The transition of main importance is the glass transition temperature (T_g) as it is characterized by sharp modifications in the materials properties. This temperature can be determined by different techniques. The most common being described in Section 2.8.

The values of the glass transition temperature depend on the definitions, testing methods and testing parameters. There is no universally accepted definition of T_g. In a DMA (see Section 2.8.3), some prefer to define T_g as the onset temperature

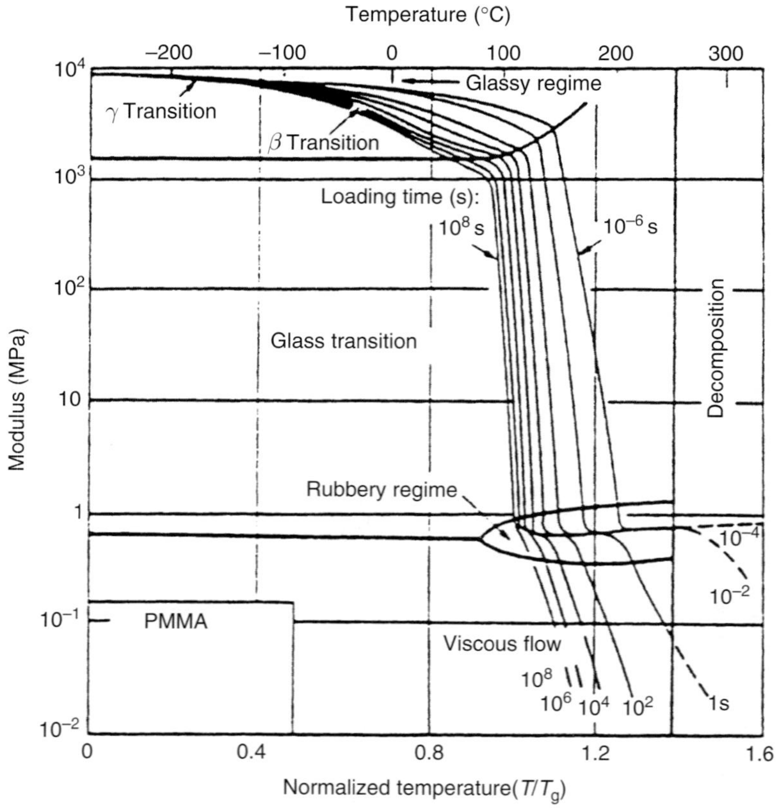

Figure 2.19. Modulus versus temperature – Combined time and temperature influence. (Copyright 1986, *Engineering Materials 2*, by M.F. Ashby and D.R.H. Jones, Pergamon Press [45].)

of the modulus drop. Others prefer to take T_g as the inflexion point in the storage modulus versus temperature curve or the temperature corresponding to the peak in the tangent delta versus temperature plot.

At the molecular level, the glass transition temperature feeds sufficient energy into the system to enable the onset of coordinated motion of large molecules. The amount of free volume *trapped* in the amorphous phase is significantly reduced (Figure 2.20).

The position of the glass transition temperature is very dependent upon the strain rate applied to the sample or part. Therefore, the glass transition temperature is not a *true* thermodynamic transition.

A common mistake is to refer to a composite undergoing the glass transition as *melting*. The glass transition only *acts* upon the amorphous phase of the polymer. Therefore, a 100% crystalline polymer (such as PE monocrystals) will never have a glass transition and a 100% amorphous polymer (such as PMMA or a rapidly quenched polymer) will never show a melting transition.

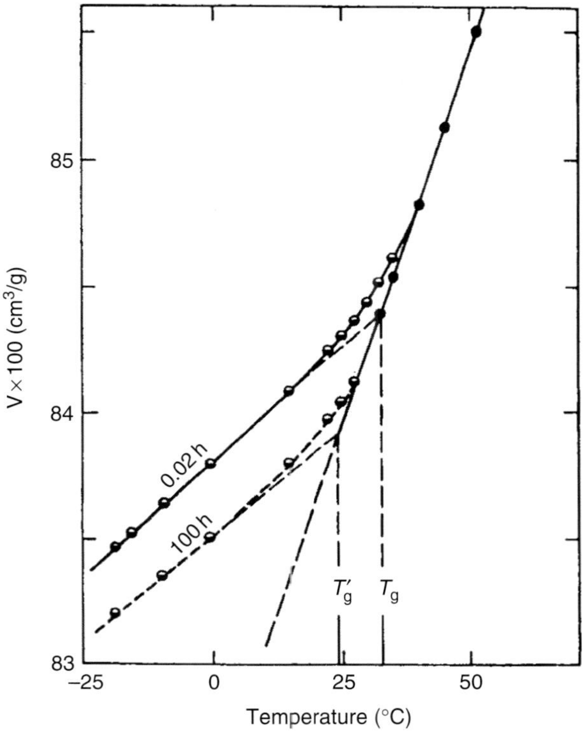

Figure 2.20. Specific volume versus temperature [11,51]. (Copyright 1983, *Introduction to Polymer Viscoelasticity*, by J.J. Aklonis and W.J. MacKnight, reprinted by permission of John Wiley & Sons.)

The glass transition is influenced by many parameters. The first parameter is the molecular nature of the polymer. Therefore, a broad range of T_g are available for the different polymers ranging from −123 to 525°C [11]. Examples of glass transition temperatures are given for various polymers in Table 2.3.

The data shown in Table 2.3 is only indicative. The glass transition temperature is very sensitive to a number of parameters: during the design process the glass transition can be influenced by tailoring the material. For example, a higher glass transition is obtained by:

- the presence of bulky side groups
- high molecular weight
- low plasticizer content
- high crystallinity content
- high filler content
- high degree of cross-linking
- high fiber content.

Table 2.3. T_g for various polymers (data from Azom [52])

Polymer	T_g (°C)	T_m (°C)
High density polyethylene (HDPE)		135
Polypropylene (PP)	−10	175
Polystyrene (PS)	100	–
Polymethylmethacrylate (PMMA)	105	
Polyvinylchloride (PVC)	65	–
Natural rubber	−75	25
Polycarbonate (PC)	50	–
Polyethylene terephtalate (PET)	70	265
Polyetheretherketone (PEEK)	145	335
Nylon 6 (PA6)	50	215
Polyamideimide (PAI)	295	
Polyphenylene sulfide (PPS)	90	285
Polyetherimide (PEI)	218	
Polytetrafluoroethylene (PTFE)	20	325

2.4.2.2 *Secondary transition temperatures*

Secondary transitions can be observed for certain polymers such as PE, PP etc. If there is only one secondary transition, the temperature will be referred to as β transition temperature. If there are two transition temperatures, the temperatures will be referred to as γ (for the lower temperature) and β transition temperatures. The spread and magnitude of such transitions is usually limited as it mainly involves the rotation of side groups in branched polymers. Therefore, no special care should be taken to avoid operation within those transitions. Modeling of the property changes can be done using the model of Section 2.4.1.5. However, modeling is only justified if the changes in properties are likely to impact the performance of the composite design (which is rather unusual).

2.4.2.3 *Melting temperature*

The melting and fusion temperatures are clearly defined: melting and fusion are *true* thermodynamic transitions. Those temperatures are independent upon experimental conditions and correspond to the peak values of exothermic (fusion) and endothermic (melting) reactions upon cooling and heating respectively.

As a rule of thumb, it was found that the glass transition and melting temperatures are related by Equation (2.35):

$$\frac{T_g}{T_m} = \frac{1}{3} \tag{2.35}$$

We previously mentioned that most polymers are used in their glassy state. It is therefore very unlikely that the melting temperature will impact the *operation* of the industrial part. However, this temperature is of prime importance for the *manufacturing* of the part.

Most thermoplastics can be recycled: the material is heated above the melting temperature then processed back into a new shape. If the degradation temperature is not reached, the recycled polymer will have excellent properties (almost no significant loss in properties is observed after re-melting of a thermoplastic).

The properties of the thermoplastic part depend strongly upon the degree of crystallization. In turn, the degree of crystallization depends on the cooling procedure. Rapid cooling, such as quenching in liquid nitrogen leads to amorphous materials. Very slow cooling leads to high crystallinity contents. Intermediate cooling rates usually lead to intermediate crystalline contents (Table 2.4).

For most industrial applications, the cooling procedure is selected to allow maximum crystallization. The degree of crystallinity in a sample can be obtained via density measurements or differential scanning calorimetry (DSC, see Section 2.8.2.2 for a detailed description).

Some polymers such as polyolefins can undergo further crystallization at temperatures below the normal (primary) crystallization temperature. Usually the rate of this secondary crystallization is much slower and can require years especially at low temperatures to become detectable. Part of the amorphous phase of the polymer trapped between the spherulites formed during the primary crystallization tend to re-organize to form further crystallites. In this new organized state, the volume occupied by the material decreases. This shrinkage results in microvoids and defects within the material. Such changes can be significant, for example, in the case of high voltage insulation. In Chapter 4 we will see that such voids can be very damaging to the insulation: they can act as ionization and eventually as tree growth sites in PE high voltage cables. Damage due to secondary crystallization in high voltage cables was observed at room temperature over several years [53]. Therefore, primary and secondary crystallization effects should be accounted for at the design state.

Table 2.4. Examples of crystallinity contents versus cooling rate [21]

Material	Cooling	Crystallinity (%)	T_g (°C)
Polyphenylenesulfide (PPS)	• Quenching (ice water)	2	92
	• In air	21	92
	• Slow, in controlled environment	52	102
Polyetheretherketone (PEEK)	• Quenching (ice water)	~0	144
	• Slow, in controlled environment	24	158

2.4.2.4 Gelation temperature

For thermoset materials, cross-linking can be obtained in different ways such as e-beam or UV radiation. The three-dimensional network can also be obtained by simple curing of the material, that is by maintaining the thermoset composite at a given temperature. Some materials vulcanize at room temperature, usually requiring storage in a cooled environment. However, most curing processes necessitate the use of a catalyst or a hardener. The cross-linking temperature (also called *gelation temperature*) varies and is not fixed for one material system. Specifically, this temperature interacts with the glass transition. A typical time–temperature-transformation diagram from [54] illustrates this concept (Figure 2.21).

The creation of cross-links has immediate effects on the materials properties, such as a stiffening of the material. The effects of gelation on the storage and loss shear modulus for a neat polymer are illustrated in Figure 2.22.

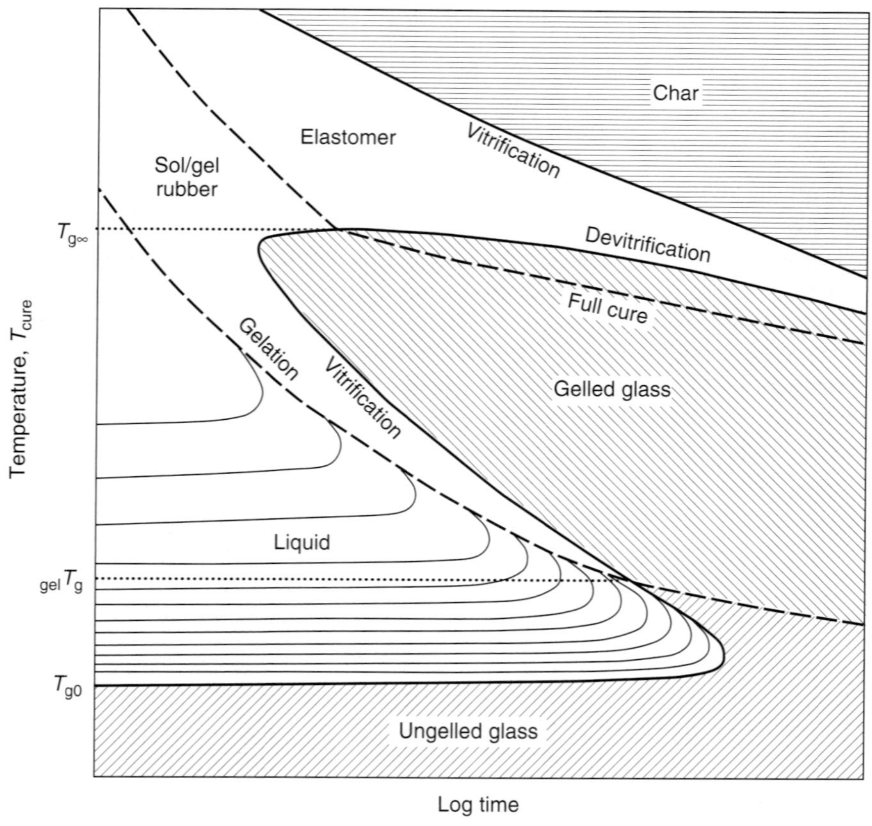

Figure 2.21. Time–temperature-transformation diagram for a thermosetting system from Gillham [54]. (Copyright 1986, *Encyclopedia of Polymer Science and Technology*, by J.K. Williams, reprinted by permission of John Wiley & Sons.)

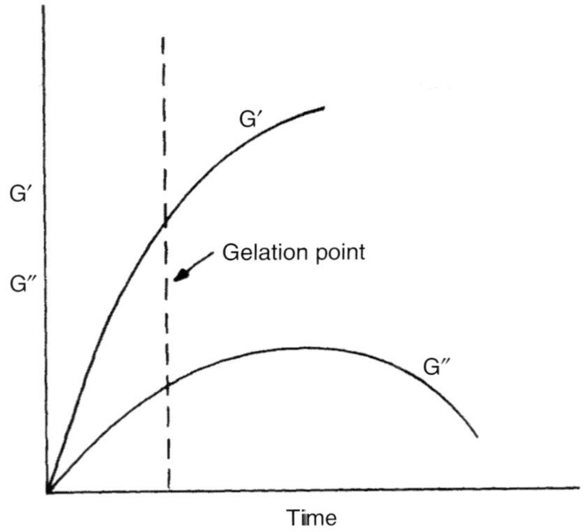

Figure 2.22. Shear moduli evolution on curing [24]. (Copyright 1992, *Introduction to Physical Polymer Science*, by L.H. Sperling, reprinted by permission of John Wiley & Sons.)

In some industrial applications the composite is voluntarily partially cured in order to allow for more flexibility during installation or to reduce production time. This process is successfully used in the case of some parts of the winding insulation of hydrogenerators (see Chapter 4) and in situ curing sewer pipes described in Chapter 3. Final curing occurs after a certain operation time. Generally speaking, in the cured state, the composite will exhibit a higher strength, higher stiffness but also a higher brittleness. Therefore, special care should be taken in this process to ensure proper consideration of the property changes during the partially cured to fully cured state transition.

2.4.2.5 Degradation temperature

Carbon and glass fibers degrade at temperatures in excess of 1000°C. The degradation temperature for polymers is well below these temperatures (between 300 and 500°C for most polymers). For example, glass melts at 1735°C [8]. Carbon fibers can be used up to 2500°C when protected from oxygen but only up to 500°C when exposed to air. Organic fibers such as aramid fibers (Kevlar) can be used only up to a 300°C [8].

This implies that in many cases the degradation temperature of the polymer matrix composite is given by the degradation temperature of the polymer. Above the degradation temperature, the material undergoes irreversible damage characterized in the case of polymers by molecular chain scission. This phenomenon will be more detailed in Section 2.7, describing the effects of fire on polymer matrix composites.

2.4.2.6 Other engineering temperatures

Materials datasheets from suppliers often mention other temperatures such as maximum operation temperature (Table 2.5). However, this definition can vary and is rather subjective. The operation time also influences a polymer's maximum use of temperature. Though considering this value in the material selection screening process, it is recommended to use real transition temperatures as references in design and lifetime calculations.

2.4.3 High Temperature Polymers

Toughness and low density often make fiber reinforced polymer matrix composites attractive candidates for many applications including high temperature parts. High temperatures for metal or ceramic matrix composites usually refer to the 1000–1500°C range. The terminology is quite different for the field of polymer-based materials where *high-temperature* polymer composites traditionally operate in the 150–400°C range.

For sustained temperatures in excess of 150°C, thermoset materials are often preferred. High degrees of cross-linking usually lead to high glass transition temperatures and high stiffness. High cross-linking degrees can also adversely affect the composite by increasing the materials brittleness. The history of high temperature commercial thermosets has been reviewed in the literature [55] and many texts focus on this specialized topic [56–58]. High temperature polymer matrix classes include polyimides, bismaleimides (BMIs), cyanate esters and phenolics. Relevant properties and characteristics of selected high temperature polymers are summarized in Table 2.6. It is worth noting that the development of those polymers was originally motivated by aerospace applications. Propagation to other industrial applications is recent and still extremely limited.

Table 2.5. Maximum operation temperature for selected polymers and composites [20]

Material	Typical recommended maximum operation temperature in air (in °C)
Glass fiber/Epoxy	200–300
Polyetheretherketone (PEEK)	154–315
Glass fiber/PEEK	188–316
Carbon fiber/PEEK	221–325
Polyimide	340–360
Graphite/Polyimide	288–480

Table 2.6. High temperature polymer composite examples

Material	Glass transition temperature (°C)	Continuous use temperature (°C)	Comments and typical applications
Polyimides			
RP-46 (Unitech Corporation) [55]	397	357	• Quartz fiber/RP-46: Radome applications
SuperImide (Goodrich Corp.) [59]	350	343 Acceptable peak temperature: 600°C	• Bulk molding powder/graphite reinforced: bushings, bearings, wear surfaces. • Bulk molding compound/graphite, quartz and glass-fiber reinforced: seals, housings, clamps. • Prepreg composites/graphite, quartz and glass-fabric reinforced: aerospace structures, electronics, spacers.
Phenylethynyl (PETI-330, UBE Corp.) [55,60]	330	288	• Fiber reinforced/PETI-330: Primary structural applications in airframes and jet engines
BMIs			
F650 (Hexcel) [61]	316		• Fiber reinforced/F650: Applications in military aircraft and helicopters
CYCOM-5250-4 (Cytec) [61]	271	204	• Fiber reinforced/ CYCOM-5250-4: wing and stabilizer spars, fuselage skins and stiffeners, low operating temperature, critical load bearing components and engine components
Cyanate esters	~350	~180	• Reinforced cyanate esters: Primary aircraft structures, radomes and space systems
Phenolics	~160		• Electrical equipment, carbon/carbon composite applications

2.5 TIME–TEMPERATURE EQUIVALENCE

Section 2.2.3 has introduced the notion of viscoelasticity. It was seen that the macromolecules in the amorphous phase of the polymer tend to re-arrange if the composite is stressed or strained over long time periods (see Sections 2.3.2.1 and 2.3.2.2). Now, if the material is simultaneously exposed to elevated temperatures, more energy is being fed into the system. The molecular re-arrangement still takes place, but this time at a higher rate. Intuition tells us that, with respect to molecular re-arrangement, long periods of time and high temperatures should lead to similar effects but at different rates. Finding the rules driving this equivalence is therefore key in allowing accelerated testing. As promising as this sounds, we should however warn the reader that the modeling of such equivalence is far from trivial. To introduce the time–temperature equivalence concept, we will first focus on the most widely used equivalence method (the WLF equation), keeping in mind that Arrhenius' equation can be found more applicable. We will then discuss its limits and reflect on the opportunities and challenges of accelerated testing.

2.5.1 Time–Temperature Superposition

If we measure the modulus of a polymer at a constant temperature T_1 (isothermal measurement), the modulus will drop with time. However, large time spans can be involved if the temperature is low (e.g. below the glass transition temperature T_g). If we now measure the modulus of the same polymer at a higher temperature T_2 ($T_2 > T_1$), the modulus for an identical period of time will be lower and show an earlier drop with time. This experiment can be performed for a constant time period (Δt) and for several temperatures T_0, $T_1, \ldots,$ T_i. Typical results of such experiments are shown in Figure 2.23. The curves can be shifted along the time axis, resulting in a continuous curve (Figure 2.23). This curve represents the modulus versus time for the polymer at the reference temperature T_3. T_3 is subjectively selected. In an industrial context, the reference temperature is often taken as the

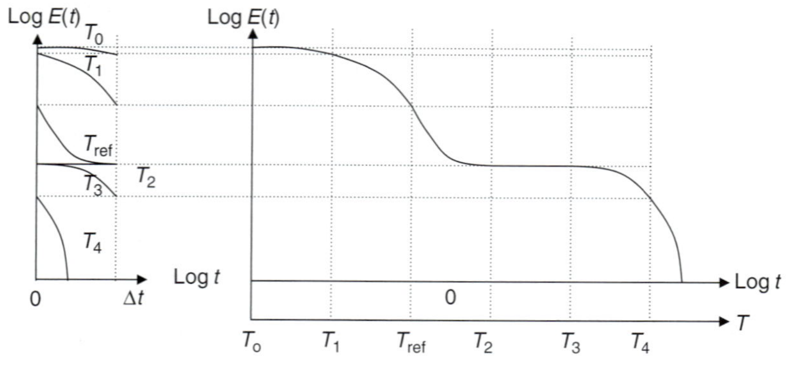

Figure 2.23. Time–temperature equivalence principle (Master curve).

glass transition temperature. Mathematically, this time–temperature equivalence can be written as:

$$E(T_1, t) = E(T_2, t/a_T) \tag{2.36}$$

where a_T is called the *shift factor* and can be estimated thanks to Equation (2.38).

Such a time–temperature superposition principle mainly holds for linear, unfilled polymers. Aklonis and MacKnight [11] propose a correction to account for the vertical shift due to the temperature variation, by introducing a normalization by the density ρ (which is proportional to the temperature). The significance of the correction factor will have to be evaluated for industrial modeling purposes that already consider large safety coefficients.

$$\frac{E(T_1, t)}{\rho(T_1)T_1} = \frac{E(T_2, t/a_T)}{\rho(T_2)T_2} \tag{2.37}$$

2.5.2 WLF Model and Limits

Williams, Ferry and Landel introduced an equation relating the shift factor a_T to the temperature and reference temperature. Original and further derivations of the WLF equation can be found in [11].

$$\log a_T = \frac{-C_1(T - T_g)}{C_2 + T - T_g} \tag{2.38}$$

The values for C_1 and C_2 are constant for a given material, but vary slightly from one polymer to the other. For linear amorphous polymers above the glass transition temperature, $C_1 \sim 17.44$ and $C_2 \sim 51.6$ [24]. However, values in excess of 100 can be found for some polymers such as polyisobutylene.

Practically, the WLF equation (Equation (2.38)) enables us to predict the modulus at any time from the modulus versus temperature curve and the shift factor of the material (keeping in mind that generally Equation (2.38) is applicable within $\pm 30°C$ about the glass transition temperature). It is also worth mentioning that the horizontal shift of Equation (2.36) that we are currently discussing holds only for linear, unfilled polymers. Not only intrinsic parameters (such as materials nature) but also extrinsic parameters (such as moisture) can introduce vertical shifts. In this case, Equation (2.37) might help obtain a better fit to the data. In any case, if the vertical shift is not negligible, time–temperature equivalence and accelerated testing should be used only if the root causes for the vertical shift are understood and properly modeled.

2.5.3 Physical Aging

Most polymer matrix composites are used in the glassy state. In the glassy state, the motion of the molecules is restricted. Amorphous materials are obtained by *freezing* molecules in a non-equilibrium state (most often by quenching the material).

However with time, the molecules will attempt to re-arrange in order to tend toward an equilibrium state (higher level of order). This process is accompanied by changes in the mechanical properties of the polymer. The density of the composite increases, which tends to corroborate the theory linking physical aging and a reduction in the free volume of the polymer (Figure 2.24).

For materials used for long-term applications, physical aging can lead to significantly altered materials properties (Figure 2.25) and its impact should therefore be systematically investigated. To do so, accelerated methods might become a necessity.

2.5.4 Accelerated Testing

Accelerated testing is common practice in the field of polymeric materials. Accelerated testing is often a necessity, as polymers and composites are designed for

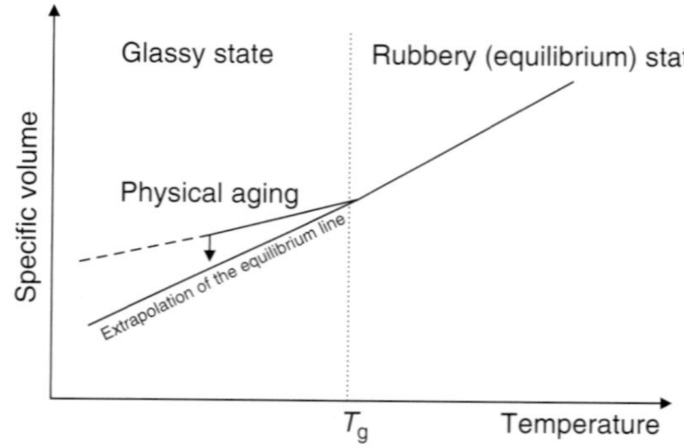

Figure 2.24. Physical aging and specific volume.

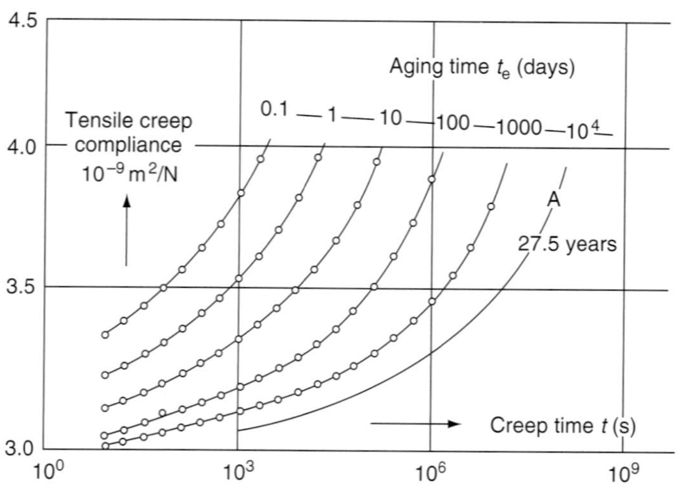

Figure 2.25. Effects of aging on creep compliance. (Copyright 1978, *Physical aging in Amorphonous Polymers and Other Materials*, by L.C.E. Struik, Elsevier Applied Science [62].)

lifetimes of several years that exceed laboratory experiment time frames. It would be unrealistic to require testing over an entire lifetime before the commercialization of such products. The literature dealing with accelerated testing is abundant [63–65] but the norms are very sparse in this field. Therefore, we should add the following guidelines which might help in establishing a valid accelerated testing program.

We have already mentioned that the molecular motion and therefore the polymer response to stresses is very different above and below the glass transition temperature. If a material is to be used in the glassy state, accelerated testing should be performed below the glass transition temperature. Especially in the case of composite materials, the damage and failure mechanisms at the various operation and test temperatures should be analyzed. *An accelerated test is only valid if no change in the damage mechanisms occurs over this temperature range.*

Gates [63] summarized the most common methods for accelerated testing of composite materials and proposes the following methodology. Gates considers three primary aging mechanisms: chemical, physical and mechanical. These mechanisms can act individually or interact. A key to accelerated testing is that the mechanisms should be consistently reproducible. While it is not possible to test all conditions and their combinations most of the time, Gates proposes to identify easily measurable key degradation mechanisms. Those mechanisms will be measured, thanks to metrics such as weight, mechanical properties etc. Real-time data provide information on the critical degradation mechanism that will serve as basis for testing. To guide those choices it is strongly recommended to use the load combination scheme and the design of experiment (DOE) method of Chapter 6.

Acceleration theories are based on superposition models such as:

(a) The time–temperature superposition model of Section 2.5.2.
(b) The Boltzmann superposition principle. The Boltzmann superposition principle (Figure 2.26) states that if we consider n stress increments σ_i applied at times t_i the stresses act independently and the resulting strain results from the linear addition of the individual strains [11].

$$\varepsilon(t) = \sum_{i=1}^{n} \sigma_i D(t - t_i) \tag{2.39}$$

Or replacing the summation:

$$\varepsilon(t) = \int_{-\infty}^{t} \frac{\partial \sigma(x)}{\partial x} D(t - x)\,\mathrm{d}x \tag{2.40}$$

with x the variable.

(c) The time-aging-time superposition. The Kohlraush Williams Watts equation (KWW) [37,38] models the modulus evolution with time:

$$E(t) = E_{t=0} \exp\left(-\left(\frac{t}{\tau^*}\right)^{1-n}\right) \tag{2.41}$$

where the parameter n indicates the degree of intermolecular coupling and τ^* is the characteristic segmental relaxation time.

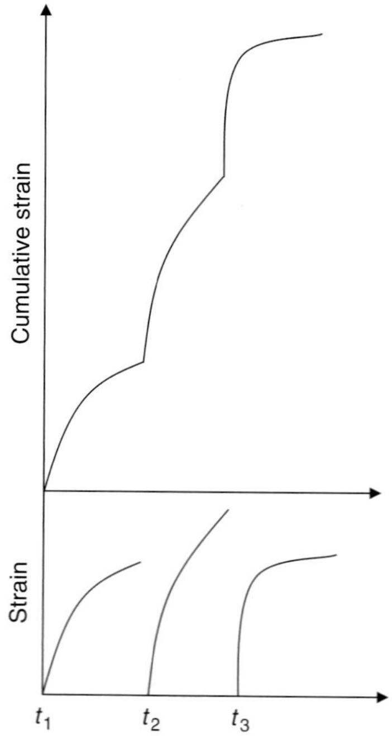

Figure 2.26. Illustration of the Boltzmann's superposition principle.

From this equation and assuming that the shift rate depends on temperature and will tend toward zero when the material approaches the glass transition temperature, Gates [63] proposes an equation for the compliance D to allow "the prediction of long-term behavior based solely on material parameters determine from short-term tests":

$$D(t) = D_{t=0} \exp\left(\frac{\int_0^t a_{t_e^0}(\xi)\mathrm{d}\xi}{\tau(t_e^0)}\right)^{\beta} \tag{2.42}$$

where $\beta = 1 - n$ and the shift factor depends on the shift rate μ:

$$a_{t_e^0}(t) = \left(\frac{t_e^0}{t_e^0 + t}\right)^{\mu} \tag{2.43}$$

where t_e^0 is the initial aging time.

To combine the effect of temperature and aging, Gates [63] proposes a time–temperature–aging-time superposition scheme based on Sullivan [66].

$$\log a = \log a_{t_e} + \log \overline{a_T} \tag{2.44}$$

with a_T being the time–temperature shift factor that depends on aging:

$$\frac{\overline{a}_{T_1/T_2}^{t_{e_2}}}{\overline{a}_{T_1/T_2}^{t_{e_1}}} = \left(\frac{t_{e_2}}{t_{e_1}}\right)^{\mu(T_2)-\mu(T_1)} \tag{2.45}$$

These equations can be directly used in the micromechanical and macromechanical models of Chapter 5. An alternate possibility for combination of aging and temperature is the use of the combination scheme of Chapter 6.

The use of such equations is scientifically sound but can lead to difficulties when used on complex systems. For example, some parameters such as aging rate and glass transition temperature are inter-dependent. When aging occurs, glass transition will vary which will in turn affect the aging rate. A further challenge is the integration of the size effect and free surfaces.

It is therefore recommended to run experiments on systems as close as possible to real operation. When such testing is not feasible based on cost or time considerations, literature, combination schemes and experience should be used to assess the limits of validity of the prediction. Engineers dealing with composite materials should always keep in mind the principal factors limiting the validity of accelerated tests, among those the fact that:

(a) A sample under a higher stress level for a short period of time generally accumulates less damage than a sample under a low stress level for a long time period.
(b) Small samples are not the real part and no global scheme that allows for sizing is available.
(c) The damage mode has to be the same in the real case and in the accelerated tests.

And to finish up with an anecdote, an egg maintained at 39°C in 50% relative humidity (RH) for 20 days will result in a chick. If we now accelerate the process by putting the egg in water at 100°C for a few minutes, we will most certainly obtain a boiled egg. Accelerated testing can definitely lead to rather unexpected results!

2.6 FURTHER TEMPERATURE EFFECTS ON COMPOSITE PROPERTIES

2.6.1 Strength and Other Properties

For *matrix*-dominated composites such as short-fiber composites, the major contribution to the temperature dependence originates from the processes described in Sections 2.2–2.5. On the other end of the spectrum, the so-called *fiber*-dominated composites are affected by the matrix behavior to a lesser extent. When a unidirectional carbon-fiber reinforced epoxy is loaded in the fiber direction, it is often assumed that its behavior is *fiber*-dominated, that is the stiffness and strength are equal to the properties of the fibers. Indeed, in the temperature range considered (polymer operation temperature), carbon fibers do not see significant variations in their elastic or strength properties. However, several experimental studies [21,23,48] clearly report an influence of temperature on the strength and stiffness of unidirectional composites.

A few examples were selected to illustrate such phenomena. Figure 2.13, for example, shows the variation of stiffness with temperature of a carbon-fiber reinforced vinyl ester composite with polyurethane interface. Figure 2.27 illustrates the strength variations of a graphite epoxy composite: the strength of the composite is halved over 200°C. Figure 2.28 in which the unidirectional carbon-fiber polyphenylene sulfide (AS4/PPS) sample strength drops by 25% when the temperature is raised from 20 to 140°C confirms this behavior. The reason for such a temperature-dependent response can be attributed to the matrix molecular creep and/or a modified matrix–fiber interface. Indeed, an increase in temperature can weaken the interface and decrease its ability to transfer stress between polymer and fibers [21]. The literature dealing with temperature effects on the matrix–fiber interface is sparse [22,23,49] and conclusions are difficult to generalize. Whether the temperature dependence of the composites response originates from matrix relaxation or interface modification, the different figures clearly indicate that strength variation with temperature may be significant and should not be neglected, even in the fiber direction.

Beyond stiffness and strength, temperature also influences other polymer materials properties, such as the Poisson's ratio or the coefficient of thermal expansion (see Figures 2.29 and 2.30 for a graphite/epoxy and a carbon/epoxy composite).

However, for properties other than stiffness and strength, very few general models are available to predict the materials response with temperature. Rosato [68]

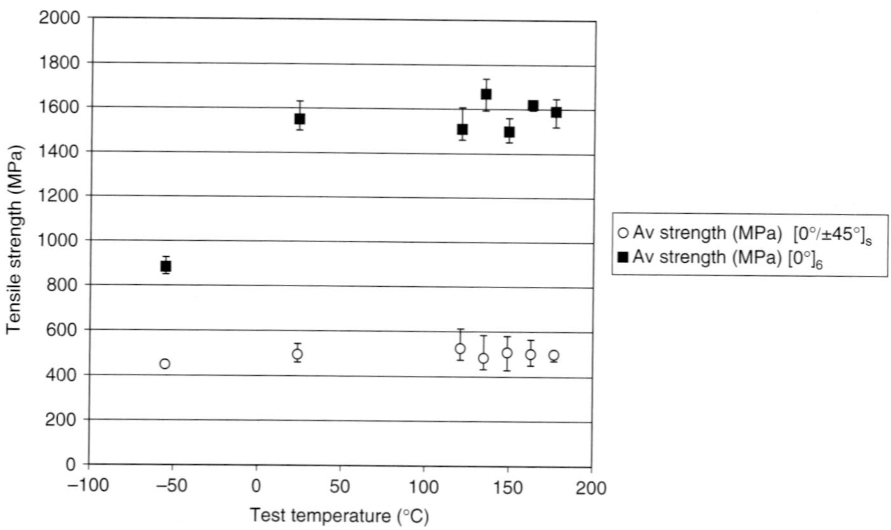

Figure 2.27. Strength of graphite epoxy composite versus temperature. (Data from Kerr and Haskins [22]).

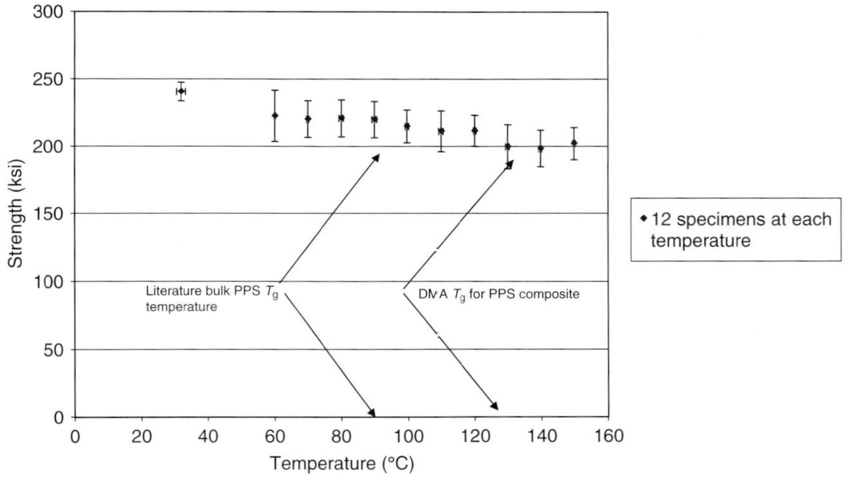

Figure 2.28. Strength versus temperature for unidirectional carbon-fiber reinforced polyphenylene sulfide (AS4/PPS) [47]. (Copyright 2001, *Applied Composite Materials*, by C.A. Mahieux et al., reprinted with kind permission of Springer Science and Business media.)

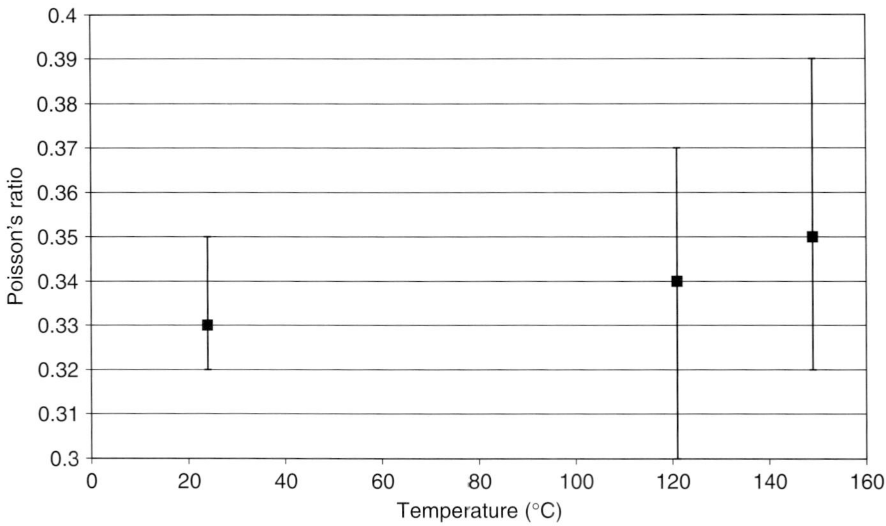

Figure 2.29. Poisson's ratio of graphite epoxy composite versus temperature. (Data from Kerr and Haskins [22].)

proposes general trends illustrating typical property variation with temperature (Figure 2.31).

Beyond the *instantaneous* effects of temperature on composites, it is necessary to consider two important additional mechanisms: residual stresses and physical aging.

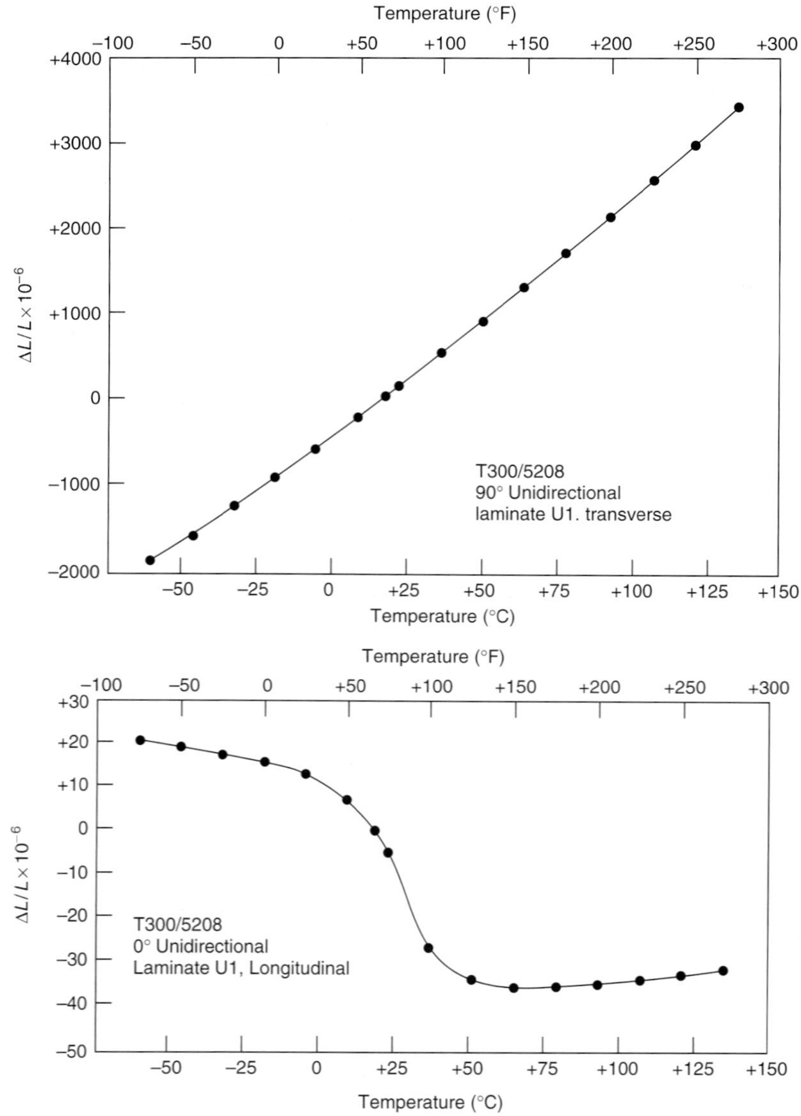

Figure 2.30. Thermal expansion coefficients of T300/5208 carbon/epoxy laminates [67]. (Copyright 1984, *Environmental Effects on Composite Materials*, Vol. 1, by G.S. Springer, reproduced by permission of Routledge/Taylor & Francis Group, LLC.)

Indeed, laminates are often manufactured by applying pressure (or vacuum) and temperature on a ply stack. When the plies have different fiber orientations, thermal stresses are created. Details on how to calculate such stresses are given in Chapter 5. Stresses remaining after manufacturing are referred to as residual stresses. Figure 2.32

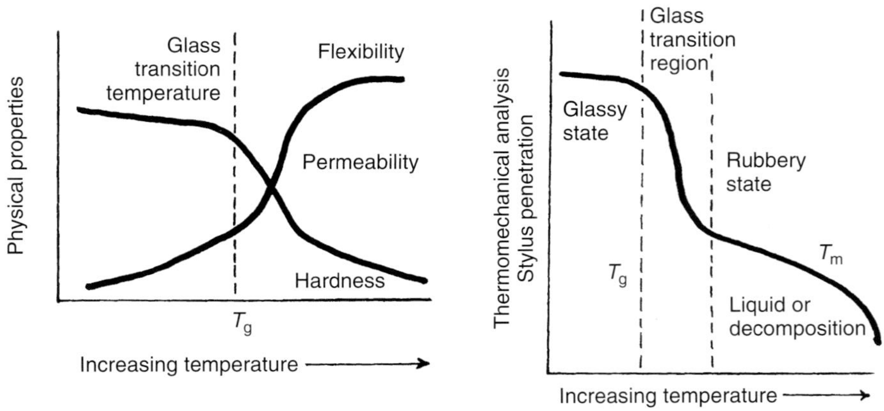

Figure 2.31. Effect of temperature on general polymer matrix composite properties. (Copyright 1991, *Designing with Plastics and Composites*, a handbook, by D.V. Rosato and D.P. DiMattia, Van Nostrand Reinhold [68].)

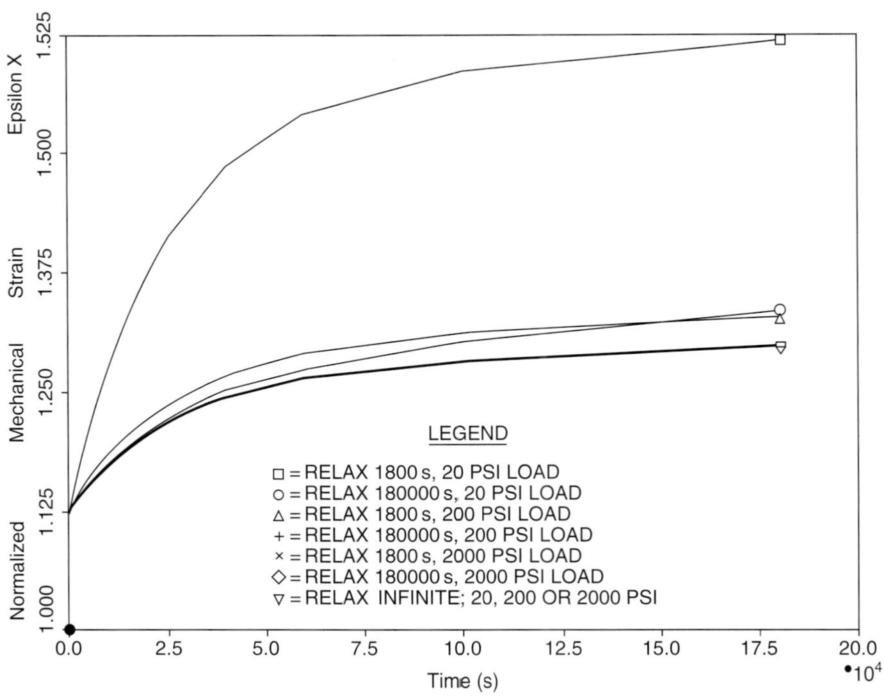

Figure 2.32. Effect of residual thermal stress relaxation on creep behavior of $[\pm 45°]_s$ GY70/339 graphite composite laminates at different mechanical load levels [69]. (Copyright 1984, *Environmental Effects on Composite Materials*, Vol. 1, by G.S. Springer, reproduced by permission of Routledge/Taylor & Francis Group, LLC.)

from [69] shows the effects of residual thermal stresses on the creep behavior of a [±45°] graphite composite laminate (GY70/339) at various load levels.

Additionally, aging will affect the composite matrices according to the laws described in Section 2.5.3. These property modifications will of course impact the global response of the composite. Indeed, aging can lead to drastic changes in the strength of materials. Figures 2.33 and 2.34 illustrate the impact of aging time on a $[0°]_6$ and a $[0/\pm45°]_s$ carbon-fiber/epoxy laminate. The laminates were aged under two distinct stress levels: $0.1\,MN/m^2$ and $0.014\,MN/m^2$. For the unidirectional graphite fiber/epoxy, Kerr and Haskins [23] report a 30% strength monotonous drop over a 100 000 hour aging time period at 450 K (177°C). The results evidence the interaction between aging conditions and laminate type (lay-up). For example, an increase in the aging stress increases the strength degradation rate for the $[0°]_6$ sample but does not significantly influence the response of the $[0/\pm45°]_s$ laminate. The failure mode can also be affected by the aging time as illustrated by Figure 2.35. With no or short aging times, the samples break via the propagation of a crack through the width of the sample (across the fibers). After a long exposure time, however, the samples exhibit splitting along the fibers, a failure characteristic of a weakened matrix and interface.

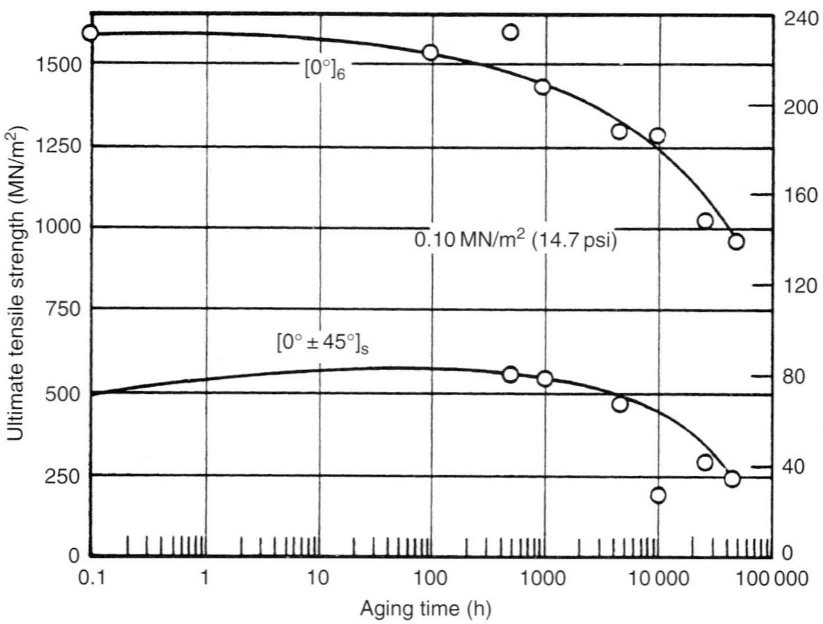

Figure 2.33. Effect of aging on strength of a graphite epoxy composite at 450 K (177°C) after thermal aging in $0.1\,MN/m^2$ air at the same temperature. (Data and graphs from Kerr and Haskins [22].)

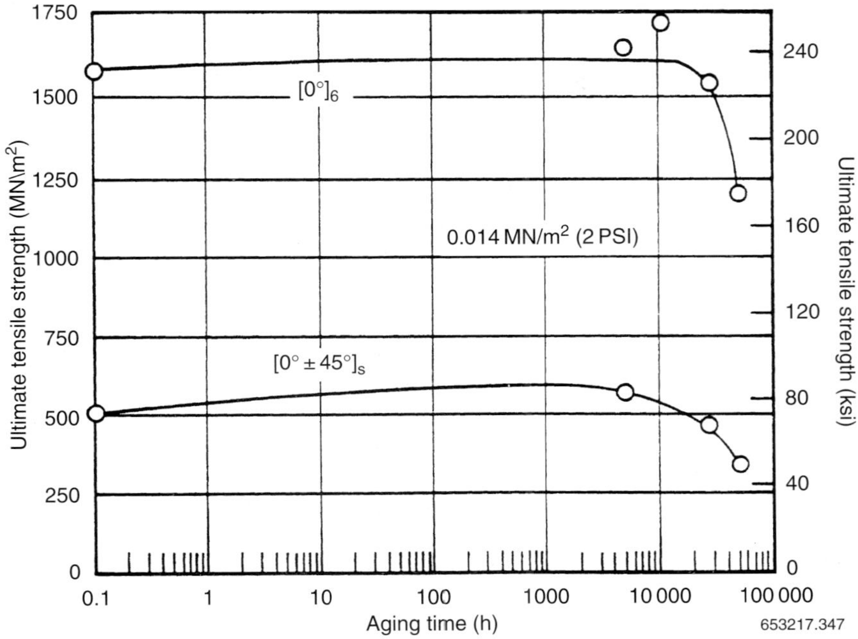

Figure 2.34. Effect of aging temperature on strength of a graphite epoxy composite at 450 K (177°C) after thermal aging in 0.014 MN/m^2 air at the same temperature. (Data and graphs from Kerr and Haskins [22].)

2.6.2 Composites, Time and Temperature – Common Pitfalls and General Precautionary Rules

Most traditional models (especially FE models) rely on Hook's law and linear elastic relationships. A careful check should be performed before applying those to composites in terms of linearity of the response and time dependency. We need to clearly separate these two concepts. To do this, Schapery et al. [70] proposes an approach. That is, the material should be tested at different stress levels. "If the material is linear but viscoelastic there will be no dependence on stress level and the compliance will only be a function of time". For example, Figure 2.36 clearly shows a non-linear (stress-level dependent, (b)) as well as a time dependent (a) response of the selected composite (a continuous AS4C carbon-fiber reinforced rubber-toughened epoxy resin Fiberite E719T).

Dealing with a time-dependent response can be challenging. For example, within a composite, a polymer matrix in its glassy state at a time $t = 0$ (such as design time) can switch into its rubbery state after a few years. The modulus may drop by a noticeable amount in this transition. Therefore such changes should be anticipated and design checked in both conditions. Rupture mechanisms may also

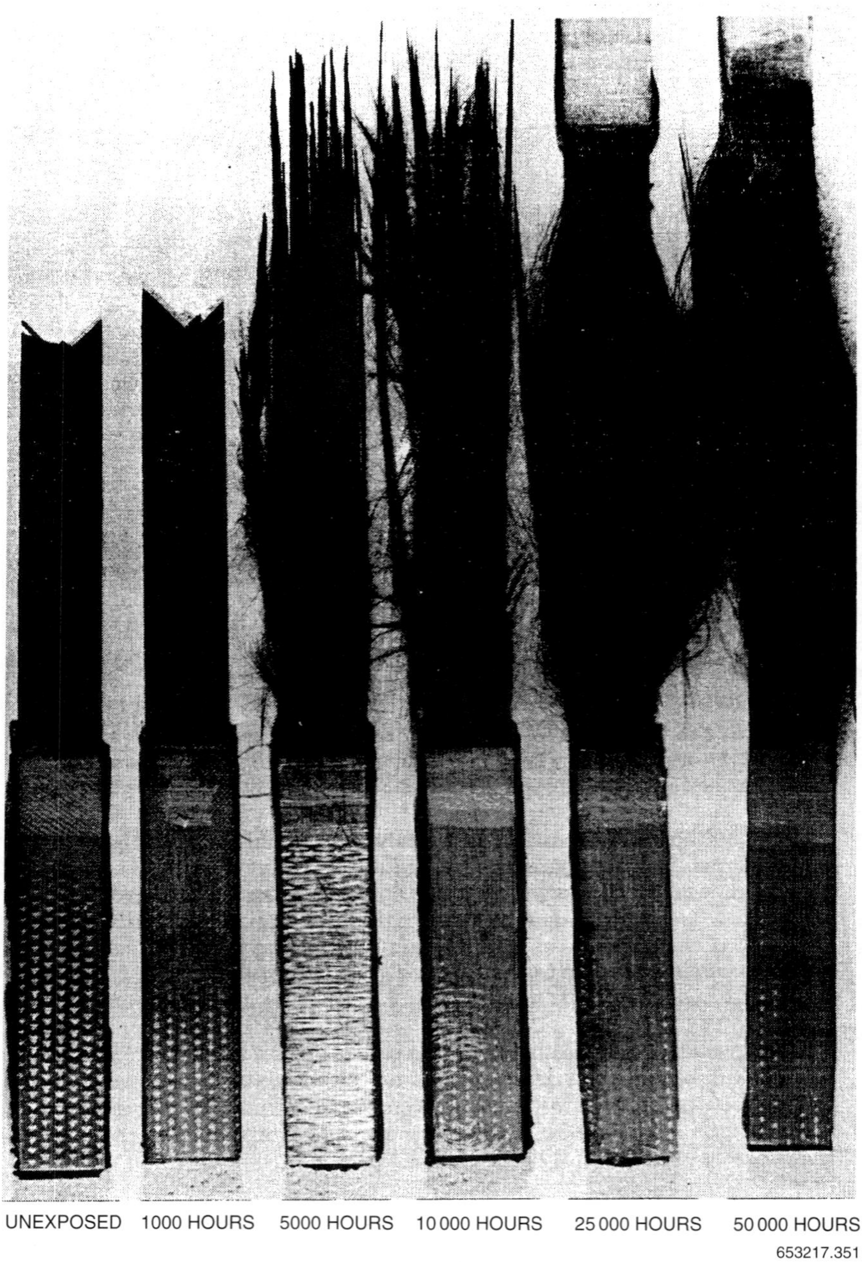

Figure 2.35. Various failure modes for aged specimens. Graphite epoxy specimens before and after thermal aging at 450 K (177°C) in one atmosphere air and tensile testing at the same temperature. (Picture from Kerr and Haskins [22].)

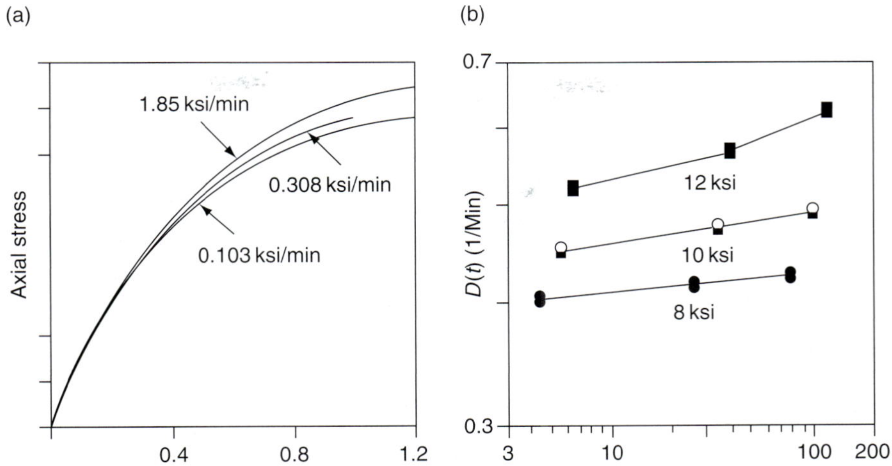

Figure 2.36. Non-linear viscoelastic behavior. (a) Axial stress–strain response of three 30° off-axis carbon-fiber reinforced rubber-toughened epoxy specimens loaded with three different loading rates. (b) Axial compliance versus time for constant levels of stress as taken from three 30° off-axis specimens tested at different stress levels. (Copyright 1999, *Composite Materials for Offshore Applications*, by S.S. Wang et al. [70] reproduced by permission of the American Bureau of Shipping.)

be fundamentally different in both states and should be investigated separately if the risk exists to move from one region to the other.

Accelerated testing is only applicable if the same damage or failure mechanisms apply over the operating range and lifetime of the composite part. Therefore, as a general rule, accelerated testing for a material used in the glassy state should not be performed above the glass transition temperature.

2.7 COMPOSITE EXPOSURE TO EXTREME TEMPERATURES

Reported investigations of the effects of low temperatures on composite materials are limited [71–75]. At low temperature, the polymer becomes more brittle and exhibits enhanced stiffness values. For example, the carbon black filled rubber of Figure 2.37 has a modulus of 251 MPa at 130 K(−143°C) and only 5 MPa at room temperature (Figure 2.37).

Figure 2.38 shows a detrimental effect of low temperature (−50°C) on the strength of a graphite epoxy composite (from Kerr and Haskins [22]). The brittleness of the material at such temperatures is a challenge to be solved for cryogenic industrial applications. Such applications range from the storage and transportation of liquid oxygen to fuel cells and space-launch technology. Recent developments at Northrop Grumman Corp and Lokheed Martin in cooperation with NASA lead to the development of cryogenic tanks obtained by innovative manufacturing processes and surviving pressure cycles in the laboratory simulating the life of a reusable tank.

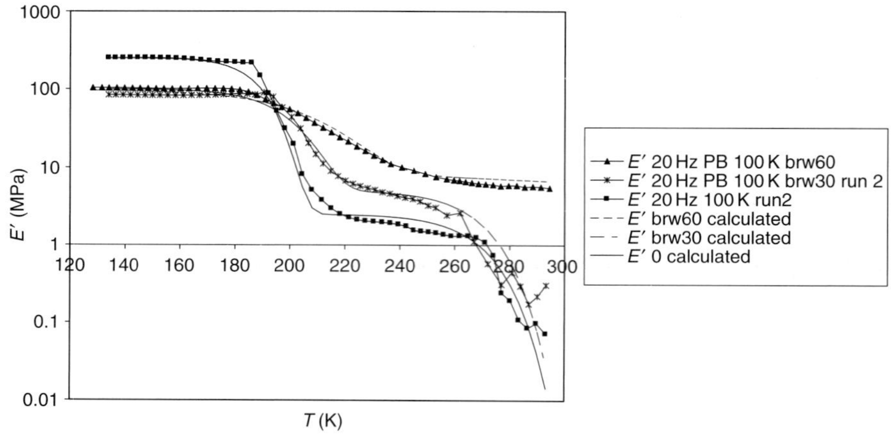

Figure 2.37. Experimental and theoretical results for polybutadiene with different contents of carbon black. PB 100 K is an unfilled polybutadiene, PB 100Kbrw60 is a polybutadiene filled with 37.5% carbon black by weight, PB 100Kbrw30 is a polybutadiene filled with 23% carbon black by weight. The continuous lines were obtained from Equation (2.31). (Copyright 2002, *Journal of Elastomers and Plastics*, by C.A. Mahieux and K.L. Reifsnider [50], reproduced by permission of Sage Publications.)

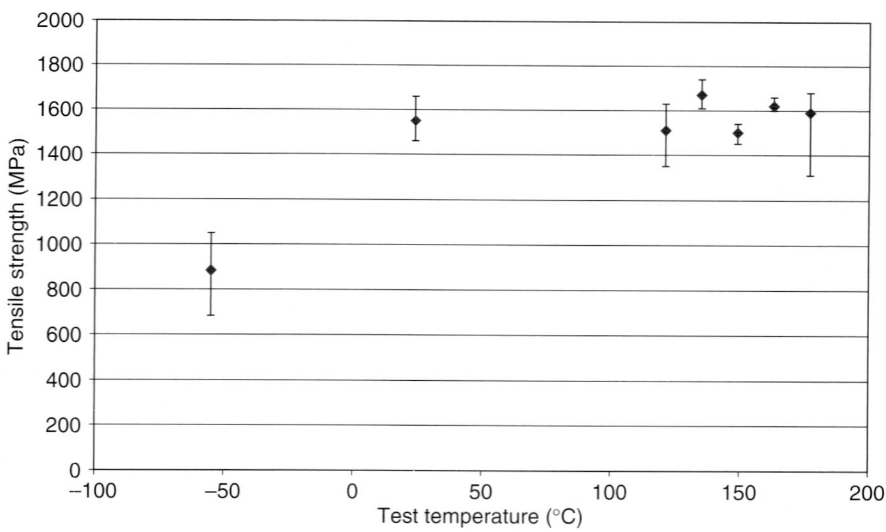

Figure 2.38. Temperature effects on graphite epoxy. (Data from Kerr and Haskins [22].)

Industrial case study: Cryogenic tanks for space re-launchable vehicles

At the time of writing, cryogenic fuel storage in launch vehicles relied on single-use metallic (aluminum) tanks. The challenge for such a product is for the material to maintain its full integrity at a temperature of −253°C required for liquid hydrogen storage. Additionally, the diffusion of the fuel molecules characterized by a fine molecular structure should be negligible over the mission's period.

Northrop Grumman Corp under a Next Generation Launch Technology contract option from NASA's Marshall Space Flight Center successfully developed a 1.8 m diameter, 4.6 m long tank produced by ultrasonic tape lamination. This new manufacturing process allowed cost reduction and large part manufacturing with out-of-autoclave curing. To avoid permeation of small fuel molecules (such as hydrogen), a large number of thin unidirectional prepreg layers were used as inner skin. An additional thin epoxy-coated aluminum shield provided enhanced diffusion protection. Finally, a honeycomb structure was used as core component. The use of composites enabled a 25% weight saving translating into an 8% decrease in vehicle acquisition cost and a 6% decrease in vehicle operation cost over the current state-of-the-art aluminum–lithium LH2 tanks for launch vehicles [76]. Most importantly, the composite tank was characterized by a possibility of multiple uses. The acceptable operation temperature range for the material was −253 to 121°C. The major challenge to solve was to prevent microcracking in the composite matrix resin at cryogenic temperatures. This condition has always occurred previously for lightweight, high performance tanks, resulting in hydrogen leaks through the tank wall. Previous tanks have prevented leaks by using increased wall laminate thickness and operating the tank at reduced strain levels. The drawback of this approach was to work against the weight saving benefits. The new development route taken by Northrop Grumman Integrated Systems was based on using special lamination techniques, redundant permeation barrier material layers and perforated honeycomb core. These allowed the tank to operate with the thin facesheet laminates at high strains, retaining the 25% weight savings over aluminum. This performance was validated via a rigorous test program at NASA MSFC [77]. In November 2003 a subscale tank manufactured by Autoclaving was tested at NASA's Marshall Space Flight Center (Figures 2.39 and 2.40). The tank was filled with liquid hydrogen then submitted to combined axial load (635 kg simulating gravitational stresses experienced during launch) and internal pressure (786 Pa simulating the fuel pressure load). During the year, the tank was drained and the loads were cycled around 40 times. No leaks were observed during this period.

In parallel with the tank development, a structural health monitoring system was thoroughly tested in which fiber-optic sensors were used to map the temperatures and strains of the structure in order to detect potential damage and leakage in an early stage. This success motivated the prototyping of a half tank using the more cost-efficient ultrasonic tape lamination process (Figure 2.41).

The half tank was produced with intentional, well-defined defects in order to validate the non-destructive evaluation techniques for large components used in this project, namely thermography where infrared energy was flashed on a composite structure creating a thermal map and laser shearography involving the scanning of the surface with a laser and comparing zero-load and loaded visual responses.

Considering the importance and size of the challenge, NASA also teamed up with Lockheed Martin to produce a liquid oxygen tank. The tank was manufactured

Figure 2.39. Figure Subscale CryoTank Test subjected to 40 simulated launch cycles including axial loads comparable to what it would experience in a typical launch vehicle *stack*. (Courtesy of Northrop Grumman.)

using automated fiber placement. The tank underwent 52 pressure cycles to the limit loads proving the possibility of multiple uses.

On the other end of the temperature scale, polymer matrix composites can be exposed to extreme heat in case of fire. A fire depends on the nature and amount of material as well as on the environment (such as amount of oxygen involved). Reproducibility and scaling are real issues for laboratory fire testing of

Figure 2.40. Composite Subscale cryo tank. (Courtesy of Northrop Grumman.)

composites. In addition, fire regulations vary from country to country. Generally, thermoplastic composites might melt and drip, while thermosets tend to retain their shapes until degradation. Ignition usually occurs around 400°C but can be delayed to temperatures as high as 500°C for high temperature polymers such as polyimide. The literature [78–81,68] relates the complexity of testing and degradation mechanisms for composites under fire.

Case study: Fire resistance for mass transportation and civil applications
For obvious economic reasons, weight reduction is highly desirable for air, ground and underground mass transportation systems. Therefore, composite materials become prime material candidates for airplane or rail transportation. However, stringent fire regulations restrict materials selection.

Fire regulations generally focus on the following aspects of the materials response:

- flame spread
- smoke density
- toxic gas emission
- ignition point
- dripping (which is a serious limiting factor for thermoplastic materials)
- retention of mechanical properties for structural elements.

Unfortunately, the different standard institutes (US ASTM, French AFNOR, German DIN etc.) all established different fire regulations. To complicate matters,

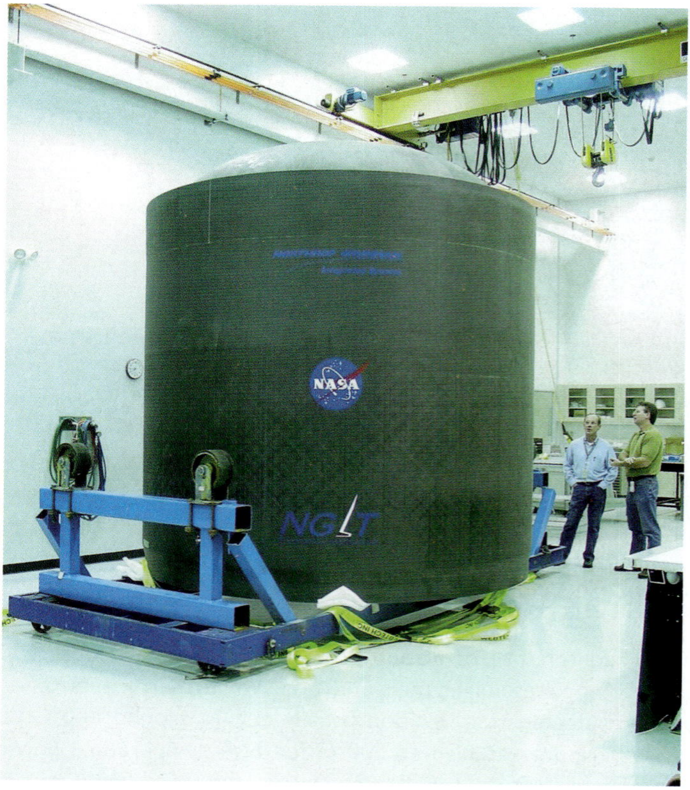

Figure 2.41. Photo of the ultrasonic tape lamination process manufactured cryogenic composite half hank. (Courtesy of Northrop Grumman.)

each fire is different depending on the amount of heat, oxygen and fuel available.

Material selection is driven by the search of the optimum compromise between the above aspects. For example, smoke emission is the major criterion for underground transportation, while the flashover occurrence (which strongly influences the escape time from a burning cabin) is key to the US federal aviation association (FAA). A fire flashover is characterized by the ignition of the hot smoky layer present in the closed area followed by rapid fire spread (non survivable condition). The limit is closely related to the heat release rate of the cabin material. Flammable components include ceiling, stowage bins, lavatory, upper and lower sidewall, windows, floors and most importantly seat cushions. Aircraft cabins are filled with several tons of polymer-based materials representing a fire load comparable to the equivalent weight of aviation fuel (Figure 2.42) [82]. In most fire accidents, fatalities actually occur as a result of smoke and toxic fumes and not directly from fire.

Figure 2.42. Full-scale aircraft fire test. (Courtesy of the FAA.)

Three possibilities are offered in the material selection of fire-resistant composite materials.

(1) *Selection of intrinsic fire-resistant matrices*: This is the case, for example, of phenolic matrices widely used in the transportation industry. A study performed by the FAA [82] reports the peak heat release rate versus cabin escape time for different materials (Figure 2.43). Clearly, the glass-fiber reinforced polyetheretherketone (PEEK)-polyimide composite is superior to the other investigated polymers (no flash-over observed after 10 min). However, high costs and aesthetical considerations do not allow aircraft manufacturers to use such an ultra-resistant composite. Phenolics are still within an acceptable range of fire performance and are more cost effective. The polyvinyl fluoride (PVF)-epoxy/glass composite on the other hand leads to cabin escape time well below the FAA requirements.

(2) *Filler addition*: Halogen fillers are widely used as fire retardant fillers. They improve ignition resistance and slow down fire spread. However, after ignition, toxic acid gases such as HCl might be emitted causing respiratory problems. Suppliers are therefore trying to develop filler substitutes such as phosphorus, melamine, aluminum trihydrate (ATH), antimony trioxide (ATO), magnesium hydroxide, zing stanate and zinc borate [83]. The selection of the appropriate fillers depends on the target response: minimizing smoke release or increasing ignition point, for example, will lead to the selection of different fillers. In under-ground transportation, smoke should be limited at any price, leaving little chances for halogenated resins.

Unfortunately, the addition of fillers is often detrimental to materials mechanical properties and processability. This can lead to weakened

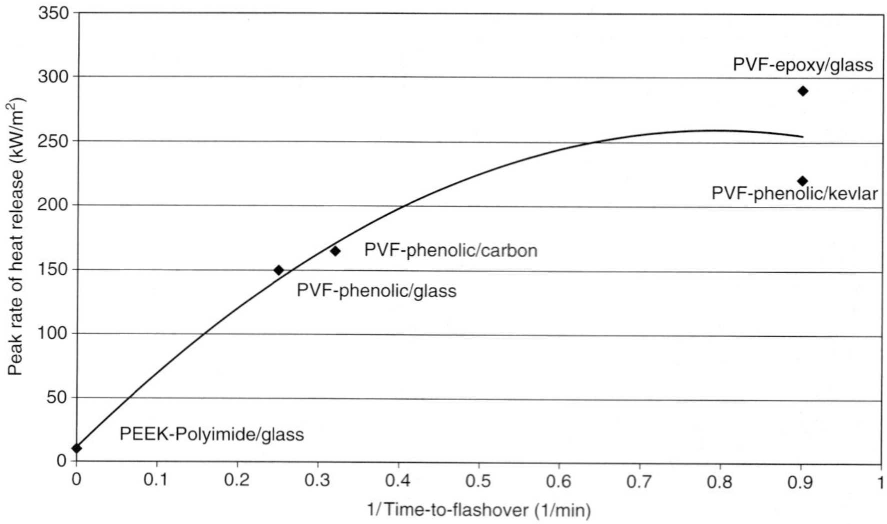

Figure 2.43. Peak heat release rate versus cabin escape time of different panel materials in a full-scale, post crash fire simulation. (Data from Lyon [82].)

structures and falling panels in case of fire. The use of fire-retardant filled composites is not restricted to transportation applications. Companies such as Clariant Exolit at the time of writing developed non-halogenated adhesives applicable to thermoplastic materials such as polyamide 66 targeting primarily the electrical and electronic industries.

(3) *Fire screen*: The last option to retard fire spread is the use of a fire screen covering the composite parts. This can be a thin material layer or a protective paint. This is, for example, the case in airplane seats, typically made of a polyurethane foam covered by a fire-blocking layer and an upholstery fabric [82].

These three technologies, especially in sandwich form, are currently being used for mass transportation applications. The Shinkansen E4 train in Japan, for example (Figure 2.44), uses a polymethacrylimide (PMI) core inside carbon epoxy skins when the exterior panels of the Refio Shuttle train from ADtranz use a sandwich structure of vacuum-infused phenolic skins and PVC foam core [83].

Flame retardant composites are also useful for building applications. Holland Composites developed student *box* housings made of a foam core covered with a fireproof sheet of glass/polyester [84] (Figures 2.45 and 2.46). The wall panels are molded and glued together. Finally, pigments are directly added during the molding phase avoiding a costly painting step.

Figure 2.44. Sandwich structures in Japan's Shinkansen E4 train utilize a near-aerospace combination of PMI structural foam core with epoxy prepreg [83]. (Picture courtesy of Razvan Photography.)

Figure 2.45. Space box installation. (Courtesy of Holland Composites.)

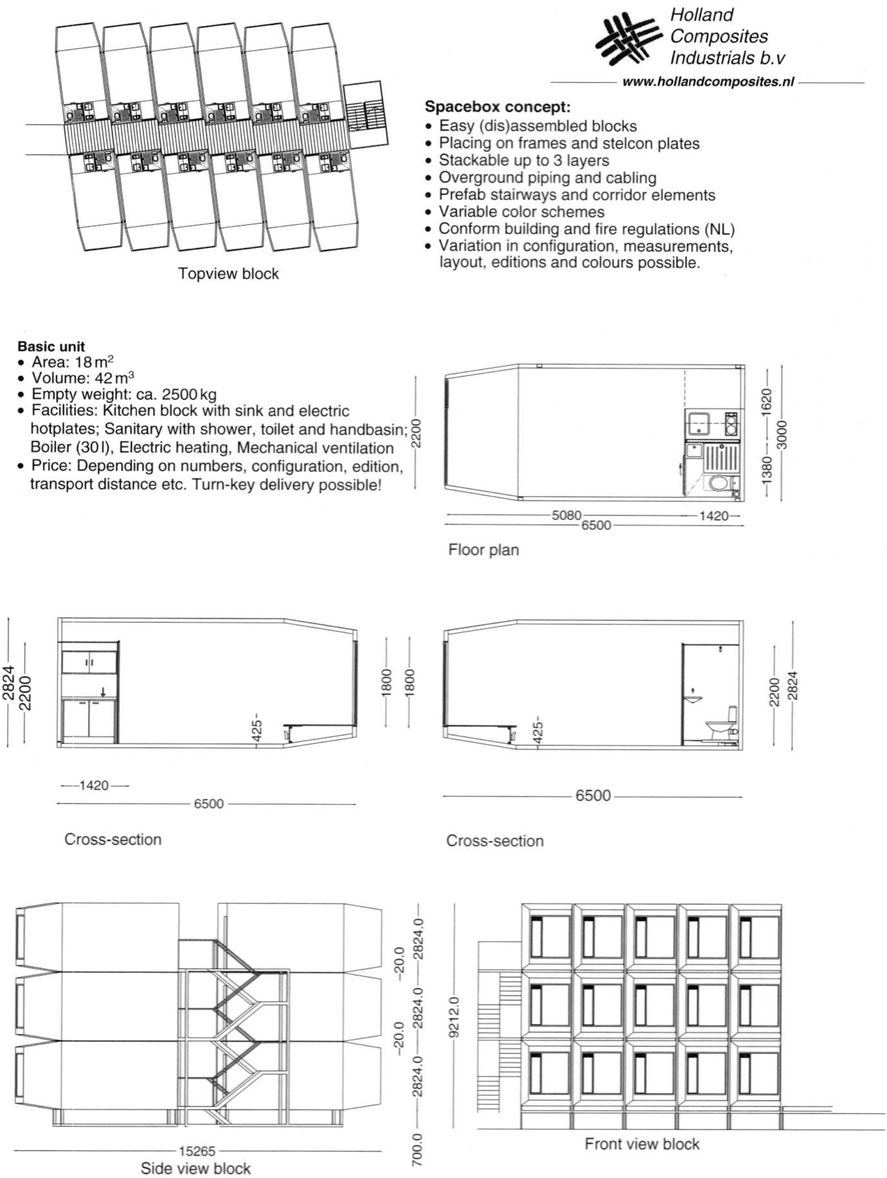

Figure 2.46. Composite housing boxes concept. (Courtesy of Holland Composites.)

2.8 TESTING

This chapter has illustrated the complexity and diversity of the degradation processes in polymer matrix composites. Due to the specificity of the degradation processes, extensive experimental work is required for all composite parts such as creep and relaxation experiments, aging, thermomechanical tests and transition temperature measurements. The test processes are numerous [85]. Only the main and most relevant methods are summarized thereafter. More details can be found in the literature [85].

2.8.1 Dilatometry Methods

The principle of dilatometry methods is to measure the materials volume variation as a function of the change in temperature. The equipment used for such measurements is usually a mercury-based dilatometer [86]. A sharp change in slope in the curve indicates the glass transition temperature. A discontinuity in the curve indicates a *true* thermodynamic transition (first order) such as melting. It is interesting to note that the glass transition temperature value obtained by this method (intersection of the extrapolation of the two linear portion of curves) measured with a temperature increase rate of 1°C/min generally corresponds to a 10-second mechanical measurement [86].

2.8.2 Thermal Methods

2.8.2.1 Differential thermal analysis (DTA)

The DTA method compares the heating rate of a reference sample (inert material such as Al_2O_3) and the sample to be evaluated (Figure 2.47). The DTA relies upon the temperature difference between the samples, to determine the presence of endothermic and exothermic peaks: the heating rate is slowed down during endothermic reactio ns and accelerated during exothermic reactions. The output of a DTA typically shows the exotherms and endotherms ($\Delta T = T_{\text{initial}} - T_{\text{final reaction}}$) versus time.

DTA methods are traditionally used for identifying initial temperatures of thermal processes, exothermic and endothermic reactions, reversibility and order of

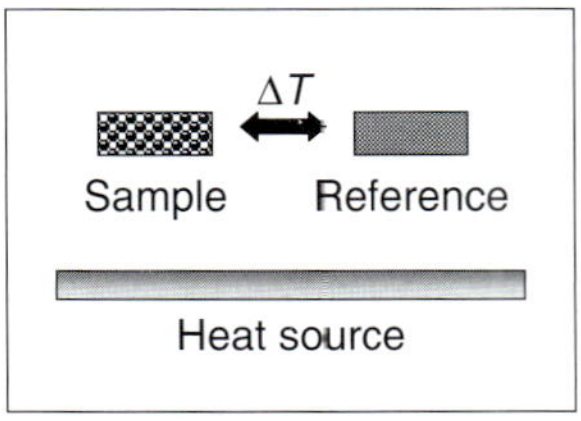

Figure 2.47. DTA principle.

the transition and finally for establishing phase diagrams. However, DTA cannot be used to perform quantitative evaluations, such as the measurement of energy amounts involved in reactions.

2.8.2.2 *Differential scanning calorimetry*

The DSC method relies on a principle similar to the DTA, but compensates some of the shortcomings of the previous method (such as compensations of the changes in the thermal transport properties or detector sensitivity [85]). The more advanced DSC method will supply energy to the two materials in order to maintain equal temperatures (Figure 2.48). The DSC method provides not only the location of the exothermic or endothermic peaks but also quantitative measurements of the enthalpy of the transition. Energy is supplied to either the sample (endothermic reaction) or the reference material (exothermic reaction) to maintain an identical temperature. Two main types of DSC are being used today [85]: *power compensating* DSC and DSC operating in *heat flux* mode. Both methods are appropriate for polymers and composites and yield excellent quantitative results. The DTA curves provide the initial and final temperatures of thermal processes, the characteristics of the reaction (exothermic, endothermic, reversibility, order of the transition) and finally the enthalpy of the reaction (ΔH). This is of particular importance as the amount of energy during the fusion process can be used to establish the amount of crystallinity of a semi-crystalline composite. The exothermic and endothermic sign conventions can vary from one instrument to the other. Usually for a DSC during an endothermic reaction, it is necessary to bring in more energy into the sample in order to keep the temperature equal to the reference material. Therefore, the endotherm will be associated with a positive electrical power in a DSC (opposite to DTA convention). For example, crystallization requires energy to be fed into the system and by convention will show an exothermic peak (energy needed in the system). Melting on the other hand will show an endothermic peak (energy being led out of the system). This concept is illustrated in Figure 2.49.

DSC experiments also evidence exotherms during cross-linking (curing process) of the material (Figure 2.50).

A final steep exothermic increase at high temperature can hint toward an oxidation process and the irreversible degradation of the material.

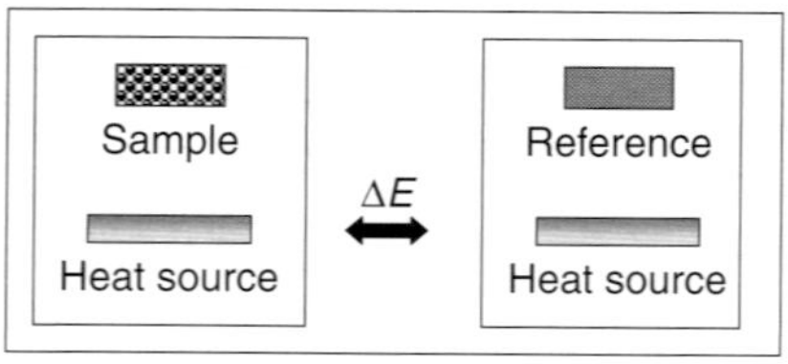

Figure 2.48. DSC principle.

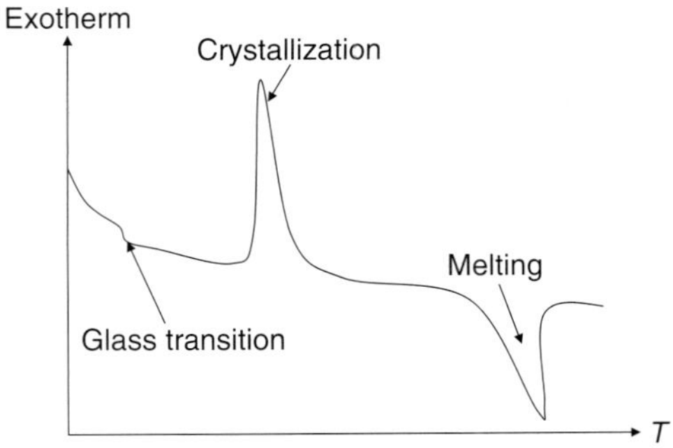

Figure 2.49. Typical DSC for a semi-crystalline material.

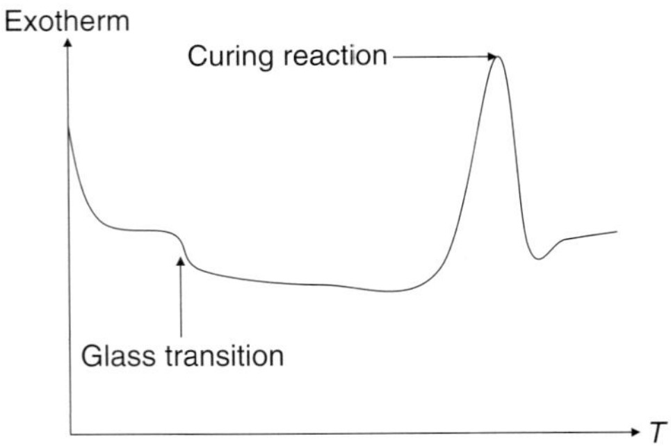

Figure 2.50. Typical DSC of a thermoset system undergoing cross-linking.

Differential scanning calorimetry is a cost-effective, rapid and reliable technique. It is recommended to systematically perform a DSC experiment on any polymer or composite system being studied in order to determine the number of transitions, transition temperatures, extent and type of transitions and crystallinity content. A DSC can also be modified to measure thermal conductivity [85].

2.8.3 Mechanical Methods

There are numerous types of mechanical methods including:

- Dynamic mechanical spectroscopy (DMS)
- Torsional braid analysis (TBA)
- Thermogravimetric analysis (TGA)
- Dynamic mechanical analysis (DMA)
- Dynamic mechanical thermal analysis (DMTA).

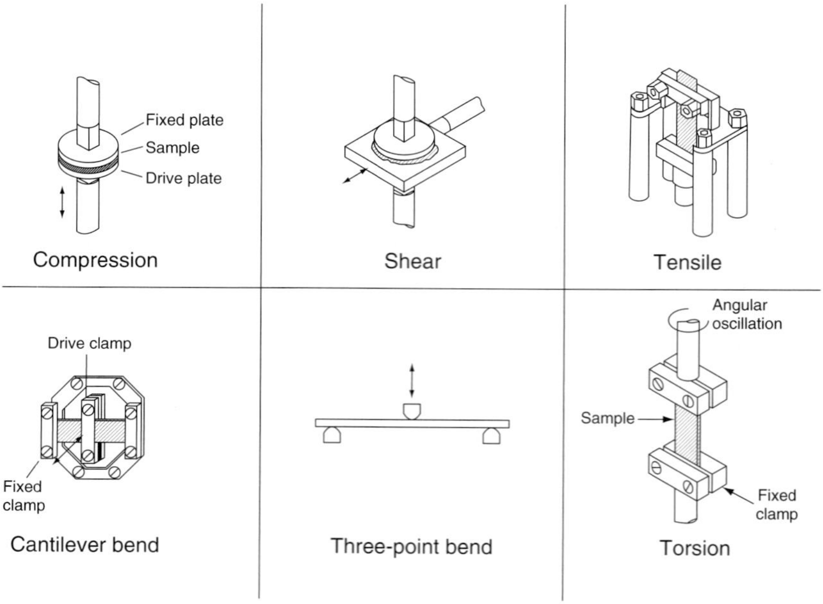

Figure 2.51. DMTA apparatus examples. (Copyright 1993, *American Laboratory*, by R.E. Wetton et al. [87], Elsevier.)

We will not go into the detail of each testing method. Instead, we will focus on the method of prime importance, DMA. The DMA is an extension of the TMA where the temperature is increased in parallel to periodical stresses and strains (Figure 2.51). The mechanical response of the sample (strain) is recorded. The transition temperatures as well as moduli variations result from the experiment. The glass transition temperature can be taken as the onset of the storage modulus drop or as the inflection point shown on the E' (tensile storage modulus) or G' (shear storage modulus) curve. An alternative is to consider the temperature corresponding to the peak value of $\tan\delta$. In any case, the measured values of the glass transition is not unique for a given material system, but is dependent upon experimental conditions such as strain rates.

2.8.4 Electric and Magnetic Methods

2.8.4.1 Conduction: Direct current (DC)

At constant current or voltage, the change in resistance of the sample is measured. Variation in the resistance can indicate major modifications in the material such as phase changes. These tests are often performed simultaneously with a DTA.

2.8.4.2 *Conduction: Alternating current (AC)*

(a) Dielectric analysis (DEA): This method is used to study the dielectric properties of the material (see Chapter 4). Dielectric analysis detects the re-orientation of dipoles under the effect of the AC. Results of DEA and DMA reflect the molecular motion in the polymer. However, DEA is more sensitive to local motions [85] and DMA more sensitive to coordinate molecular motion such as the glass transition.

(b) Thermally stimulated currents (TSC): TSC methods are used to study the dielectric response of polymer films [85]. This method is mainly of interest to the food packaging industry. In this method, the temperature is raised until all molecular motion of interest occur. A DC voltage is applied and the sample is cooled down to a temperature at which the molecular motion of interest does not occur anymore. The electrical field is then removed and the depolarization current is measured as the temperature is gradually increased [85]. A relaxation map can finally be drawn from the measurements.

(c) Spectroscopic methods: Spectroscopic methods enable direct visual observation while temperature is changed. Spectroscopic methods include reflectance spectroscopy, transmission infrared spectroscopy (FTIR), Raman spectroscopy, stimulated thermal emission, thermoluminescence and oxyluminescence [85].

(d) Other methods: Other methods include nuclear magnetic resonance (NMR) and temperature programmed electron spin resonance (EST). Magnetic transitions can be observed in polymer matrix composites. When temperature is increased and the sample becomes paramagnetic the transmission decreases. "Without magnetic splitting the absorption is greater at zero velocity", in the "magnetically aligned phase, the adsorption is spread over a wide range of energies. Consequently, the transmission decreases abruptly as the temperature is raised through the transition temperature and the spectrum collapses to a single or quadrupole split pair of absorption lines" [85].

2.8.5 Standard Test Methods

Common norms related to the temperature testing of polymers and polymer matrix composites are listed in Table 2.7.

Table 2.7. Common normalized testing methods for temperature effects

Organism	Designation	Title
ASTM	E2254-03	Standard Test Method for Storage Modulus Calibration of Dynamic Mechanical Analyzers
ASTM	C1181-00	Standard Test Methods for Compressive CREEP of Chemical-Resistant POLYMER Machinery Grouts
ASTM	E1867-01	Standard Test Method for Temperature Calibration of Dynamic Mechanical Analyzers
ASTM	E1640-01	Standard Test Method for Assignment of the Glass Transition Temperature by Dynamic Mechanical Analysis
ASTM	D4591-01	Standard Test Method for Determining Temperatures and Heats of Transitions of Fluoropolymers by Differential Scanning Calorimetry
ASTM	D3418-03	Standard Test Method for Transition Temperatures of Polymers by Differential Scanning Calorimetry
	E794-01	Standard Test Method for Melting and Crystallization Temperatures by Thermal Analysis
ASTM	D3417-99	Standard Test Method for Enthalpies of Fusion and Crystallization of Polymers by Differential Scanning Calorimetry (DSC)
ASTM	E2253-03	Standard Method for Enthalpy Measurement Validation of Differential Scanning Calorimeters
ASTM	E1356-03	Standard Test Method for Assignment of the Glass Transition Temperatures by Differential Scanning Calorimetry
ASTM	E793-01	Standard Test Method for Enthalpies of Fusion and Crystallization by Differential Scanning Calorimetry
ASTM	E1953-02	Standard Practice for Description of Thermal Analysis Apparatus
ASTM	E1952-01	Standard Test Method for Thermal Conductivity and Thermal Diffusivity by Modulated Temperature Differential Scanning Calorimetry
ASTM	E1269-04	Standard Test Method for Determining Specific Heat Capacity by Differential Scanning Calorimetry
ASTM	E794-01	Standard Test Method for Melting and Crystallization Temperatures by Thermal Analysis
ASTM	E2039-04	Standard Test Method for Determining and Reporting Dynamic Dielectric Properties
ASTM	E2038-99(2004)	Standard Test Method for Temperature Calibration of Dielectric Analyzers

2.9 TOOL KIT

Topic	Equation	Assumptions	Importance
Creep	Maxwell $D(t) = D_o + t/\eta$ Voigt $D(t) = D_o(1 - \exp(-t/\tau))$ Voigt–Kelvin $D(t) = \sum_{i=1}^{z} D_{o_i}(1 - \exp(-t/\tau_i))$	$\sigma = cst$	Engineering models to fit viscoelastic response of material
Relaxation	Maxwell $E(t) = E_o \exp(-t/\tau)$ Maxwell–Wiechert $E(t) = \sum_{i=1}^{z} E_i \exp(-t/\tau_i)$	$\varepsilon = cst$	Engineering models to fit viscoelastic response of material
Dynamic	Maxwell $E' = \frac{E_o\omega^2\tau^2}{1+\omega^2\tau^2}$ and $E'' = \frac{E_o\omega^2\tau^2}{1+\omega^2\tau^2}$ Voigt $D' = \frac{D_o}{1+\omega^2\tau^2}$ and $D'' = \frac{D_o\omega\tau}{1+\omega^2\tau^2}$ Voigt–Kelvin $D' = \sum_{i=1}^{z} \frac{D_{o_i}}{1+\omega^2\tau_i^2}$ and $D'' = \sum_{i=1}^{z} \frac{D_{o_i}}{1+\omega^2\tau_i^2}$ Maxwell–Wiechert $E' = \sum_{i=1}^{z} \frac{D_{o_i}}{1+\omega^2\tau_i^2}$ and $E'' = \sum_{i=1}^{z} \frac{D_{o_i}}{1+\omega^2\tau_i^2}$	$\sigma(t) = \sigma_o \exp(i\omega w)$ or $\varepsilon(t) = \varepsilon_o \exp(i\omega w)$	Engineering models to fit viscoelastic response of material
Instantaneous Modulus	$E = (E_1 - E_2)\exp\left(-\left(\frac{T}{T_1}\right)^{m_1}\right) + (E_2 - E_3)\exp\left(-\left(\frac{T}{T_2}\right)^{m_2}\right) + E_3\exp\left(-\left(\frac{T}{T_3}\right)^{m_3}\right)$	High loading rate	Engineering model to fit E versus T across the transitions
WLF Time–temperature equivalence	$\frac{E(T_1, t)}{\rho(T_1)T_1} = \frac{E(T_2, t/a_T)}{\rho(T_2)T_2}$ and $\log\, a_T = \frac{-C_1(T - T_g)}{C_2 + T - T_g}$	Only valid for linear amorphous polymers near T_g	Gives time–temperature equivalence
Boltzmann superposition principle	$\varepsilon(t) = \sum_{i=1}^{n} \sigma_i D(t - t_i)$		Strain superposition principle

REFERENCES

1. Mahieux, C.A., Materials influence on hydrodynamic behavior of thrust bearings: A comparison of Babbitt, PTFE and PFA. in print, *Journal of Tribology*, July 2005, 127, 568–575.
2. Lal, J. and J.E. Mark (Ed.) *Advances in Elastomers and Rubber Elasticity*. Plenum Pub. Corp., 1986.
3. Mark, J.E. and J. Lal (Ed.), Elastomers and Rubber Elasticity. ACS symposium series, American Chemical Society, July 1, 1982.
4. Tobolsky, R.V., Rubber Elasticity (Technical report). Frick Chemical laboratory, Princeton University, 1968.
5. Pearson, R.A. and A.F. Yee, Toughening mechanisms in elastomer-modified epoxies, Part 3: The effect of cross-link density, *Journal Materials Science*, 24, 2571, 1989.
6. Lee, H. and K. Neville, *Handbook of Epoxy Resins*, McGraw-Hill, New York, 1967.
7. Zhang, M., A Review of the Epoxy Resin Toughening. Department of Chemical Engineering and Materials Science, Syracuse University, New York, April 29, 2003.
8. Matthews, F.L. and R.D. Rawlings, *Composite Materials: Engineering and Science*. Chapman & Hall, London, 1994.
9. Tobolsky, A.V., *Properties and Structure of Polymers*, 3rd ed. Wiley, New York, 1980.
10. Pipkin, A.C., *Lectures on Viscoelasticity Theory* (Applied Mathematical Sciences), 2nd ed. Springer, 1986.
11. Aklonis, J.J. and W.J. MacKnight, *Introduction to Polymer Viscoelasticity*, 2nd ed. John Wiley & Sons, 1983.
12. Findley, W.N., J.S. Lai and K. Onaran, *Creep and Relaxation of Nonlinear Viscoelastic Materials: With an Introduction to Linear Viscoelasticity* (Dover Books on Engineering). Dover Publications, 1989.
13. Drozdov, D., *Finite Elasticity and Viscoelasticity: A Course in the Nonlinear Mechanics of Solids*. World Scientific Publishing Company, 1996.
14. Riande, E., R. Diaz-Calleja, M.G. Prolongo, R.M. Masegosa and C. Salom, *Polymer Viscoelasticity: Stress and Strain in Practice*. Plastics Engineering. Marcel Dekker, Inc., 1999.
15. Junisbekov, T.M., V.N. Kestelman and N.I. Malinin, *Stress Relaxation in Viscoelastic Materials*. Science Publishers Inc., 2002.
16. Rittel, D., An investigation of the heat generated during cyclic loading of two glassy polymers, Part I: Experimental. *Mechanics of Materials*, 2000, 32, 131–147.
17. Weitsman, Y., Private communication.
18. Ferry, J.D., *Viscoelastic Properties of Polymers*. Wiley, New York, 1980.
19. Kreyszig, E., *Advanced Engineering Mathematics*, 7th ed., John Wiley & Sons, 1993.
20. www.matweb.com.
21. Mahieux, C.A., A Systematic Stiffness – Temperature Model for Polymers and Applications to the Prediction of Composite Behavior, ETD, Virginia Polytechnic Institute and State University, February 1999.
22. Kerr, J.R. and J.F. Haskins, Time-Temperature-Stress Capabilities of Composite Materials for Advanced Supersonic Technology Application. NASA Contractor Report 178272, May 1987.
23. Walther, B.M., An Investigation of the Tensile Strength and Stiffness of Unidirectional Polymer-Matrix, Carbon-Fiber Composites under the Influence of Elevated Temperatures. Master of Science, ETD-5298-22449, Virginia Polytechnic Institutes and State University, 1998.

24. Sperling, L.H., *Introduction to Physical Polymer Science*, 2nd ed. John Wiley & Sons, New York, 1992.
25. Aklonis, J.J. and W.J. MacKnight, *Introduction to Polymer Viscoelasticity*, 2nd ed. John Wiley & Sons, 1983.
26. Schatzki, T.F., Statistical computation of destruction functions of dimensions of macromolecules. *Journal of Polymer Science*, 1962, 57(165), 337–356.
27. Boyer, R.F., The relation of transition temperatures to chemical structure in high polymers. *Rubber Chemistry and Technology*, 1963, 36(5), 1303–1422.
28. Krishnaswamy, R.K. and D.S. Kalika, Dynamic relaxation properties of poly(ether ether ketone). *Polymer*, 1994, 35(6), 1157–1165.
29. Bartolotta, A., G. Di Marco, M. Lanza and G. Carini, Charge Effect of the Cation on the Structure and the Molecular Mobility of Polymer Electrolytes. *Journal of Polymer Science: Part B: Polymer Physics*, 1995, 33, 93–104.
30. Wang, X. and J.K. Gilham, Tg-temperature property (TgTP) diagram for thermosetting systems: anomalous behavior of physical properties vs. extent of cure. *Journal of Applied Polymer Science*, January 1993, 47(3), 425–446.
31. Krishnaswamy, R.K. and D.S. Kalika, Dynamic mechanical relaxation properties of poly(ether ether ketone). *Polymer*, March 1994, 35(6), 1157–1165.
32. Muzeau, E., J. Perez and G.P. Johari, Mechanical spectrometry of the β relaxation in poly(methyl methacrylate). *Macromolecules*, 1991, 24, 4713–4723.
33. Williams, M.L., R.F. Landel and J.D. Ferry, The temperature dependence of relaxation mechanisms in amorphous polymers and other glass-forming liquids. *Journal of the American Chemical Society*, 1955, 77(14), 3701–3707.
34. Matsuoka, S., *Relaxation Phenomena in Polymers*. Carl Hanser Verlag, 1992.
35. Gibbs, J.H. and E.A. Di Marzio, Nature of the class transition and the glassy state, *Journal of Chemical Physics*, 1958, 28(3), 373–383.
36. Fox, T.G. and P.J. Flory, Second-order transition temperatures and related properties of polystyrene. I. Influence of Molecular Weight. *Journal of Applied Physics*, 1950, 21(6), 581–591.
37. Kohlrausch, R., *Pogg. Ann. Phys.* 1847, 12, 393–.
38. Williams, G. and D.C. Watts, Non-symmetrical dielectric relaxation behaviour arising from a simple empirical decay function. *Transactions of the Faraday Society*, 1970, 66, 80–85.
39. Rouse, P.E., A theory of the linear viscoelastic properties of dilute solutions of coiling polymers. *Journal of Chemical Physics*, 21(7), 1953, 1272–1280.
40. Zimm, B.H., Dynamics of polymer molecules in dilute solution: Viscoelasticity, flow birefringence and dielectric loss. *Journal of Chemical Physics*, 1956, 24(2), 269–278.
41. Bueche, F., The viscoelastic properties of plastics. *Journal of Chemical Physics*, 1954, 22(4), 603–609.
42. De Gennes, P.G., *Scaling Concepts in Polymer Physics*. Cornell University Press, 1979.
43. Khanna, Y.P., Estimation of polymer crystallinity by dynamic mechanical techniques. *Journal of Applied Polymer Science*, May 1989, 37(9), 2719–2726.
44. Murayama, T. and J.P. Bell, Relation between the network structure and dynamic mechanical properties of a typical amine-cured epoxy polymer. *Journal of Polymer Science*, 1970, 8(3), 437–445.
45. Ashby, M.F. and D.R.H. Jones, *Engineering Materials 2*. Pergamon Press, Oxford, 1986, 226–227.

46. Mahieux, C.A. and K.L. Reifsnider, Property modeling across transition temperatures in polymers: A robust stiffness-temperature model. *Polymer*, 2001, 42(7), 3281–3291.
47. Mahieux, C.A., K.L. Reifsnider and S.W. Case, property modeling across transition temperatures in PMC's, Part I: Tensile properties. *Applied Composite Materials*, 2001, 8(4), 217–234.
48. Reifsnider, K.L. and S.W. Case, *Damage Tolerance and Durability of Material Systems*. Wiley, New York, 2002.
49. Mahieux, C.A. and K.L. Reifsnider, Property modeling across transition temperatures in polymers: Application to thermoplastic systems. *Journal of Materials Science*, 2002, 37(5), 911–920.
50. Mahieux, C.A. and K.L. Reifsnider, Property modeling across transition temperatures in polymers: Application to filled and unfilled polybutadiene. *Journal of Elastomers and Plastics*, 2002, 34(1), 2002, 79–90.
51. Kovacs, A.J., La contraction isotherme du volume des polymeres amorphes. *Journal of Polymer Science*, 1958, 30(121), 131–147.
52. www.Azom.com.
53. Horvath, D.A. and R.L. Steinman, Advent Engineering Services, Inc., Microstructure Assessments for Determining Electric Cable Insulation Remaining Life, http://www.adventengineering.com/Publications/ICWSA2002_A.pdf.
54. Gillham, J.K., Curing. In *Encyclopedia of Polymer Science and Technology*, 1986, 4, 519.
55. Black, S., Are high-temp thermosets ready to go commercial. *High Performance Composites*, November 2004.
56. Mittal, K.L., *Polyimide and Other High Temperature Polymers: Synthesis, Characterization and Applications*. Brill Academic Publishers.
57. Luise, R.R., *Applications of High Temperature Polymers*, CRC Press, 1996.
58. Serfafini T.T., *High Temperature Polymer Matrix Composites*, Noyes Data Corporation/Noyes Publications 1987.
59. Goddrich Aerospace De-Icing and Specialty Systems. SuperImide datasheet.
60. UBE PETI-330 datasheet.
61. Hexcel F650 Product Data Sheet.
62. Struik, L.C.E., *Physical Aging in Amorphous Polymers and Other Materials*. Elsevier Applied Science Publishers, Ltd., 1978.
63. Gates, T.S., On the Use of Accelerated Test Methods for Characterization of Advanced Composites Materials. NASA/TP-2003-212407, 2003.
64. Arnold-McKenna, C. and G.B. McKenna, Workshop on Aging, Dimensional Stability and Durability Issues in High Technology Polymers, National Institute of Standards and Technology, Gaithersburg, July–August 1993, 98(4), 523–533.
65. Bank, L.C., A. Barkatt and T.R. Gentry, Accelerated test methods to determine the long-term behavior of FRP composite structures: Environmental effects. *Journal of Reinforced Plastics and Composites*, 1995, 14(6), 559–587.
66. Sullivan, J.L., Creep and physical aging of composites. *Composite Science and Technology*, 1990, 39, 207–232.
67. Springer, G.S., *Environmental Effects on Composite Materials*, Technomic Publishing Company, 1984, vol. 1, 17.
68. Rosato, D.V., D.P. DiMattia and D.V. Rosato, *Designing with Plastics and Composites: A Handbook*. Van Nostrand Reinhold, 1991.
69. Flaggs, D.L. and F.W. Crossman, in G.S. Springer (Ed.), *Environmental Effects on Composite Materials*. Technomic Publishing Company, 1984, vol. 2.

70. Bocchieri, R.T. and R.A. Schapery, Nonlinear viscoelastic constitutive equations for carbon/epoxy and glass/epoxy composites and their comparison through micromechanics, in *Composite Materials for Offshore Operations*, 2nd ed. S.S. Wang J.G. Williams and K.H. Lo (Eds.), American Bureau of Shipping, 1999.
71. Bechel, V.T. and R.Y. Kim, Damage trends in cryogenically cycled carbon/polymer composites. *Composites Science and Technology*, September 2004.
72. Bechel, V.T., J.D. Camping and R.Y. Kim, Cryogenic/elevated temperature cycling induced leakage paths in PMCs. *Composites Part B: Engineering*, March 2005.
73. Roy, S. and M. Benjamin, Modeling of permeation and damage in graphite/epoxy laminates for cryogenic fuel storage. *Composites Science and Technology*, October 2004.
74. Takeda, T., Y. Shindo and F. Narita, Three-dimensional thermoelastic analysis of cracked plain weave glass/epoxy composites at cryogenic temperatures. *Composites Science and Technology*, November 2004.
75. Mahieux, C.A. and K.L. Reifsnider, Property modeling across transition temperatures in polymers: Application to filled and unfilled polybutadiene. *Journal of Elastomers and Plastics*, 2002, 34(1), 79–90.
76. Press Release El Segundo, Calif, Jul 14, 2003/PRNewswire-First Call via COMTEX.
77. McKinney, B., Northrop Grumman Integrated Systems, Private communication.
78. Nelson, G.L., Flammability: Understanding Requirements and Test Procedures. Proceedings, SPI Structural Plastics Div. Annual Conference, 1989.
79. Hilado, C.J., *Flammability Handbook for Plastics*. Fire and Flammability Series, 2nd ed. Technomic, 1974.
80. Gibson, A.G. (Ed.), Composites in Fire, *Proceedings of the International Conference on the Response of Composite Materials to Fire*. University of Newcastle Upon Tyne, U.K., September 1999, Woodhead Publishing Ltd, 2001.
81. Technical and Marketing Issues Impacting the Fire Safety of Electrical, Electronic and Composite Applications. Fire Retardant Chemical Assn. Technomic Pub Co, 1992.
82. Lyon, R.E., Fire-Resistant Materials: Research Overview, DOT/FAA/AR-97/99, http://www.tc.faa.gov/its/worldpac/techrpt/ar97-99.pdf, 1997.
83. Fire-safe composites for mass transit vehicles. *Reinforced Plastics*, September 2002, 26–30.
84. Flame retardant composites solve temporary housing crisis. *Reinforced Plastics*, November 2004, 4.
85. Gallagher, P., Instrumentation, Techniques, and Methodology, in E.A. Turi (Ed.), *Thermal Characterization of Polymeric Materials*, Volumes 1 & 2, 2nd ed. Academic Press, New York, 1997, 1–192.
86. Meares, P., *The Second-order Transition of Polyvinyl Acetate*, Transactions of the Faraday Society, 1957, 53, 31–41.
87. Wetton, R.E., J.S. Fisher, K.E. Pettitt, A. Evans and J.C. Duncan, *Third Generation DMTA Instrumentation, American Laboratory*, January 1993, 15–20.

3

LIQUIDS AND GAS EXPOSURE

3.1 INTRODUCTION

We have described in Chapter 2 the complex multilayer barrier structure used by Northrop Grumman for the cryogenic fuel tanks of re-launchable space vehicles, specially developed to avoid the displacement of the small hydrogen gas molecules through the composite wall. The diffusion of liquids or gases through composite materials is indeed a complex process depending on many parameters such as the nature of the polymer, the nature and geometry of the reinforcement, and the environment (concentrations, temperature, pressure etc.).

Most studies available in the literature deal with composite exposure to water under its various forms. All composites are exposed during their life cycle to moisture present in the ambient air and many composites will experience contact with water at some point – during storage or operation, for example. However, composites can also be exposed to more aggressive liquid and gaseous environments. This is the case, for example, of composites used as implants in the human body. Body liquids can contain enzymes, which can accelerate the degradation rate of the polymer composite (see Chapter 1). Simple addition of salt to water can also induce high-rate corrosion damage. The following case study was chosen to illustrate the complexity of corrosion protection for house ducting in coastal areas.

Industrial case study: House ducting

Building houses in coastal areas can lead to extreme requirements on the materials. Piping work such as air return ducting systems need to be able to withstand the Hundred-Year Storm in which the foundations of the house might be flooded inducing unusual hydraulic forces and sea water corrosion. The heating and air conditioning system must also comply with building fire resistance (Class 1) and low smoke-emission requirements. Traditional thin gauge metallic materials and thin wall composite ducting are ruled out due to the extremely corrosive environment and the possibility of flooding (flotation) which would cause tremendous strain and damage on traditional type materials.

This challenge was solved by the Environmental Heating and Air Conditioning company who used RAM Fiberglass Vipel® K022-AAA vinyl ester to build the air return ducting system for a California coastal habitation [1]. Once material selection was performed based on the stringent corrosion and temperature requirements, RAM engineers focused on designing a thick wall, which would withstand the external forces of burial and flotation. This effort was supported by the selection of the appropriate earth anchoring system by the architect. Figure 3.1 is a picture taken just prior to completion of the underground 14 in. (35.56 cm) to 30 in. (76.2 cm) return air piping (duct). The whole area under the house is backfilled with pea gravel. The house foundation is supported on piles driven approximately 70 ft into the ground.

Despite the omnipresence of diffusion, no systematic model was available at the time of writing to predict gas and liquid absorption into a polymer composite and its consequences on the materials properties. This chapter will focus on describing elements to help understand the basic phenomena: after an introduction to the diffusion principles, the effects of liquid and gaseous environment on the polymer and fibers will be presented. The impact of composite polymer/fiber mixture and common testing procedures are then described. Finally, freeze thaw and cavitation erosion are introduced as special cases of water environment.

Figure 3.1. Composite air ducting return system. (Courtesy of RAM Fiberglass.)

3.2 THE DIFFUSION PHENOMENON

Foreign molecules constantly migrate in and out of the composite. For atoms to migrate within a solid body, the presence of *holes* is required. Polymers are characterized by the presence of free volume (see Chapter 2). It is assumed (but not proven!) that water infiltrates this free volume. This assumption is supported by the fact that diffusion in the amorphous phase of the material occurs at a faster rate than in the crystalline phase.

The rate at which this migration occurs can be very different depending upon the nature of the permeating specie. For example, the diffusion of *small* molecules (e.g. water) is easier than of *large* molecules (e.g. oil). Diffusion is further influenced by the nature of the composite. It is also well recognized that the rate of diffusion can be altered by chemical degradation and other forms of damage of the polymer during exposure. For example, exposure to a dilute solution of hydrochloric acid could lead to chain scission of the polymer, leading to microcracks which could then accelerate the diffusion of water in the solution.

Generally speaking, parameters that restrict molecular motion such as cross-linking, crystallinity or fillers influence permeation leading to a broad range of diffusivity coefficients for polymers and composites.

Diffusion is also greatly influenced by temperature, pressure, permeant concentration, polymer viscoelasticity and material damage (either due to chemical degradation, mechanical stresses or a combination of both). The multitude of parameters adds to the complexity of the process and to date diffusion is still not fully understood. As mentioned in Chapter 1, polymers in the glassy state are not in equilibrium. This leads to significantly different diffusion mechanisms for polymers above and below the glass transition temperature [2]. Mainly, vibrational motions are authorized in the glassy state. This relative *stiffness* of the molecular structure leads to low diffusivity. In the rubbery state however, the polymer chains move in a coordinated manner, facilitating the *acceptance* of foreign molecules. Therefore, diffusivity coefficients above the glass transition temperature are significantly higher than below the glass transition temperature [3]. Practically, this means that a polymer composite exposed to a liquid or gaseous environment over a large temperature range (comprising the glass transition) requires the experimental establishment of at least two diffusivity coefficients.

Mechanical loads are not to be neglected in diffusion studies. For example, a sample under tensile stress might have a higher diffusivity than an unstressed counterpart. Environmental factors might interact strongly and it is recommended to perform immersion experiments with load condition combinations when allowed by time and budget (see Section 6.4).

Gas and liquid permeation into a polymer follow similar rules, with the exception that the matrix swelling associated with liquid absorption might be much greater than for gases. Molecular reorganization and local stresses induced by the swelling can in turn modify the diffusivity. Therefore, in the case of liquids, the diffusivity D is dependent upon the permeant concentration present in the sample (typically exponentially) [3]. Gas permeation on the other hand is usually independent of the permeating specie concentration in the material.

Gas and liquid diffusion can both be modeled by solution-diffusion theory. More elaborate theories such as irreversible thermodynamics (sorption-capillary flow theory etc.) are sometimes required for liquids and water vapor [2,4,5].

In the present text, we will focus on understanding the basic concepts and practical models widely used in industrial contexts.

3.2.1 Fickian Diffusion

The properties (i.e. performance) of the material are related to the amount of foreign atoms present in the sample. It is therefore of prime importance to model the progression of gas or liquids within the composite.

Diffusion is a time- and temperature-dependent process. Under general conditions (non-steady state), diffusion can often be described by Fick's second law:

$$\frac{\partial c}{\partial t} = D\frac{\partial^2 c}{\partial x^2} \tag{3.1}$$

where c is the concentration of the diffusing species, D is the diffusion coefficient or diffusivity expressed in m^2/s, and x is the space coordinate measured perpendicular to the section. Solving this equation becomes quickly complex. Initial and boundary conditions play a deciding role in the possibility to establish a solution for the equation. If the concentrations are held constant, the steady-state flux (J) is controlled by Fick's first law:

$$J = -D\frac{\mathrm{d}c}{\mathrm{d}x} \tag{3.2}$$

Fick's first law leads to a linear concentration profile inside the sample (Figure 3.2). This simple model is very rarely applicable to real cases. A simplification of

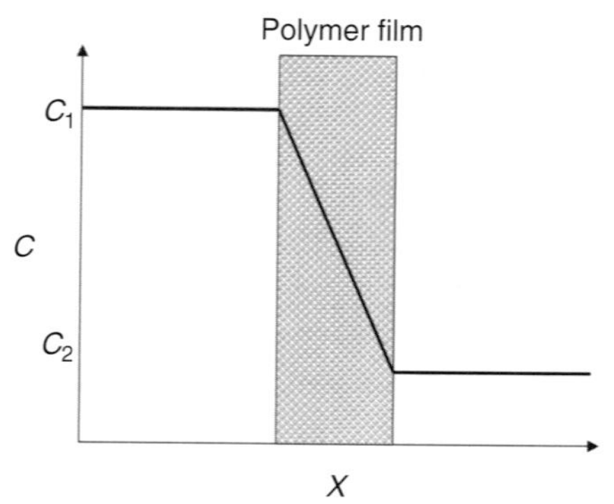

Figure 3.2. One-dimensional steady state Fickian diffusion through a polymer film.

Fick's second law which is more applicable to thick composites is presented later (Equation (3.4)).

It is interesting to note that Equations (3.1) and (3.2) apply to permeants in both gaseous and liquid states.

3.2.2 Practical Implications of Fick's Laws

The first step in any diffusion study is the calculation of the diffusing species uptake. The diffusing species content varies as a function of time. Assuming that no degradation occurs in the sample, an increase in the weight of the composite part reflects an absorption or adsorption of the diffusing specie.

Generally, the solvent content (M) in the composite is expressed as a percent. M refers to moisture; however, Equation (3.3) can be used for all diffusing species and solvents.

$$M_{\text{sample at time t}}(\%) = \left(\frac{\text{Weight}_{\text{sample at time t}} - \text{Weight}_{\text{sample initial}}}{\text{Weight}_{\text{sample initial}}} \right) \times 100 \qquad (3.3)$$

Early in a diffusion study, it is necessary to determine the saturation level characterized by the maximum weight uptake. This is done by immersing materials samples into the appropriate environment. The samples are weighed regularly and the content of diffusing species in the sample is calculated according to Equation (3.3). In most cases, the weight eventually levels out tending toward a saturation value M_∞. In the case of water vapor, M_∞ is independent of temperature but is also related to the moisture content (concentration level). Typical saturation values for polymers and composites are shown in Table 3.1.

Table 3.1. Moisture absorption at equilibrium [6]

Material	Moisture absorption at equilibrium (%)
50% Glass fiber/liquid crystal polymer (GF/LCP)	0.01
High density polyethylene (HPDE), injection molded	0.01–0.05
Polyester film (PE)	0.2
20% Glass fiber/acrylonitrile butadiene styrene (GF/ABS)	0.3
Polyetheretherketone (PEEK)	0.2–0.5
Glass fiber/polyetheretherketone (GF/PEEK)	0.14–0.5
Carbon fiber/polyetheretherketone (CF/PEEK)	0.3–0.5
30% Glass fiber/polyethersulfone (GF/PES)	0.5
Epoxy	0.05–0.5
Mica filled polytetrafluoroethylene (Mica/PTFE)	2–3
Kevlar® aramid fiber	3.5–4.5
Nylon 46	2.4–3
Nylon 6 (Cast)	1–7

Table 3.1 clearly indicates that the saturation level M_∞ depends on the nature of the material. Thermoplastics usually absorb less moisture than epoxies [7]. In case of water diffusion, hydrophobic polymers are characterized by low values. Hydrophilic materials such as Nylon 6 on the other hand have high values.

M_∞ also depends upon the thermal history of the sample. For some polymers such as cyanate ester modified epoxies [8] or epoxy vinyl esters [9], the amount of water sorbed increases paradoxically when a water-saturated specimen is transferred from a high temperature to a lower temperature water bath. This phenomenon illustrated by Figure 3.3 is referred to as reversed thermal effect or RTE.

Inversely, applying a high temperature for a short period of time (e.g. 60 s) is referred to as *thermal spiking*. This thermal treatment, even of very short duration, can have significant effects on the moisture absorption. Thermal spiking can be of particular importance for given applications, such as aircraft composite components that can experience temperatures around 100°C for a few minutes followed by a rapid cool-down.

The moisture content typically increases with increasing spiking temperature up to a threshold value above which the moisture content decreases again [10]. Hough et al. investigated the thermal spiking behavior of three carbon-fiber epoxy composites:

(1) Narmco rigidite 5245C (BASF) – a $[0]_{16}$ carbon-fiber reinforced bismaldeide modified epoxy resin.
(2) Fibredux 927 (Ciba Geigy) – a $[0]_8$ carbon-fiber reinforced high temperature, modified epoxy blend.
(3) Fibredux 924 (Ciba Geigy) – a $[0]_8$ carbon-fiber reinforced thermoplastic/epoxy blend.

The samples were conditioned in a sealed humidity chamber at 96% RH. The RH is defined as the amount of water vapor in the air expressed as the ratio between the measured amount and the maximum possible amount (the saturation point at which water condenses as dew) [11]. Temperature spiking was performed in an air-circulating oven. The specimens were left in the oven for periods varying from 3.5 to 8 minutes depending upon the sample thickness and spiking conditions. The calculated times allowed for a full 60-second exposure at the chosen temperature.

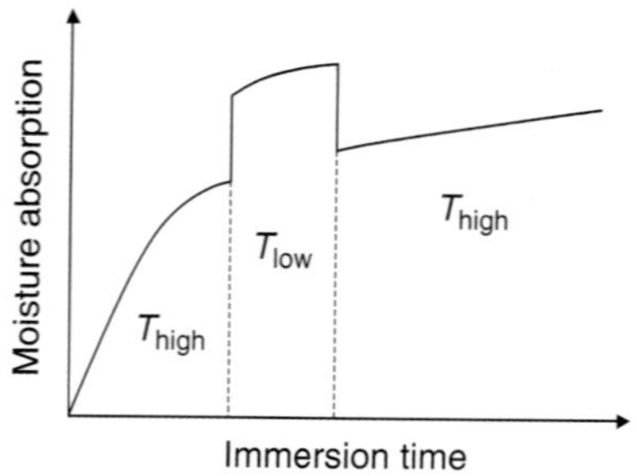

Figure 3.3. Reverse thermal effect.

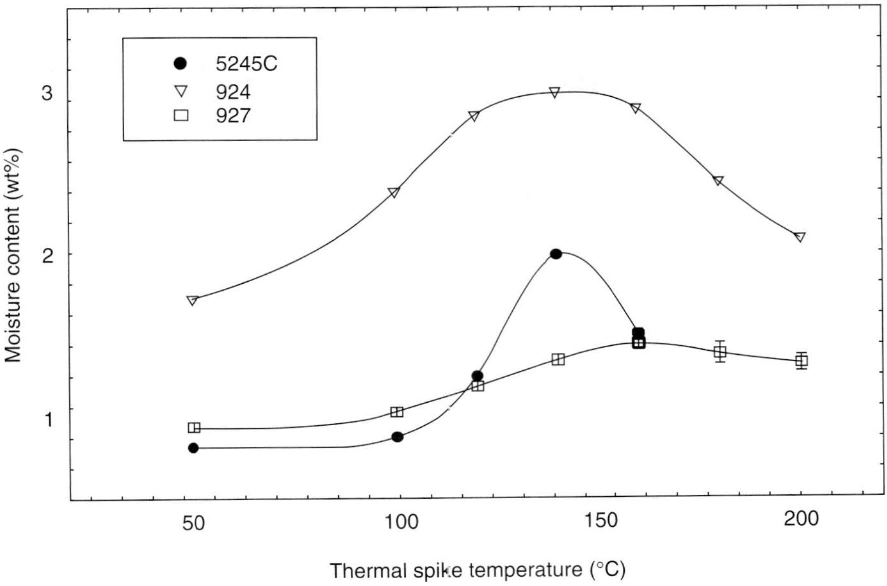

Figure 3.4. Moisture concentration in the carbon-fiber epoxy (5245C, 927, 924) laminates, after thermal spiking and conditioning at 96% RH for 10 000 h (5245C/927) and 5100 h (924). (Copyright 1998, *Key Engineering Materials*, by J.A. Hough et al. [10], reproduced by permission of Trans Tech.)

All three laminates showed a pronounced increase in moisture absorption when spiked (Figure 3.4).

Such increased moisture content of course influences the thermo-mechanical response of the composite. Figures 3.5 and 3.6 show the expected response in terms of loss tangent and storage modulus as a function of temperature for a dry laminate, a laminate conditioned under static conditions and a thermally spiked sample. The responses of the composite are significantly different in the three cases.

The flexural strength of the various laminates was also found to be strongly affected by the spike temperature (Figure 3.7). The strength of the carbon-fiber reinforced thermoplastic/epoxy blend laminate and of the carbon-fiber reinforced bismaldeide modified epoxy resin composite decreased by around 30% when the spiking temperature was increased from 50 to 160°C. Inversely, the average flexural strength of the carbon-fiber reinforced high temperature, modified epoxy blend composite increased by around 50% in the same conditions. This increase however coincides with the experimental scatter and its significance should be further investigated [10].

Studies have shown [12] that temperature equilibrates much faster than moisture. This statement holds for most solvents. It is therefore reasonable to assume that the temperature gradient in and outside the material is null. Assuming a diffusivity independent of the diffusing species content, it is possible to simplify Equation (3.1).

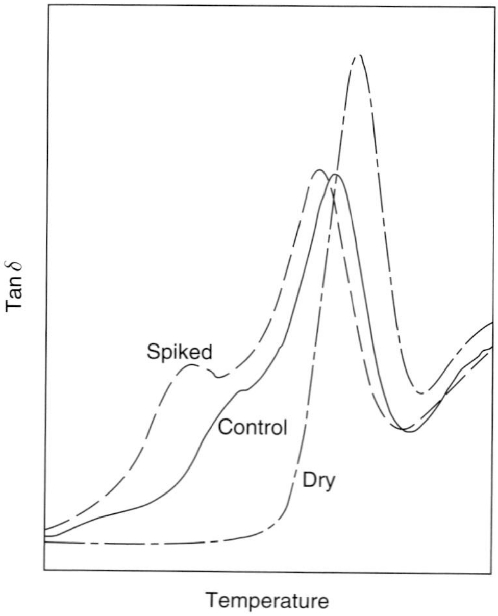

Figure 3.5. Schematic illustration of the effect of moisture absorption and thermal spiking on the relaxation spectra of the resin matrix. (Copyright 1998, *Key Engineering Materials*, by J.A. Hough et al. [10], reproduced by permission of Trans Tech.)

Considering that the initial concentration of the diffusing species is c_i within the material of thickness h for $t \leq 0$ and assuming that the concentration of the diffusing species is maintained constant and equal to c_s at the surface sample for $t > 0$, Jost [13] proposes a very useful exact solution for Equation (3.1):

$$\frac{c - c_i}{c_s - c_i} = 1 - \frac{\pi}{4} \sum_{j=0}^{\infty} \frac{1}{(2j+1)} \sin \frac{(2j+1)\pi x}{h} \exp\left[-\frac{(2j+1)^2 \pi^2 D \cdot t}{h^2}\right] \tag{3.4}$$

The practical consequences of this equation are twofolds. The moisture content at a time t as well as the time to equilibrium can be predicted using the more *friendly* Equations (3.5) and (3.6) [12].

$$M = G(M_m - M_i) + M_i \tag{3.5}$$

where

$$G = 1 - \frac{8}{\pi^2} \sum_{j=0}^{\infty} \frac{\exp\left[-(2j+1)^2 \pi^2 \left(\frac{D \cdot t}{h^2}\right)\right]}{(2j+1)^2} \tag{3.6}$$

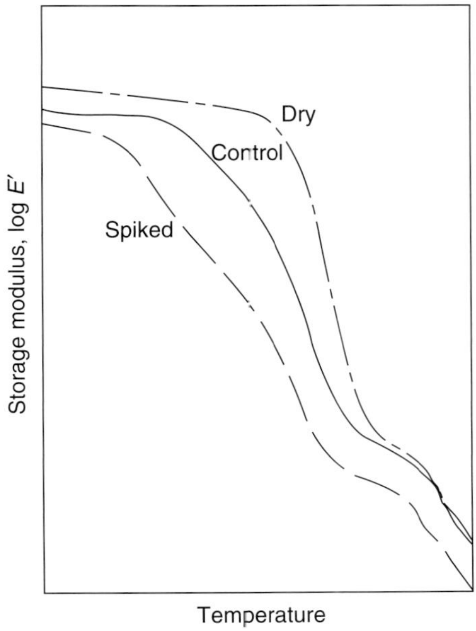

Figure 3.6. Schematic illustration of the effect of moisture absorption and thermal spiking on the DMTA storage modulus. (Copyright 1998, *Key Engineering Materials*, by J.A. Hough et al. [10], reproduced by permission of Trans Tech.)

These equations are valid for adsorption as well and for samples exposed on one side or both sides.

Springer [12] simplifies the equations further for a unidirectional problem leading to:

$$G = 1 - \exp\left[-7.3\left(\frac{D_x t}{s^2}\right)^{0.75}\right] \tag{3.7}$$

where $s = h$ if the sample is exposed on both sides and $s = 2h$ if only one side is exposed and the other side is insulated.

Parameter of importance, the diffusivity D can be measured or calculated. Note that anisotropic materials such as laminated composites have different diffusion coefficients along the various x, y, z directions. Springer [12] proposes to calculate the diffusion coefficients parallel (D_{11} and D_{22}) and normal to the fibers using equations analog to Springer and Tsai [14] (for $v_f < 0.785$ and for $D_f << D_r$, where f and r stand for fiber and resin respectively):

$$D_{11} = (1 - v_f)D_r \tag{3.8}$$

$$D_{22} = \left(1 - 2\sqrt{\frac{v_f}{\pi}}\right)D_r \tag{3.9}$$

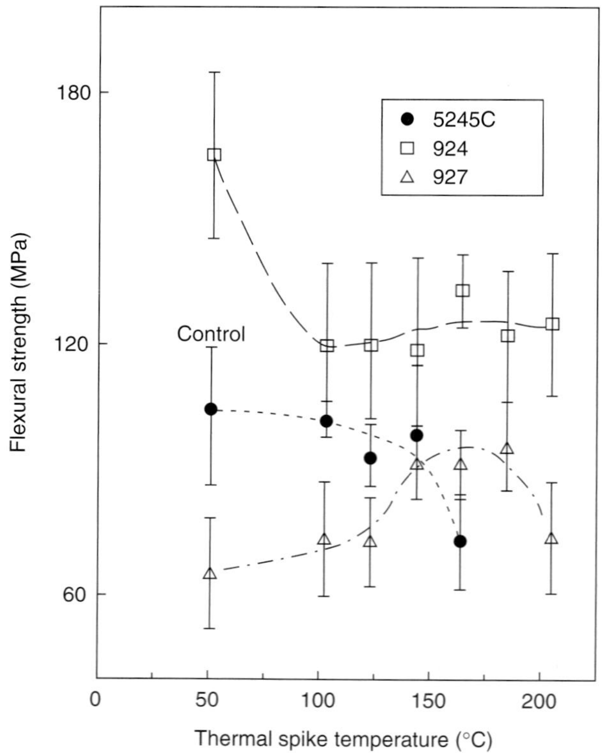

Figure 3.7. Transverse flexural strengths of wet 5245C laminates after spiking and 10 000 h conditioning (5245C/927) and 5100 h (924). (Reproduced from Hough et al. [10] by permission of Transtech.)

It is clear that in order to predict the amount of diffusing specie at a time t, it is necessary to have reliable M_∞ and D values. Immersion experiments are often required to qualify composite candidates. In such laboratory tests, the weight of the sample is measured and the specie content, for example the moisture content (M), is plotted as a function of the square root of time. The asymptote to the curve gives the M_∞ value and the slope of the linear portion gives the diffusivity coefficient D according to Equation (3.10) (Figure 3.8). Edge effects can also be accounted for and the procedure is described in the literature [12].

$$\text{Slope} = \frac{4M_\infty}{h\sqrt{\pi}}\sqrt{D} \tag{3.10}$$

It happens that the measured value of M reaches a maximum and starts decreasing after a certain time. This late decrease is related to the material loss (degradation) due to prolonged exposure to the liquid or gaseous environment. Such a phenomenon is particularly critical for marine applications: the boating industry, for example,

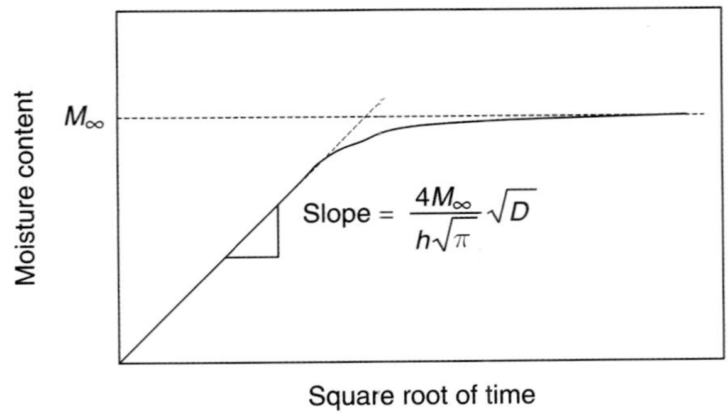

Figure 3.8. Obtainment of saturation level and diffusivity from experimental data.

strongly relies on composite materials for structural parts of the vessel, such as the boat hull. The hull is constantly immersed in lake or sea water. Exposure to salt water is known to accelerate polymer composite degradation. Furthermore, water absorption can lead to swelling and significant internal stresses build-up. Therefore, diffusion into the composite should be avoided at any price. The following industrial case study illustrates the achievement of appropriate diffusion resistance for a large composite vessel hull mainly achieved via the use of corrosion resistant materials such as vinyl esters combined with carefully selected gel coats.

Case study: Boat

Large sailboat and yacht manufacturers are preponderant consumers of polymer composites. The requirements on boat materials are common to other marine applications such as oil and gas offshore products: salt water resistance, light weight combined with high stiffness, fire resistance and no UV degradation.

It is no surprise that the boat builders heavily rely on glass reinforced epoxy vinyl ester resins. Foam sandwich structures provide most of the stiffness required for deck and hull. Carbon fibers are sometimes used to reinforce highly loaded locations and Aramid fibers can be used to improve potential impact resistance. However the Kevlar® fibers are usually used away from the surface where glass fibers are preferred in case of accidental surface abrasion. UV resistance, moisture barrier and surface finish are most often ensured by a carefully selected gel coat. Indeed, we should add to the stringent requirement list, a surface finish standard close to the automotive industry (Class A) required on very large surfaces.

This is typically achieved by spraying or painting gel coats onto the molds. The challenge is to apply the gel coat within a tolerance of 0.25 to 0.5 mm thickness. Below this thickness, the gel coat is not efficient against weathering and above this thickness, the gel coat sags with time. The boating industry is facing an additional challenge in which the increasing demand for lower emissions is guiding gel coat materials toward lower styrene contents that usually translate in a reduced product lifetime.

Figure 3.9. Yacht hull lamination. (Courtesy of Bavaria Yachts.)

Boat production rates are ranging from unique parts (Figure 3.9) to semi-mass production. The size of the hulls and the geometry similitude lead to manufacturing processes close to the windmill blade industry (see Chapter 1). Hand lamination of prepregs in a female mold and vacuum bagging are traditional methods for manufacturing. In order to reduce manufacturing costs however, boat builders are trying to apply more automated processes such as resin infusion or RTM for smaller parts.

3.2.3 Gas Permeation

The basic principles introduced in Section 3.2 can be used for both liquid and gaseous environments with the general rule that the smaller the diffusing species, the higher the transmission rate. The rate at which a gas penetrates into the material is referred to as permeation.

The importance of permeation in materials selection is traditionally illustrated by the example of polymer sparkling water bottles. However, permeability is also key for other industrial applications such as contact lenses or blood platelets storage [3,15–18]. Indeed, contact lenses require high oxygen permeability combined with excellent wettability and flexibility. Developments at the time of writing have lead to the replacement of silicone by polymethylmethacrylate (PMMA for hard lenses), poly (2-hydroxyethyl methacrylate) (pHEMA) or polypeptide films.

Selecting blood platelet storage material is an equally challenging task regarding diffusion: platelets are living organisms requiring oxygen to survive. First, they create CO_2. Second, the gas should be eliminated in order to maintain a constant pH of the solution. Third, water should be retained in the recipient. The polymer should therefore allow oxygen and carbon dioxide to permeate through the package, while hindering water molecule diffusion. The material of choice today is plasticized polyvinylchloride (PVC). This material introduced half a century ago for blood bags is still being used while other candidates were under evaluation at the time of writing [19].

Permeation typically occurs in three steps [20]:

(1) Absorption of the gas into the composite.
(2) Diffusion of the gas through the composite.
(3) Desorption from the sample surface.

In the case of gases, the diffusion coefficient (D) is directly related to the solubility (S) and the permeability (P) (volume of vapor passing through the polymer):

$$P = D \cdot S \tag{3.11}$$

The permeability P is in turn related to temperature via an Arrhenius equation of the type:

$$P = P_o \cdot \exp\left(-\frac{\Delta E}{RT}\right) \tag{3.12}$$

where ΔE is the activation energy for permeation. In SI units, P is expressed in $\dfrac{\text{cm}^3(273.15\,\text{K},\ 1.013\times10^5\,\text{Pa})\times\text{cm}}{\text{cm}^2\times\text{s}\times\text{Pa}}$, S in $\dfrac{\text{cm}^3(273.15\,\text{K},\ 1.013\times10^5\,\text{Pa})}{[(\text{cm}^3)\,(\text{atm})]}$, D in cm^2/s and ΔE in kJ/mol.

Diffusivity, permeability and solubility parameters can be of great practical importance for applications such as drug and food packaging. In particular, designing polymer molecules to have higher selectivity (oxygen to carbon dioxide) in diffusivity is the key objective these days in the food packaging industry. Table 3.2 presents values of permeability and activation energy for permeation for different polymers and diffusing species.

Permeant atoms can move inside the polymer only if spaces are available within the material. Therefore, reducing free volume and mobility will increase the diffusion barrier characteristics [22,23]. Specifically, this can be done by increasing:

- crystallinity content
- crystallite size
- molecular weight
- degree of cross-linking
- packing density
- glass transition temperature
- solubility of the diffusing species in the polymer
- by adding inorganic fillers, provided that voids do not multiply and that the fillers do not absorb the diffusing species.

Table 3.2. Permeability coefficient and activation energy for various polymers and permeants [20–22]

Polymer	Permeant	$P \times 10^{13}$ (SI units)	$P_o \times 10^7$ (SI units)	ΔE (kJ/mol)
Low density polyethylene	O_2	2.2	4.62	42.7
	N_2	0.73	329	49.9
	CO_2	9.5	62	38.9
	H_2O	68	48.8	33.5
Poly(vinylidene chloride)	N_2	0.000 70	900	70.3
	O_2	0.003 83	825	66.6
	CO_2	0.021 8	24.8	51.5
	H_2O	7.0	863	46.1
Poly (dimethylsiloxane)	He	4.429		
	N_2	2.635		
	O_2	5.862		
Poly(styrene)	He	0.168		
	N_2	0.004		
	O_2	0.021		

Permeability (P), diffusivity (D) and solubility (S) values are therefore very sensitive to the nature of the polymer, elastomers being the most permeable and semi-crystalline polymers the least. In the case of composite materials, little data is available from the literature: permeability and diffusivity have to be determined experimentally. To do so, a film is generally placed between two chambers – one containing the diffusing species and a second chamber where the diffusing species concentration is monitored [24]. The test procedures – absolute pressure method, isostatic method and quasi-isostatic method – are described in the ASTM methods of Section 3.8.

For membrane, filtering or food packaging applications, another property of interest is the selectivity. The first reported application of selective permeation through polymers was for the extraction of H_2 gases from a gas stream by the Monsanto Company in 1977 [22]. This process is still used today for the separation of H_2, CO_2, H_2S and N_2. Pervaporation (liquid separation) was introduced at an industrial scale in the early 1980s by Le Carbone-Lorraine/GFT Co. to separate industrial alcohols, methyl ether ketone and ethyl acetate from diverse mixtures [22,25]. The materials selection for such applications requires the comparison of the permselectivity. This coefficient is defined as the ratio of the materials permeability toward two different species (e.g. a and b):

$$\mu_{\mathrm{a/b}} = \frac{P_{\mathrm{a}}}{P_{\mathrm{b}}} = \frac{S_{\mathrm{a}} D_{\mathrm{a}}}{S_{\mathrm{b}} D_{\mathrm{b}}} \tag{3.13}$$

Equation (3.13) can be used for gases as well as for liquids.

Diffusion is a central issue in hydrogen storage applications. Hydrogen storage is in turn key to enabling fuel cell technologies. This matter is discussed in the following case study dealing with recent static and transportation fuel cell developments.

Case study: Fuel cells

Environmental consequences of mass energy production are a center topic in today's society. Global warming accelerated (if not caused) by carbon dioxide (CO_2) emissions has consequences of increasing impact on the earth and its population: melting of ice at poles, flooding and accelerated desertification of arid areas (Figures 3.10 and 3.11). Most developed countries have agreed to implement measures aiming at reducing CO_2 emissions.

However, the issue is complex. The need for power and electricity is constantly rising. If in developed countries consumption could be slowed down by greater awareness on energy saving possibilities, developing countries do not have this option at a time when clean water and electricity access are recognized requirements to fighting hunger and malnutrition. The use of fossil fuel, however, cannot be increased without catastrophic impact for the next generation.

Along with water, wind and solar power, fuel cells constitute today a significant hope in sustainable development. Fuel cells are electrochemical energy conversion devices. Hydrogen and oxygen are combined to produce water, heat and electricity. Under the condition of using pure hydrogen, the emissions from such a process are null.

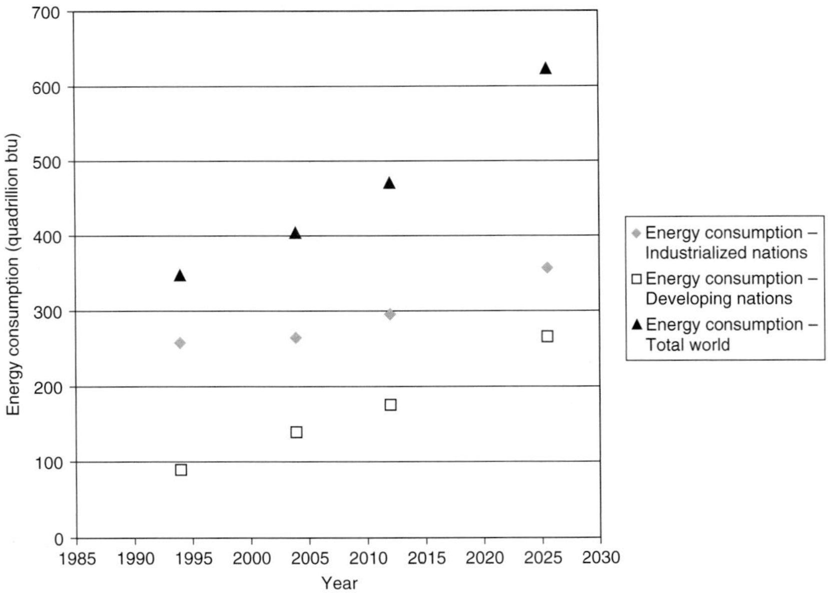

Figure 3.10. World energy consumption, 1990–2025 [26].

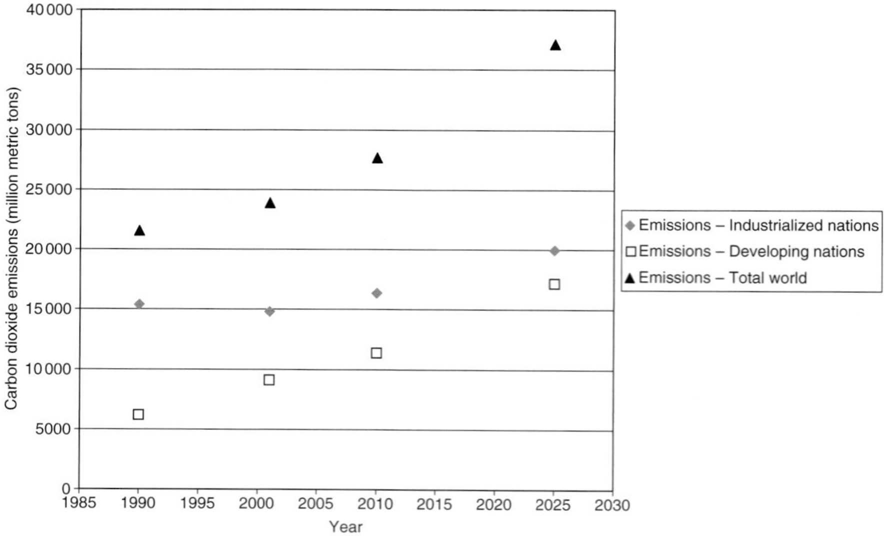

Figure 3.11. World CO_2 emissions, 1990–2025 [26].

From a technology perspective, composites have an important role to play in fuel cell technology development. Polymer-based materials are prime candidates for membranes, end-plates or storage tanks. However, hydrogen supply is a major challenge still hindering the development of fuel cells. Indeed, hydrogen can be stored as a gas, liquid or solid matrix. The type of hydrogen supply then determines the type of technology to be used. At the time of writing, no consensus was reached on the best hydrogen storage alternative and possibilities are still under heavy discussion [27]. This lack of agreement leads to a great diversity in few cell concepts.

Fuel cells are usually classified by the type of electrolyte: polymer (direct methanol fuel cell), phosphoric acid (phosphoric acid fuel cell), fast polymer (proton exchange membrane, PEM), molten carbonate (molten carbonate fuel cell) and solid oxide (solid oxide fuel cells).

Proton exchange membranes are attractive candidates as they operate within a reasonable temperature range (50–80°C). The company Plug Power, for example, has commercialized a stationary fuel cell system operating on natural gas with 2.5, 4 and 5 kW at 120 and 240 V AC [28] (Figure 3.12). Such outputs are obtained by stacking cells together. Plug Power uses bipolar plates to separate the cells, each plate acting as the anode of one cell and the cathode of the next cell. The two main functions of the composite plates are reaction containment within a given cell and current conduction. The combined functions lead to complex requirements including high electrical conductivity, high thermal conductivity and corrosion resistance. Additionally, the material should be inert to eliminate the risk of contaminant emission. Last but not least, an excellent machinability is required as the distribution of oxygen and hydrogen is achieved by intricate flow channels in the plates. Facing

Figure 3.12. Uncovered view of Gensys stationary fuel cell system. (Courtesy of Plug Power.)

Table 3.3. PEM fuel cell bipolar plates properties

Fuel cell composite materials	BMC 940 bipolar plates
Specific gravity	1.82
Flexural strength	46.9 MPa
Flexural modulus	11.7 GPa
Compressive creep, 1000 h, 80°C	0.3% at 1.4 MPa
Conductivity (in x, y directions)/ (in z direction)	95 S cm^{-1}/70 S cm^{-1}
Thermal conductivity	16 W/m K

those complex materials specifications, Plug Power selected a highly conductive graphite filled vinyl ester bulk molding compound supplied by Bulk Molding Compounds (Table 3.3).

Though successfully applied to static applications, the major potential for composites probably lies in low-weight liquid hydrogen tanks. Liquid hydrogen, with a high liquid density of 70.8 kg/m^3 is a competitive candidate for the automotive industry where fuel tank volume should be minimized [27]. Liquid hydrogen fuel tanks are subjected to difficult environmental conditions such as cryogenic temperatures and/or high pressure. Indeed, hydrogen remains in the liquid state under normal atmospheric pressure for a temperature of −252°C. At ambient temperature however, the pressure should be increased to 1 GPa to maintain the liquid state! [29]. In addition to low temperature and high pressure resistance, the material if used for transportation purposes should comply to stringent safety regulations (fire and crash resistance).

Fiber reinforced materials and hybrid composites are thought as excellent candidates for structural and environmental resistance. Quantum Technologies Worldwide Inc, for example, has successfully designed for this purpose a high-pressure composite hydrogen tank combining a metal alloy and a polymer liner resistant to embrittlement and stress corrosion cracking.

3.3 LIQUID AND GASEOUS ENVIRONMENT EFFECTS ON THE MATRIX

Diffusion can have many (positive or negative) effects on polymers. Some diffusion effects such as molecular degradation by hydrolysis or microcracking are irreversible and lead to permanent materials degradation. On the other hand, some of the moisture-induced changes such as swelling or plasticization are reversible and disappear with desorption. Diffusion rules and equations are generally applicable to desorption. However, for highly polar polymers such as polyamides polymers, the strong polarity might prevent loss of moisture for temperatures up to the glass transition.

The exact effects of diffusion on the polymer depend on the nature of both material and solvent. For example, it is established that water plasticizes epoxies significantly [30]. A good metric for plasticization is not only the stiffness (static or storage modulus, see Chapter 2) but also the glass transition temperature. Indeed, plasticization typically translates into an increased compliance and a broader glass transition temperature happening at lower temperatures [31,32]. Wright indicates that as a rule of thumb a polymer's glass transition temperature decreases by 20°C for every 1% of moisture absorbed [33].

There is no need to convince the reader of the potential impact of such a drop for polymer composite parts designed to be operated (and supposed to remain) in the glassy state. The influence of water absorption on transition temperatures in polymers is therefore detailed thereafter.

3.3.1 Influence of Water Absorption on Transition Temperatures in Polymers

The Kelley-Bueche [34] equation relates the glass transition temperature to the properties of the *polymer-diluent system*:

$$T_{\mathrm{g}} = \frac{\alpha_{\mathrm{p}} V_{\mathrm{p}} T_{\mathrm{gp}} + \alpha_{\mathrm{d}} (1 - V_{\mathrm{p}}) T_{\mathrm{gd}}}{\alpha_{\mathrm{p}} V_{\mathrm{p}} + \alpha_{\mathrm{d}} (1 - V_{\mathrm{p}})} \tag{3.14}$$

with

T_{gp} as the glass transition temperature of the polymer
T_{gd} as the glass transition temperature of the diluent
α_p as the coefficient of thermal expansion of the polymer
α_d as the coefficient of thermal expansion of the diluent and
V_p as the volume fraction of the polymer.

Unfortunately, the coefficient of thermal expansion is often unknown and Equation (3.14) is difficult to apply. Luckily, Equation (3.14) can be simplified for specific cases such as for moisture absorption in Epoxies [31]:

$$T_g = V_p T_{gp} - (1 - V_p) T_{gd} \tag{3.15}$$

Equation (3.15) still requires knowledge of the diluent glass transition temperature. This information can be found in the literature for common solvents; for example, the glass transition temperature of water ranges between 128 and 165 K [31,35]. For other substances, Nielsen [36] proposes an experimental equation mainly based on free volume considerations:

$$\frac{T_g}{T_m} \approx \frac{1}{2} \ to \ \frac{1}{3} \tag{3.16}$$

where T_m is the material melting point.

Equation (3.16) does not consider features such as the hydrogen bonding capabilities of water and can lead to calculated glass transition temperatures greater than actual values [31,37]. It is therefore strongly recommended to perform thermomechanical tests (such as a DMA described in Chapter 2) to determine the actual glass transition temperature after immersion.

Moisture can also induce modifications to the secondary, β and tertiary, γ relaxations. For example, in the case of Nylon 66, the β-relaxation magnitude is increased (i.e. water confers more mobility to the polymer branches), when the γ-relaxation temperature is decreased (a phenomenon referred to as *antiplasticization*). Such transition temperature shifts with moisture content for Nylon 66 are illustrated in Figure 3.13.

Secondary relaxation amplification phenomena are not restricted to water. High density polyethylene, for example, shows none or a very weak relaxation peak from −30 to 50°C in the dry state. After CCl_4 exposure, however, a distinct relaxation appears from −60 to −80°C [38].

3.3.2 Polymer Swelling

Absorption of liquid or vapor generally leads to matrix swelling. In addition, the fibers themselves might swell upon moisture uptake (case of Kevlar reinforcement). Swelling modifies the state of stresses within the composite and therefore influences

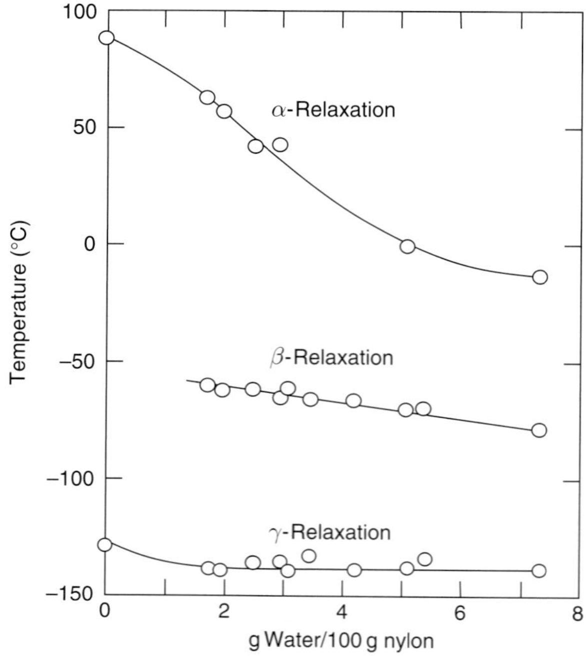

Figure 3.13. α- (or T_g), β- and γ-Relaxation shift with moisture content. (Copyright 1980, ACS symp., by Starkweather [38] reproduced by permission of ACS.)

the moisture uptake. The state of stress and resultant strain is particularly sensitive to the part geometry since water diffusion/uptake is a slower process as compared to thermal heat transfer. This in turn results in moisture gradients which then set up stress gradients, specially early on in the uptake process. It can be a good indicator of the materials integrity. Sudden swelling can indeed indicate a change in the degradation phenomenon such as delamination or fiber debonding.

In a Fickian diffusion process, the absorption coefficient is related to the material and water densities. For a homogeneous material [39]:

$$\beta = \frac{1}{3}\frac{\rho}{\rho_{\mathrm{w}}} \tag{3.17}$$

where β is the strain per unit change of moisture content, ρ is the density of the material and ρ_{w} is the density of water.

As a first approximation for fiber-dominated composite (unidirectional carbon- or glass-fiber reinforced plastic), the swelling in the fiber direction can be neglected. In the transverse direction (matrix dominated), we can write [39]:

$$\beta_2 = \frac{\rho_{\mathrm{c}}}{\rho_{\mathrm{m}}}(1 - V_{\mathrm{m}})\beta_{\mathrm{m}} \tag{3.18}$$

where the subscripts m and c refer to matrix and composite and V_m is the matrix volume fraction.

Equation (3.18) was derived for a perfect system and leads to approximate values that can be used as indicators. The calculation results, however, should always be validated by experimental values. Swelling data can be obtained during moisture uptake experiments by simultaneously measuring the composite sample dimensions and weight.

3.3.3 Changes in the Thermo-mechanical Properties

Measuring the effect of moisture on thermo-mechanical properties of composites is not trivial. Indeed, most methods such as DMA and DSC require heating of the samples, during which some water may evaporate.

Moisture was, however, found to consistently influence the properties of thermosets by reducing modulus, glass transition temperature and coefficient of thermal expansion. Moisture was also reported to induce creep and lower fatigue life at room temperature [40,31]. Indeed, the presence of a solvent in the material modifies the viscoelastic response of the polymer composite. Analogies between fluid content and temperature were even drawn in various studies [41] and the possibility to introduce a moisture related shift factor established. A master curve much alike the one presented in Chapter 2 for time–temperature equivalence can be drawn, where the shift factor is now a combination of the temperature- and the moisture-dependent parameters. Much like temperature, aging in moist environments can lead to changes in the ductile–brittle transition and therefore dictate damage and failure mode.

The consequences of moisture absorption naturally extend onto the polymer dielectric properties [40] and insulators loss factors (see Section 4.2.2.2) can increase dramatically with increased moisture contents.

3.3.4 Limits of the Model

It is not uncommon to see experimental data, even in the case of water diffusion in simple polymers, deviating significantly from the predictions obtained from Equations (3.1) and (3.2) – the diffusion is then said to be *non-fickian*.

Water is probably the most basic and aggressive solvent. Even then, different types of water can be identified influencing the absorption and desorption phenomena in different ways. Bulk water (I) is free or pure water. On the other hand, bound water molecules (II) are closely attached to the polymer hydroxyl groups after intermolecular hydrogen bond scission. Bound water can only be found in polar polymers. A portion of this bound water never crystallizes and constitutes a third potential type of water called non-freezing bound water.

Water diffusing in polyester or vinyl ester can be found in both free and bound water forms (Figures 3.14 and 3.15). The diffusion rates appear to depend upon the

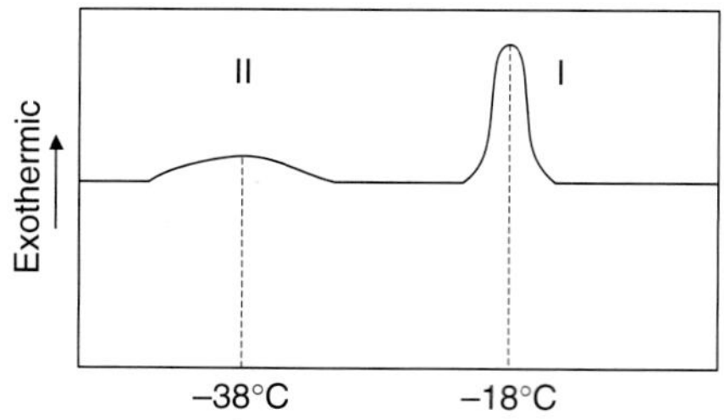

Figure 3.14. Cooling curve.

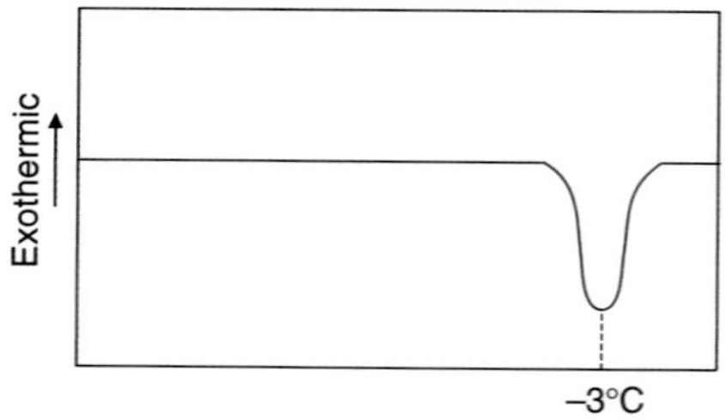

Figure 3.15. Heating curve.

interaction between the two water phases. The Langmuir model [30,42] is applicable to non-fickian diffusion cases involving two interacting phases in the solvent and can allow to distinguish between the two different types of water.

Osmosis is another case of non-Fickian diffusion occurring in materials. Indeed, water-soluble impurities contained in polymers can further pull water inside the sample, resulting in extensive plasticization of the resin and eventually blistering around the inclusion [7]. Osmosis should therefore be minimized at all costs in industrial applications. Minimizing the impurity level in the composite and avoiding abrupt polymer/gel coat property transitions usually limit the osmosis phenomenon.

Macroscopic voids will also influence diffusion mechanisms and rates. It is sometimes assumed (but not proven) that moisture can be transported by capillary action along cracks and fiber–matrix interfaces. To model such phenomena is unrealistic for industrial products as it would require a complete knowledge of the material state including defect location and size. In the frame of industrial development projects, parts as close as possible to the final composite product should therefore be tested. Indeed, results on the part (and not samples) encompass all effects and are reflected by the empirically determined diffusivity coefficients.

3.4 LIQUID AND GASEOUS ENVIRONMENT EFFECTS ON THE FIBERS

Separating the diffusion-induced damage mechanisms in the fibers from the global processes in the material is a challenging task. Diffusion in the polymer is time dependent and often involves swelling. This swelling translates into localized time-dependent stresses on the fibers. Calculating those stresses requires a perfect knowledge of the microscopic composites state (presence of voids, location of the amorphous and crystalline phase etc.) and is generally unrealistic. Those local stresses can, however, not be overlooked: indeed the failure mechanisms can be significantly altered by water exposure combined with mechanical stresses. For example, the matrix plasticization can modify the stress transfer to the fibers, relaxing stresses around fiber cracks and slowing down damage mechanisms. Due to the complexity of the matter, only the most important concepts directly related to fiber degradation are introduced thereafter.

Advanced polymer matrix composites often involve a ceramic phase such as glass or carbon. Due to their different chemical nature, glass or carbon have significantly distinct responses to water exposure. Carbon fibers are rather inert and resistant to gases and solvents. Glass fibers, on the other hand, are far more sensitive and react to water exposure especially under load. Stress corrosion usually refers to the degradation process in the case of combined water and load exposure.

Standard ceramic models for damage mechanism can be used to model the crack propagation in the fibers. The crack growth velocity is related to the stress intensity factor by [30]:

$$V = AK_{\mathrm{I}}^{n} \tag{3.19}$$

Figure 3.16 illustrates typical crack growth rates in ceramics and indicates the presence of three regions with different governing mechanisms.

The stress intensity factor is defined as the ratio of the maximum stress at the crack tip divided by the nominal applied tensile stress [43]:

$$K_{\mathrm{I}} = \frac{\sigma_{\mathrm{m}}}{\sigma_{\mathrm{o}}} \tag{3.20}$$

From Equation (3.20), we see that the intensity factor, that is the crack growth velocity, strongly depends upon the state of stresses at the crack tip which in turn is influenced by the moisture content. Indeed, the presence of water – or as a matter of fact ammonia, hydrazine and formamide [44] – was shown to significantly increase the rate of crack growth in glass fibers. Small variations in the stress intensity factor can lead to a change in the region (Figure 3.16). Moisture absorption or desorption can therefore trigger a passage from a stable crack growth (Region II) to a very rapid crack growth (Region I or III).

With this in mind, glass-fiber reinforced polymers still remain prime candidates for corrosion-resistant applications. Degradation processes of E-glass in neutral

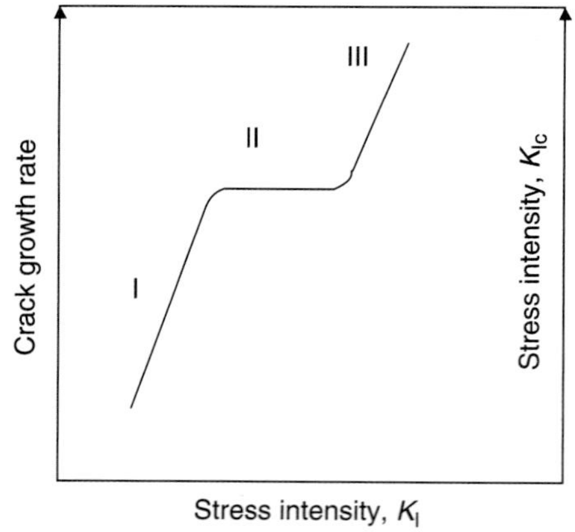

Figure 3.16. Three regions for crack growth in ceramics.

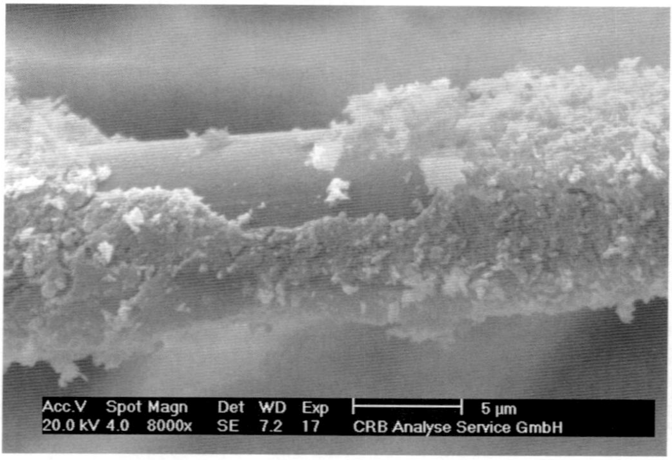

Figure 3.17. Micrograph of E-glass fiber in 5% NaOH at 23°C after 28 days. (Courtesy of DBW fiber Neuhaus.)

aqueous solutions are generally slow, but quickly become unacceptable in highly acidic or alkaline solutions. Micrographs of E- and S-glass fibers after prolonged exposure (28 days) to 5% NaOH are shown in Figures 3.17 and 3.18.

To better resist such harsh environments different variations of glass fibers such as C-glass (chemical glass), E-CR glass (electrical-corrosion resistant glass) or AR-glass (alkali-resistant glass) were developed at the time of writing, resulting in slower degradation rates in specific aggressive environments. Table 3.4 compares typical properties of E-glass and E-CR glass fibers.

Figure 3.18. Micrograph of S-glass fiber in 5% NaOH at 23°C after 28 days. (Courtesy of DBW fiber Neuhaus.)

Table 3.4. E-glass and E-CR glass property comparison [45]

Properties	E-glass (direct roving)	E-CR glass (direct roving)
Tensile strength	1500 MPa	1650 MPa
Tenacity	83 GPa	88 GPa
Acid resistance: 24 h soak at 96°C	43%	7%
Electrical properties	0.4 Ω cm	1.16 Ω cm

Good acid resistance and high dielectrical strength make E-CR glass fibers attractive candidates for electrical applications such as insulator rods. Indeed, in the presence of moisture and ozone (see Section 4.2.3), nitric acid can be generated leading to embrittlement of the rods and premature failure. The use of E-CR glass reinforcement can drastically increase the insulator lifetime. For example, in an experiment, 24 mm E-glass and E-CR glass reinforced polymer samples were dipped in nitric acid. The E-glass-based composite failed after 100 min while the time to failure exceeded the 5700 min for E-CR glass [45]. In addition to the excellent E-CR glass fiber corrosion resistance, the dielectric strength of the E-CR glass fibers is slightly in excess of the one of E-glass fibers (320 and 310 V/mil respectively) making them perfectly adapted to medium and high voltage applications.

Recent advances in the field of alkali exposure resistance also provide end-users with more durable fibers. An example of a micrograph after alkali exposure of newly developed Powertex® fibers is shown in Figure 3.19 and can be directly compared with Figures 3.17 and 3.18.

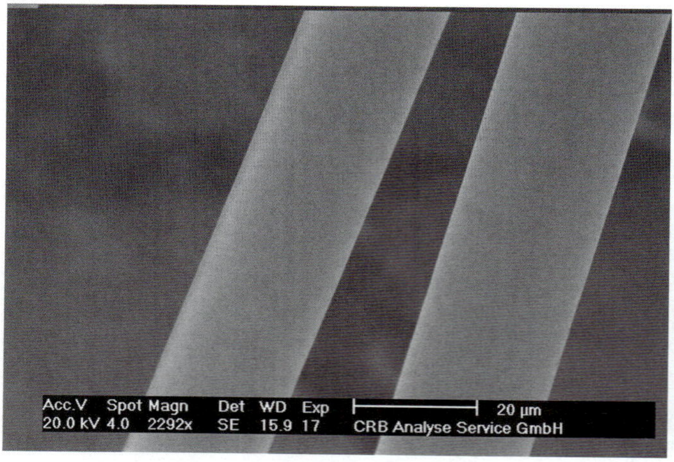

Figure 3.19. Micrograph of Powertex® fiber in 5% NaOH at 23°C after 28 days. (Courtesy of DBW fiber Neuhaus.)

Though limited and despite such advances, degradation in the modified glass fibers still occurs in most cases: indeed in the later case of AR-glass, for example, the alkaline environment leads to limited silica leaching from the fiber surface, leaving a ZrO_2-rich surface which inhibits the corrosion process. Unfortunately, the porous ZrO_2 progressive growth can also result in 50% strength loss after extended exposures within a magnitude of ten years [39].

3.5 LIQUID AND GASEOUS ENVIRONMENT EFFECTS ON THE COMPOSITE

We are now aware that moisture absorption can have dramatic effects on the fiber crack growth rate. The complexity of the problem further increases when the fibers are imbedded in a matrix due to fiber–matrix interactions and the presence of a fiber–matrix interface. In addition, the role of this interface has clearly been shown to be crucial in affecting the properties of the bulk composite [46–48]. Numerous composite level responses can be dramatically influenced by the correct choice of this region. This includes moisture diffusivity, saturation moisture levels, post aging property reduction, unaged fatigue lifetimes as well as other key quasi-static properties such as compression, tensile and flexure strength.

Diffusion and degradation phenomena are complex and multiple due to the variety of environments, polymers and reinforcement (types and shapes). Though difficult to generalize, water exposure degradation in composites often occurs in phases [30]. The first step is characterized by a matrix modification such as creep or microcracking. Further, moisture uptake then induces matrix plasticization, followed by stress corrosion of the fibers and possibly debonding at the fiber–matrix interface.

After having reviewed the diffusion effects on the isolated matrix and fiber constituents, we will now consider the composite *as a whole* in order to characterize the composite global response to liquid and gaseous exposure.

3.5.1 Diffusion in Composites

The most comprehensive model to date for diffusion in composites [30] is probably Marom's model. Marom et al. [49–51] introduced a comprehensive model that describes the diffusion of water in composites. This model includes the effects of stresses by relating them to the free volume (Doolittle equation). Moisture progression is obtained via Fick's laws. Finally, classical lamination theory (CLT, see Chapter 5) is used to calculate stresses and strains. Though capillary water motion was not considered, model and experimental data showed good correlation. The authors of this study also pointed out that the moisture saturation level M_∞ was decreasing with increasing stress. Fiber orientation was found to influence diffusion as well and matrix dominated 90° plies suffered greater moisture exposure consequences than 0° plies.

Practically, the experimental procedure for diffusion assessments in composites and polymers is identical. Figures 3.20 and 3.21 show examples of experimental absorption curves for composite materials thought as candidates for oil and water bearing applications [52]. After immersion in 80°C water, carbon-fiber reinforced Teflon (CF/PFA) showed almost no sign of absorption. Under identical conditions, moderate weight uptake could be observed for carbon-fiber reinforced polyetheretherketone (AS4/PEEK) and polyphenylene sulfide (AS4/PPS).

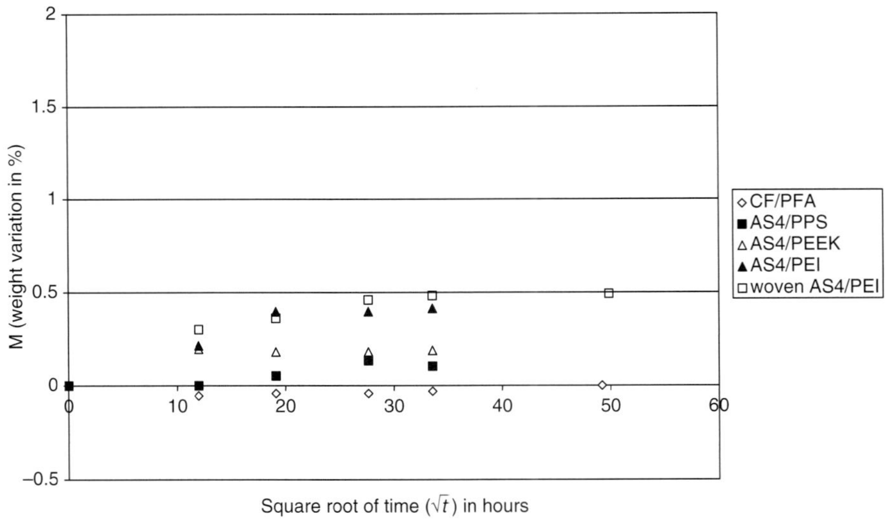

Figure 3.20. Weight variations (M) versus $\sqrt{t}$ for samples immersed in water at 80°C [52]. (Copyright 2002, *Polymer Testing*, by C.A. Mahieux et al., Elsevier.)

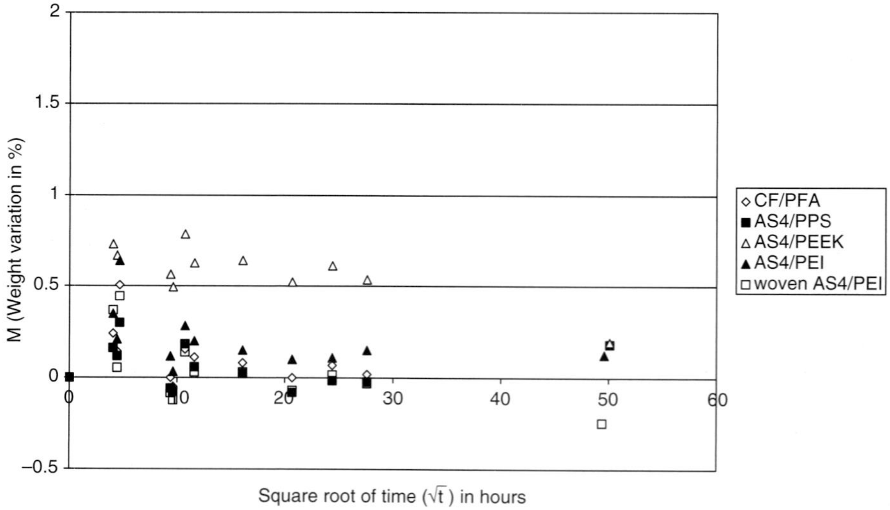

Figure 3.21. Weight variations (M) versus $\sqrt{t}$ for samples immersed in oil at 80°C [52]. (Copyright 2002, *Polymer Testing*, by C.A. Mahieux et al., Elsevier.)

Woven and unidirectional carbon-fiber reinforced polyetherimide (AS4/PEI) exhibited larger moisture uptakes. These results clearly illustrate that absorption rate and saturation level depend on the nature of both matrix and fibers.

Figure 3.21 presents the weight variations of the same composite samples, this time immersed in 80°C industrial oil. The diffusion of large oil molecules in the matrix-free volume is rather unlikely. The samples were therefore expected to exhibit null mass variation. This assumption was validated for the AS4/PPS, CF/PFA and AS4/PEI samples. Surprisingly, the PEEK-based composite exhibited a rapid and significant weight uptake. Though no explanations were given for this phenomenon, the weight uptake is thought to be related to the presence of macroscopic voids in the composite. This concrete example reveals the complexity of environmental exposure and evidences the systematic need for performing experiments as close as possible to operating conditions, even when theory does not anticipate problems or changes [52].

3.5.2 Effects of Exposure on Composite Properties

The potential effects of liquid and gas diffusion in composites are numerous and include changes in the glass transition temperature, alterations of the elastic constants, modifications of the materials ductility and strength. All reversible and irreversible effects on the matrix listed in Section 3.3 apply to composites. Additionally, and in the case of composites, we should add to this long list potential irreversible fiber damage.

3.5.2.1 Changes in transition temperatures

The glass transition temperature T_g is greatly influenced by the presence of solvents in the material. This was shown for the polymer in Section 3.3.2. For an equivalent weight uptake however, the change in the T_g of the pure polymer might not be the same as of the composite. The magnitude of the change in the transition breadth and magnitude will be determined experimentally.

Dynamic mechanical analysis performed on the AS4/PPS samples of Section 3.5.1 after oil immersion at elevated temperature show no significant difference in the transition temperatures (Figure 3.22). CF/PFA samples, however, exhibit a consistent glass temperature reduction of 30°C and a 50°C shift in the rubbery flow region (Figure 3.23).

Composite secondary relaxations (such as β relaxations) can also appear or be strongly modified as a result of solvent exposure. This was shown by a series of studies investigating the effect of boiling water on epoxies with various reinforcements [53–58]. The changes in magnitude and temperature of the β transition after exposure were found to depend upon the chemical nature of the epoxy, static and cyclic mechanical and thermal stresses and moisture content. For example, when the pure polymers did not exhibit any changes after exposure to boiling water, the composites clearly showed a new β transition with larger magnitude for glass-fiber reinforced epoxies than for carbon-fiber reinforced samples [30].

3.5.2.2 Changes in mechanical response

Solvent absorption generally results in matrix plasticization. Short and random fiber composites therefore suffer most from moisture exposure. Unidirectional reinforced

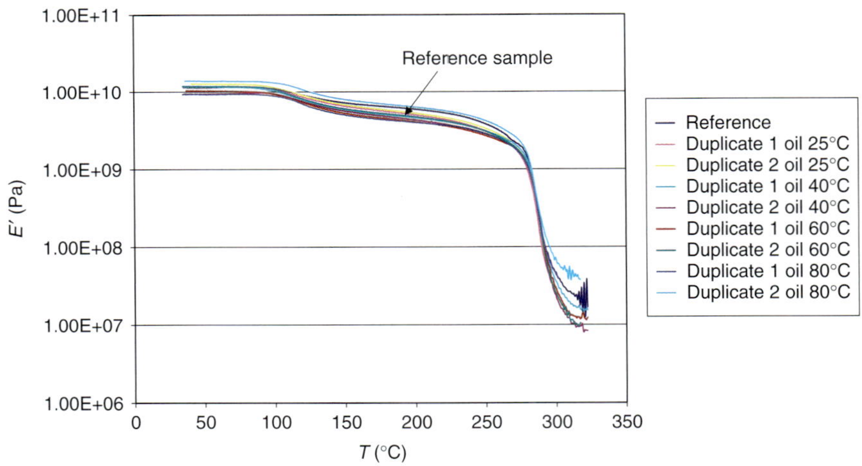

Figure 3.22. DMA AS4/PPS after immersion in oil at room temperature, 40°C, 60°C and 80°C [52]. (Copyright 2002, *Polymer Testing*, by C.A. Mahieux et al., Elsevier.)

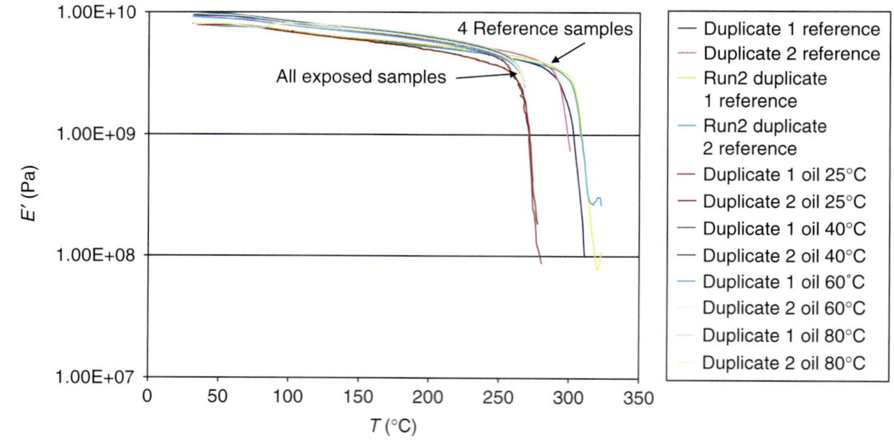

Figure 3.23. DMA AS4/PFA after immersion in oil at room temperature, 40°C, 60°C and 80°C [52]. (Copyright 2002, *Polymer Testing*, by C.A. Mahieux et al., Elsevier.)

fiber dominated composites also generally experience a drop in their transverse elastic moduli. Figure 3.24 shows a clear plasticization of the AS4/PFA composite samples of Sections 3.3.2 and 3.5.2.1 aged at 80°C in water. The storage modulus E' was measured by DMA. The modulus exhibits a spectacular drop with temperature. Aged samples show storage moduli values at 200°C almost 10 times lower than non-aged samples. Between 200 and 300°C, the storage modulus of the aged samples experiences an increase to finally reach non-aged sample values at 300°C.

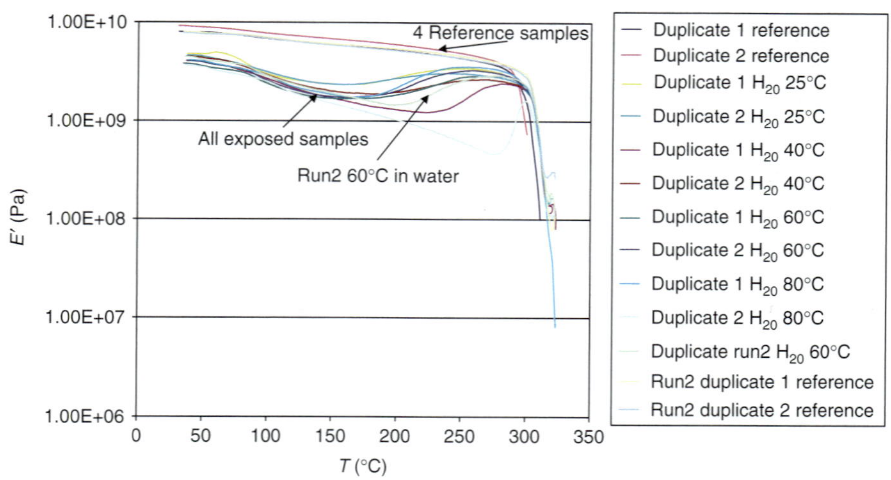

Figure 3.24. DMA AS4/PFA after immersion in water at room temperature, 40°C, 60°C and 80°C [52]. (Copyright 2002, *Polymer Testing*, by C.A. Mahieux et al., Elsevier.)

This modulus increase is due to water desorption during testing under the effect of temperature and illustrates the reversibility of the process. Thermomechanical testing is not trivial. Indeed, the solvent can evaporate and microscopic phenomena such as re-crystallization can occur under the effects of the rising temperature.

During a composite design process, moisture uptake and related effects on mechanical properties should be considered at a very early stage. Indeed, the performance of a composite product made of the AS4/PFA material of Figure 3.24 should be calculated in the original state with a storage modulus equal to 10^{10} Pa at room temperature and with a reduced $0.5.10^{10}$ Pa modulus after a three-month water exposure at room temperature.

Such elastic property variations in the composite influence local damage and global failure modes. Solvents may significantly modify the threshold for sudden failure (Figure 3.25). Studies on polyester and glass epoxy composites [30,59] revealed wet specimen behaviors significantly different under static and cyclic mechanical loads. Indeed, fiber debonding resulting from moisture absorption under static loads was found to slow down crack propagation by relieving stresses at the crack tips. On the other hand, crack growth rates were significantly increased by moisture absorption under cyclic load.

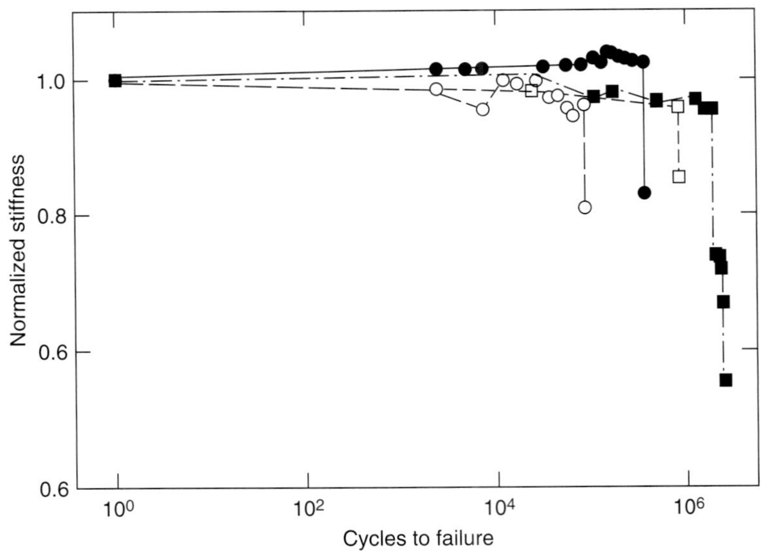

Figure 3.25. Typical stiffness reduction curves for samples cyclically loaded at 65% UTS, for both all-glass-fiber and hybrid samples tested under dry and wet conditions. Unidirectional E-glass composites (● (dry), ○ (wet)) are compared to hybrid composites with 25% of the fibers by volume of Type A carbon fiber (PAN HTA 6000 Asahi Nippon) (■ (dry), □ (wet)). Fiber volume fraction (V_f) was 30%. (Copyright 2003, *Fatigue in Composites*, by B. Harris [7], reproduced by permission of Woodhead publishing Ltd.)

3.5.2.3 Changes in the failure mechanisms

The behavior of composites can be significantly altered when subjected to an acidic or alkali environment. Damage extent and failure mechanisms vary as a function of the materials nature and the exposure conditions. Schutte [30] illustrates the specificity of the rupture process by the following example. Glass polyester composites exposed to sulfuric acid generally fail in the exposed region. On the other hand, glass epoxy samples break in the region that is not immersed in the acid [60]. If the acid is now switched to hydrochloric acid, the glass epoxy composites also fail in the immersed region.

The final failure of fiber-dominated composites is generally controlled by the fibers (and the interface) though the state of stress can also be significantly influenced by changes in the matrix (see Chapters 5 and 6). Carbon fibers are generally inert except for highly oxidizing acids [7]. Glass fibers on the other hand are very sensitive to environmental stress corrosion cracking (ESCC). S and R glasses usually show better chemical resistance than commonly used E-glass. Aramid fibers are in turn extremely sensitive to moisture with saturation levels up to 5% and can oxidize under UV exposure [61,62]. Kevlar fibers are also known to undergo hydrolysis irreversible damage [63].

In an effort to generalize the process of corrosion damage, Jones [7] proposes the presence of three regions describing the lifetime of E-glass composites undergoing ESCC (Figure 3.26).

The first region is controlled by standard stress rupture laws and crack propagation in composites. Region 2 is characterized by the highest degradation rate and is dominated by environmental stress corrosion failure resulting from the concomitant effects of static fatigue (crack propagation) and glass degradation. The cracks first develop in the fiber region (at the presence of a flaw) then progress in the matrix to the next fiber. The fibers are greatly weakened by the acidic environment and failure of the composite is accelerated as a function of the applied stress. Region 3 is characterized by a lower degradation rate where the crack propagation is slow enough to enable corrosion in the fiber resulting in crack tip radius increase.

Figure 3.27 is a micrograph of a stress corrosion fracture surface resulting from nitric acid exposure on an E-glass fiber epoxy composite. Figure 3.27 evidences a significant amount of fiber pull-out. Stress corrosion cracking can occur in many

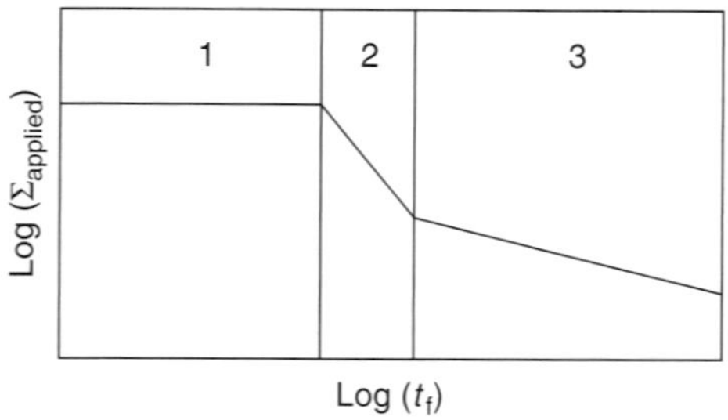

Figure 3.26. Three ESCC regions for glass-fiber reinforced plastics.

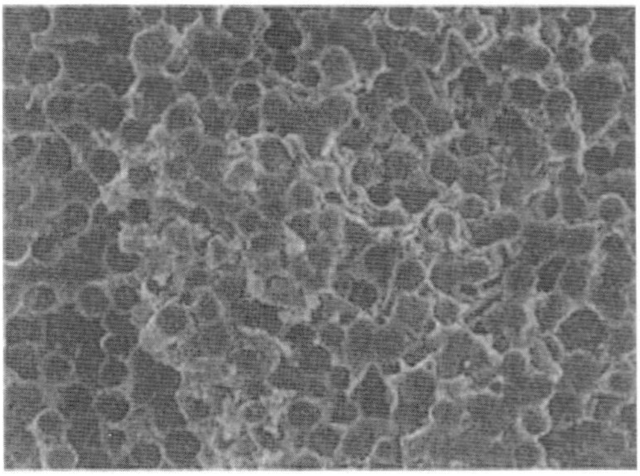

Figure 3.27. Stress corrosion fracture surface from nitric acid at 500× for an E-glass/Epoxy [64]. (Copyright 2001, *Journal of Composite Materials*, by T.D. Ely et al., reproduced by permission of Sage Publications.)

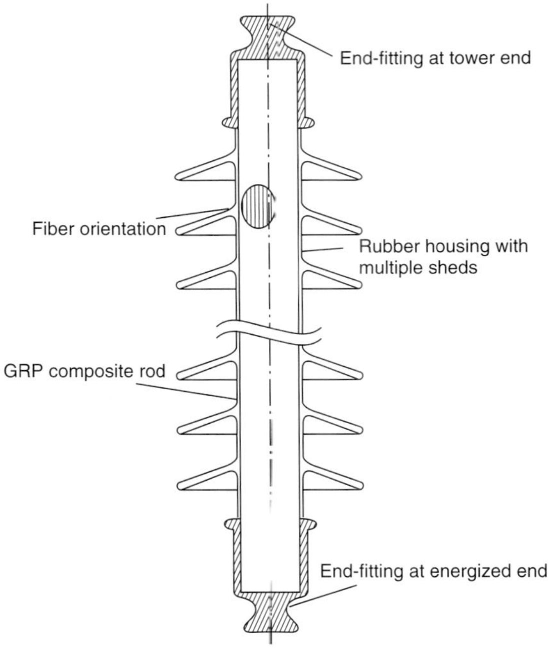

Figure 3.28. Schematic diagram of a composite suspension insulator [64]. (Copyright 2001, *Journal of Composite Materials*, by T.D. Ely et al., reproduced by permission of Sage Publications.)

industrial applications. The E-glass pultruded composite rods used in suspension insulators, for example, are known to fail under the presence of an acidic liquid despite low mechanical load levels (Figures 3.28 and 3.29). Further industrial examples are developed in the following case study.

Figure 3.29. Brittle fracture surface of a 500 kV composite suspension insulator [64]. (Copyright 2001, *Journal of Composite Materials*, by T.D. Ely et al., reproduced by permission of Sage Publications.)

Case study: Corrosion resistance – Sewer pipes

Corrosion can occur at different stages of a product life and is a main concern for many industrial applications including infrastructure, pharmaceutical, chemical, food processing, transportation, marine and utilities. The yearly corrosion costs for the US Department of Defense was estimated around US$20 billions [65] and the total direct cost for the US state in the range of US$280 billions [66].

Historically, the use of composite materials as replacement of stainless steel started in the chloro-alkali facilities of the pulp and paper industry in the early 1950s [67]. Ten years later, the metal treatment and refining industry introduced glass-reinforced plastics for electrowinning process tanks. More recently, the wastewater treatment industry started viewing glass reinforced plastic composites as prime candidates for sewer rehabilitations. Corrosion can be delayed by different measures including regular operation corrosion maintenance (such as repainting) and usage of corrosion resistance materials. Polymers are often used as corrosion-resistant coating materials and glass-fiber reinforced composites are attractive candidates for structural applications. The use of a corrosion-resistant material often increases the original part price but can offer substantial savings when considering reduced maintenance costs and extended part lifetime (Figures 3.30 and 3.31).

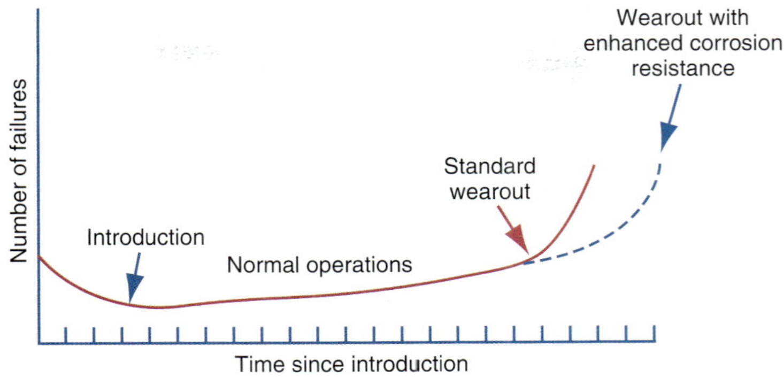

Figure 3.30. Expected life curves (reliability) for standard and corrosion resistance part [65]. (Copyright *AMPTIAC Quarterly*, by D.H. Rose, reproduced by permission of Alion Science and Technology.)

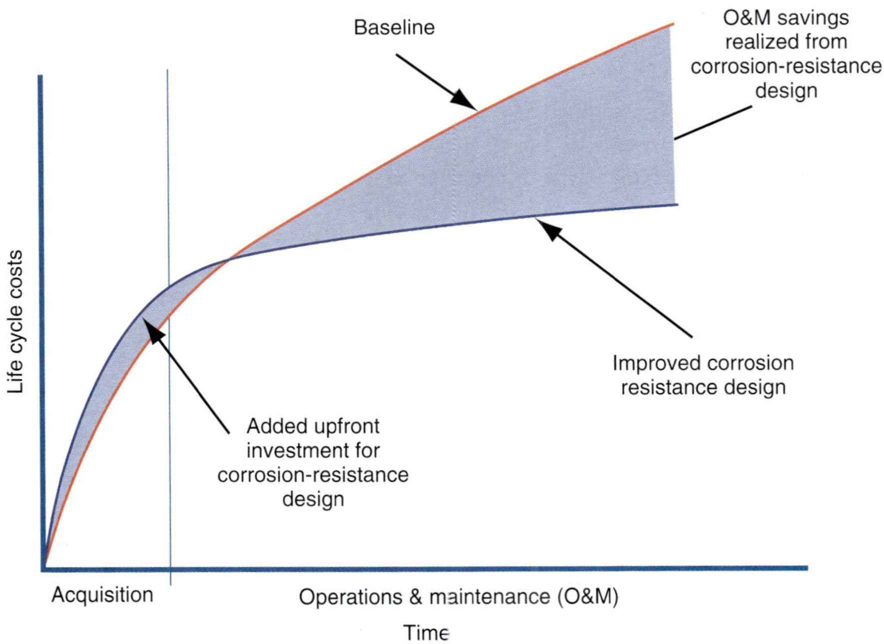

Figure 3.31. Operation costs and potential savings [65]. (Copyright *AMPTIAC Quarterly*, by D.H. Rose, reproduced by permission of Alion Science and Technology.)

Saturated, high molecular weight and highly cross-linked materials usually offer greater chemical stability and are less likely to react with environment chemicals. Polyester resins are traditionally preferred for corrosion resistant applications. However, the use of vinyl-ester-based materials (such as Bisphenol-A-based epoxy vinyl ester) is constantly increasing driven by the development of very harsh and structurally demanding oil and gas composite applications. Chemical and infrastructure industries also often add stringent fire resistance requirements. More exotic materials such as fluorinated ethylene propylene might be necessary for composite parts operating in temperatures beyond 200°C.

Material and design selection for chemical pipes and towers is therefore complex and designs differ significantly from one plant to another. The example of an 8.5 m fiberglass/vinyl ester stack liner designed for the Santee Cooper Power facility in Cross, SC, USA is detailed in the literature [66]. The assembly is complex and comprises a carbon fiber veil to ensure grounding of parasitical static electricity, a resin-rich C-glass veil for enhancing the corrosion protection on the inner surface, multiple E-glass rovings for the structural wall together with a wide stitched unidirectional glass material and chopped strand mat wound over the corrosion shield.

The use of corrosion resistant polymer composite materials sometimes appears in unexpected areas. In many urban locations, century-old underground infrastructures such as sewers need rehabilitations. Sewer pipes undergo complex environmental loading leading to stringent specifications including:

- structural strength sufficient to resist fluid normal and cyclic surge pressure and buckling resistance
- corrosion resistance with respect to waste fluid exposure leading to acid and alkali corrosion
- resistance to methane, hydrogen sulfide and other gases
- erosion resistance and minimum flow resistance
- impermeability cyclic water pressure
- dent resistance (rodent attack)
- a hundred year lifetime.

It is no surprise that vinyl-ester-based or HDPE polymers are considered prime candidates for such piping applications.

Access to sewers is sometimes restricted and the need to dig for rehabilitation or installation adds significantly to the general costs of the work. When trenching is possible, glass reinforced polymer curved panels can be placed to reinforce existing sewer parts (Figures 3.32, 3.33 and 3.34). Typically, the composite pipe comprises (from the outside inward) a gel coat, an E-CR glass fiber anti-corrosion veil, a glass reinforced thermoset layer, a polymer concrete core and another composite layer [68]. The composite matrices are usually based upon silica sand – thermoset mix.

Figure 3.32. Large glass-fiber reinforced pipes for sewer rehabilitation. (Courtesy of Hobas.)

The problem is more complex for rehabilitation works where trenching is not possible. Indeed, the pipe needs to be pushed round bends, have good fit to the original pipe and provide added structural strength. To answer this challenge, the company Insituform developed a process in the 1970s where a *soft* composite, pre-impregnated with polyester fiber felt pipe could be introduced through an existing manhole (Figure 3.35). The pipe was then forced down the hole by water pressure and finally cured *in situ* by circulating heated water [68].

Figure 3.33. Sewer pipe system re-lining with composite corrosion-resistant pipes. (Courtesy of Hobas.)

Considering the very large number and diversity of materials alternatives, the selection of optimized corrosion-resistant polymers and reinforcements might be challenging. Suppliers can usually be a good source of guidance in the material selection process. Online material selection tools are even provided by suppliers, where operating conditions and chemical exposure can be selected [69].

Figure 3.34. Inside composite sewer pipe. (Courtesy of Hobas.)

3.6 FREEZE THAW

The use of composite materials for civil engineering applications, such as bridges [70,71], is constantly rising. A standard test for concrete in those applications is the *freeze-thaw* test. Typical *freeze-thaw* tests are of two types:

(1) Rapid freezing and thawing both in water.
(2) Rapid freezing in air and thawing in water.

Typical freeze-thaw cycle times are 2–24 h including 20–25% of the time for thawing. Sample properties such as weight and modulus are measured with a periodic frequency. The test ends after a chosen number of cycles or if some specific criterion are met; for example, if the specimen has lost 40% of its initial modulus or if a 0.1% expansion is observed. Such experiments based on global sample measurements are useful general indicators but do not provide details on damage and water diffusion during heating and cooling.

During thawing, two mechanisms occur and possibly interact: thermal aging and moisture absorption. Some of the water absorbed mainly at high temperature freezes during the low temperature cycle. When water becomes ice, it undergoes a 9% volume expansion. Therefore, the cavity where the water is contained must dilate or the excess water must be expelled from it (and flow toward escape boundaries).

Figure 3.35. *In situ* curable sewer pipe. (Courtesy of Insituform.)

The pressure in this process is proportional to the coefficient of permeability of the material, the distance from the escape boundary and the rate of freezing (speed of water motion). If a point is sufficiently remote from an escape boundary, permanent damage can occur.

Several theories try to explain and model the expansion mechanism of water being expelled from a concrete cavity [72–76]: hydraulic pressure theory (air bubble), Helmuth's hypothesis (expansion mechanism is a result of the growth of ice dendrites in the capillaries), the osmotic pressure theory (as mentioned above), the Litvan's hypothesis (desorption of the gel pores toward the freezing sites) and the diffusion theory. The latest asserts that the simultaneous presence of the bulk water in the material and the ice in larger capillaries creates a non-equilibrium situation where the bulk water acquires a potential energy enabling it to move into the cavity and cause the ice crystal to grow and enlarge the cavity. The non-frozen water has a higher energy state than the ice in the cavities: the bulk water flows in the cavity to relieve the destructive hydraulic pressure.

Generally, freezing initiates the process. Flaws act as freezing sites and attract all the water from the surroundings. Growth of this ice might help propagate the flaw. If water expulsion is hindered, the internal pressure can cause fracture of the material.

If freezing in pure polymers is rather unlikely since the size of the free volume region is smaller than the minimum domain size for water clustering [77], polymer matrix composites on the other hand can exhibit a large number of flaws due to curing, mismatches in coefficients of thermal expansion or poor fiber wetting during manufacturing. All these flaws can serve as freezing sites. Water absorption in composites can induce swelling and warping, delamination and debonding. The expansion of the freezing water causes further delamination and also induces the coalescence of microscopic cracks leading to macroscopic damage. If the assumption that moisture is transported into the composite by capillary action along cracks and fiber–matrix interface is correct, then the freezing damage may allow further water absorption. Though not supported by experimental data, the previous statements tend to indicate that non-polar polymers would undergo less damage (less water absorption) during freeze-thaw experiments and should be prime candidates for such applications.

The following case study reports some highlights of freeze-thaw experiments performed on the fiber reinforced composite bridge deck from the company Creative Pultrusion installed on the Salem Avenue (see Chapter 6).

Case study: Freeze-thaw results highlights for Creative Pultrusion Bridge Deck (Salem Ave, Ohio)

The Creative Pultrusion all fiber reinforced plastic deck was installed on the Salem Avenue Bridge in Ohio (see Chapter 6 for details). Prior to installation, samples of the selected composite material underwent a series of environmental tests in the laboratory including freeze-thaw experiments. The results were communicated as a courtesy of Creative Pultrusion Inc. and can also be found in the literature [78].

The composite beams were made of pultruded hexagonal and double trapezoid profiles. The material was a combination of:

- a Reichhold (Atlac 480-05) vinyl ester
- pultruded 0° rovings
- a multiaxial [90°/±45°] stitched fabric (BTI TH4000/IHX4501).

The total volume fraction of reinforcement was around 49.2%, with 19% of axial reinforcement, 11% of transverse reinforcement and 3.6% of mat layer (in volume). The sample thickness for the freeze-thaw experiments was 2.54 cm.

The freeze-thaw equipment comprised a chest freezer and type I distilled water tanks equipped with 250 Watt heaters and 4 l/min circulator pumps. The tank water temperature for thawing was set at 38°C and the freezer temperature adjusted to −18°C.

The samples were conditioned 21 days in a water bath prior to freeze-thaw experiments. The samples were then repeatedly immersed 12 h in the warm water bath, dried on the surface with a lint-free rag then placed 12 h in the freezer.

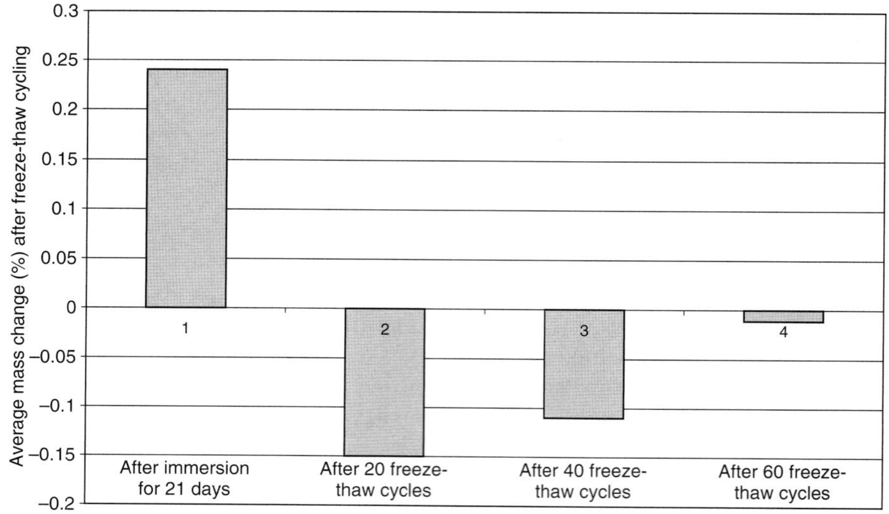

Figure 3.36. Average mass change after freeze-thaw treatment. (Data from Lopez-Anido et al. [78].)

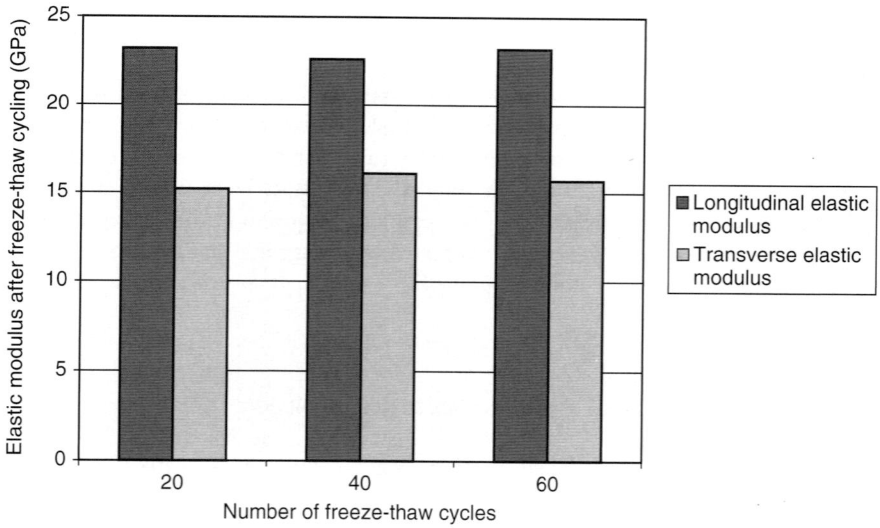

Figure 3.37. Longitudinal and transverse elastic moduli after freeze-thaw cycling. (Data from Lopez-Anido et al. [78].)

The samples weights were recorded prior to immersion, after the 21-day conditioning period and after every 20 cycles of freezing and thawing. At the end of the tests, the aged samples were stored at room temperature then tested in tension, short beam shear or cut for cross-section observations.

An expected initial weight uptake occurred during the 21 days simple exposure to distilled water. A systematic decrease in the sample mass was then observed after 20, 40 and 60 freeze-thaw cycles (Figure 3.36). This decrease was concomitant with the progressive but visually observable loss of edge sealing resin. The mass difference did not, however, reflect in an alteration of the failure mode, materials strength or elastic modulus (Figure 3.37).

3.7 CAVITATION EROSION

Cavitation erosion is another consequence of fluid exposure and generally occurs in rapidly moving fluids. Cavitation occurs in systems where the local pressure in the liquid drops below its vapor pressure. It is characterized by the formation of vapor cavities and vapor bubbles in the liquid. Cavitation typically occurs in hydraulic systems, where low fluid levels may draw air into the system. The small bubbles resulting from cavitation might expand explosively and cause irreversible damage such as materials erosion. The resistance to cavitations erosion varies as a function of the fluid velocity and surface material (Table 3.5).

Table 3.5. Cavitation erosion resistance of plastic structure materials from Kallas and Lichtman [79]

Material	Erosion rate, μl/h, at velocity, m/s (fps)		
	30.48 m/s (100 fps)	38.1 m/s (125 fps)	45.72 m/s (150 fps)
Polyamide, molded nylon, 2.5% water	–	–	0.1
Polyamide, nylon, fiber reinforced	<0.05	0.18	–
Polycarbonate	0	1.03	1.25
Polycarbonate, fiber reinforced	0.1	1.4	2.0
Poly(vinyl chloride)	<0.03	2.7	5.8
Epoxy, glass-cloth-reinforced	0	8.4	23
Poly (methyl methacrylate)	1.6	16	38
Styrene-acrylonitrile	<0.1	0.8	25
Acetal	0	0.26	0.76
Polyimide	0	0.17	0.71

Composite materials are being investigated as potential candidates for pelton water turbines. Indeed, the performance of the turbine can be drastically enhanced through wall thickness reduction. Excellent fatigue resistance and significant mass reductions are further advantages offered by composite systems. However, it is necessary to protect the composite structure from flow aggression and specifically cavitation. ALSTOM has developed a carbon-fiber epoxy bucket where *sharp edges* coating at the bucket lip and splitter was obtained by an electrolytic method (Figure 3.38). The other bucket surfaces are protected by a proprietary coating resisting abrasive erosion and cavitation damage. At the time of writing, this promising development was under experimental stage and required further investigations before commercialization.

Cavitation occurs only in specific system. Rain erosion however can occur in common outdoor equipment such as high speed vehicles. In this case, erosion mechanisms and rates are a function of speed, droplet size and time of exposure.

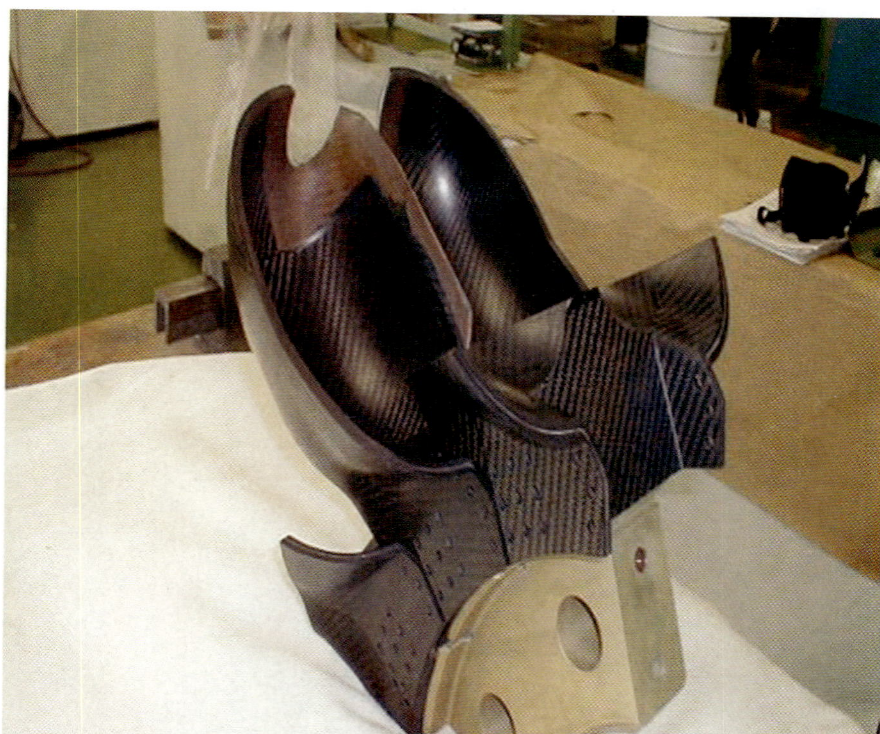

Figure 3.38. Carbon-fiber epoxy pelton turbine bucket prototype. (Courtesy of ALSTOM.)

3.8 TESTING

Test norms and models alike mainly deal with water absorption. Fortunately, most test methods can be adapted to other permeating species.

Common testing norms are listed for water in Table 3.6 and for other permeating species in Table 3.7.

High temperature accelerated testing for gas and liquid absorption should be used with caution. Such tests should be used only if the diffusion and failure mechanisms remain unchanged over the operation-testing range (see Chapter 2).

A general summary of the method for assessing the effect of gas or liquid environment on polymer matrix composites is provided in Figure 3.39 and main tools are summarized in Section 3.9.

Table 3.6. Common normalized testing methods for water absorption

Organism	Designation	Title
ASTM	D5229/D5229M-92 (2004)	Standard Test Method for Moisture Absorption Properties and Equilibrium Conditioning of Polymer Matrix Composite Materials
ASTM	F1769-97	Standard Test Method for Evaluation of Solubility, Diffusivity, and Permeability of Organic Vapor Barriers Using a Flame Ionization Detector
ASTM	F1770-97e1	Standard Test Method for Evaluation of Solubility, Diffusivity, and Permeability of Flexible Barrier Materials to Water Vapor
ASTM	D1653-03	Standard Test Methods for Water Vapor Transmission of Organic Coating Films
ASTM	F1249-01	Standard Test Method for Water Vapor Transmission Rate Through Plastic Film and Sheeting Using a Modulated Infrared Sensor
ASTM	F2298-03	Standard Test Methods for Water Vapor Diffusion Resistance and Air Flow Resistance of Clothing Materials Using the Dynamic Moisture Permeation Cell
ASTM	E398-03	Standard Test Method for Water Vapor Transmission Rate of Sheet Materials Using Dynamic Relative Humidity Measurement

(*Continued*)

Table 3.6. (*Continued*)

Organism	Designation	Title
ASTM	D6908-03	Standard Practice for Integrity Testing of Water Filtration Membrane System
ASTM	D5744-96 (2001)	Standard Test Method for Accelerated Weathering of Solid Materials Using a Modified Humidity Cell
ASTM	E2321-03	Standard Practice for Use of Test Methods E96 for Determining the Water Vapor Transmission (WVT) of Exterior Insulation and Finish Systems (EIFS)
ASTM	D4279-95 (2003)	Standard Test Methods for Water Vapor Transmission of Shipping Containers – Constant and Cycle Methods
ASTM	E96-00e1	Standard Test Methods for Water Vapor transmission of Materials

Table 3.7. Common normalized testing methods for gas and liquid absorption

Organism	Designation	Title
ASTM	D1434-82 (2003)	Standard Test Methods for Determining Gas Permeability Characteristics of Plastic Film and Sheeting
ASTM	D2684-95 (2001)	Standard Test Method for Permeability of Thermoplastic Containers to Packaged Reagents or Proprietary Products
ASTM	F778-88 (2001)	Standard Methods for Gas Flow Resitance Testing of Filtration Media
ASTM	D3902-90 (1998)	Standard Test Method for Rubber Hose for Gas Diffusion of Liquefied Petroleum Gas
ASTM	F949-03	Standard Specification for Poly(Vinyl Chloride) (PVC) Corrugated Sewer With a Smooth Interior and Fittings
ASTM	F739-99a	Standard Test Method for Resistance of Protective Clothing Materials to Permeation by Liquids or Gases Under Conditions of Continuous Contact
ASTM	F1307-02	Standard Test Method for Oxygen Transmission Rate Through Dry Packages Using a Coulometric Sensor
ASTM	D3985-02e1	Standard Test Method for Oxygen Gas Transmission Rate Through Plastic Film and Sheeting Using a Coulometric Sensor
ASTM	F1927-98 (2004)	Standard Test Method for Determination of Oxygen Gas Transmission Rate, Permeability and Permeance at Controlled Relative Humidity Through Barrier Materials Using a Coulometric Detector

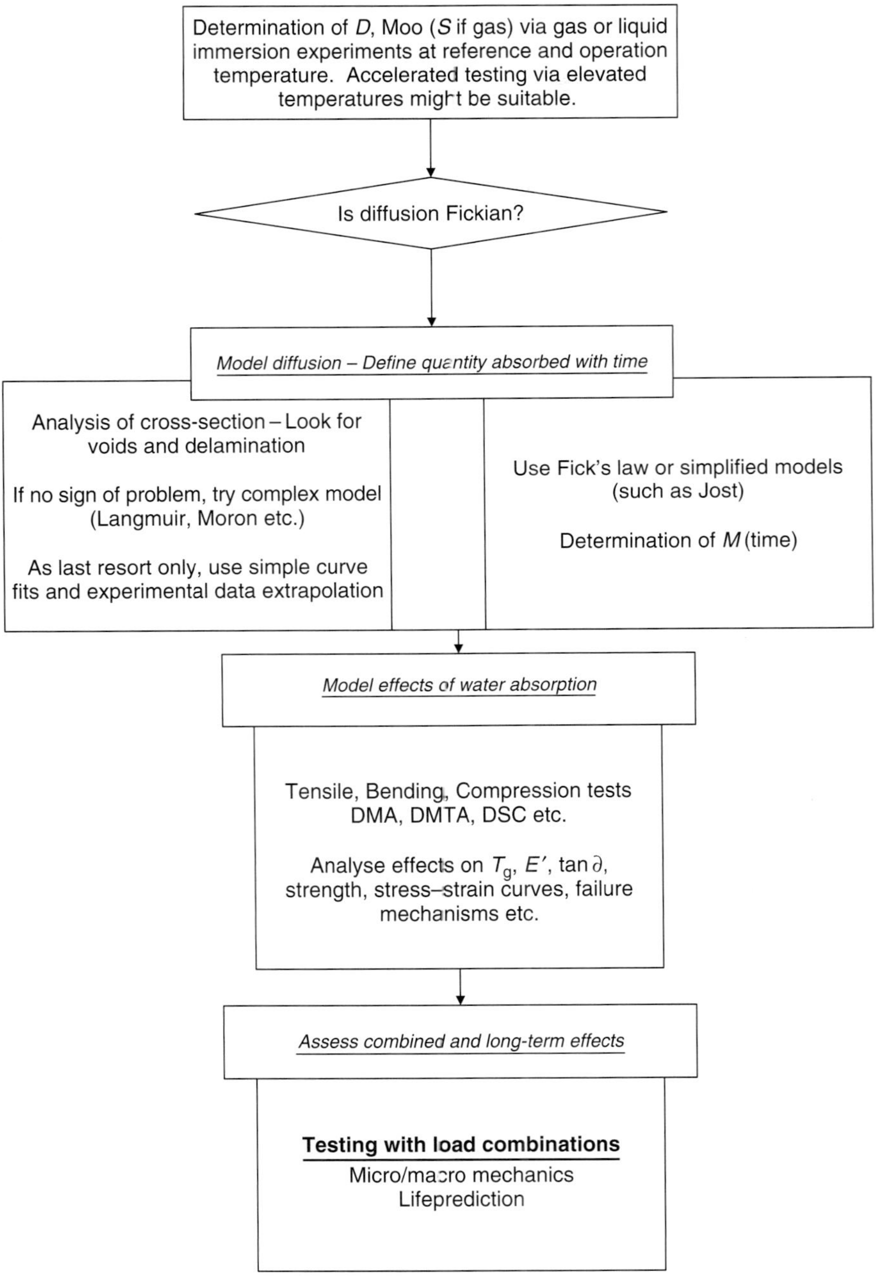

Figure 3.39. Moisture and liquid effect assessment Flow-chart.

3.9 TOOL KIT

Topic	Equation	Assumptions	Importance
General eq. for diffusion (Fick)	$J = -D\frac{dc}{dx}$	Steady state	Most widely used to predict diffusing species concentration versus time
	$\frac{\partial c}{\partial t} = D\frac{\partial^2 c}{\partial x^2}$		
Special solution (Jost)	$\frac{c - c_i}{c_m - c_i} = 1 - \frac{\pi}{4}\sum_{j=0}^{\infty}\frac{1}{(2j+1)} \sin\frac{(2j+1)\pi x}{h}\exp\left[-\frac{(2j+1)^2\pi^2 D\cdot t}{h^2}\right]$	$C_s = \text{cst}$ at $t > 0$	Provides useful solution for composites
Diffusivity in composites	$M_{\text{sample at time t}}(\%) = \left(\frac{\text{Weight}_{\text{sample at time t}} - \text{Weight}_{\text{sample initial}}}{\text{Weight}_{\text{sample initial}}}\right) \times 100$	None	To define weight uptake
	$\text{Slope} = \frac{4M_\infty}{h\sqrt{\pi}}\sqrt{D}$		Useful to calculate D from experiments
	$D_{11} = (1 - v_f)\, D_r,\ D_{22} = \left(1 - 2\sqrt{\frac{v_f}{\pi}}\right) D_r$	Assumes *Perfect composite* (no voids, etc)	To estimate D in longitudinal and transverse directions if needed in finite element method (FEM)
Gas permeation	$\tilde{P} = \tilde{D}\cdot\tilde{S}$	None	General equation for gases
	$\tilde{P} = \tilde{P}_o\cdot\exp\left(-\frac{\Delta E}{RT}\right)$	This equation does not always hold. Empirical result	Dependence to temperature
	$\mu_{a/b} = \frac{P_a}{P_b} = \frac{S_a D_a}{S_b D_b}$	None	Permselectivity definition
T_g versus solvent quantity	$T_g = \frac{\alpha_p V_p T_{gp} + \alpha_d(1 - V_p)T_{gd}}{\alpha_p V_p + \alpha_d(1 - V_p)}$	Assumes *perfect polymer*	Predicts T_g of the polymer after exposure
	$T_g = V_p T_{gp} - (1 - V_p)T_{gd},\ \frac{T_g}{T_m} \approx \frac{1}{2} to \frac{1}{3}$	Approximate, assumes Epoxy	Simplified, gives order of magnitude of T_g after exposure

Topic	Equation	Assumptions	Importance
Polymer swelling	$\beta = \frac{1}{3}\frac{\rho}{\rho_w}$	Assumes homogeneous material	Gives order of magnitude of swelling
Composite swelling	$\beta_1 = 0, \beta_2 = \frac{\rho_c}{\rho_m}(1 - v_m)\beta_m$	Assumes *perfect composite*	Gives order of magnitude in longitudinal and transverse directions if needed in FEM

REFERENCES

1. A dream home – thanks to corrosion resistant and fire-retardant resins. *JEC-Composites*, January 2004, no. 6.
2. Stern, S.A. and S. Trobolaki, Barrier Polymers and Structures, W.J. Koros (Ed.), ACS Symp. Ser., Americal Chemical Society, Washington DC, 1990, 22–59.
3. Sok, R.M., Permeation of small molecules across a polymer membrane: A computer simulation study. University of Gorningen, Netherlands, 1994. www.ub.rug.nl/eldoc/dis/science/r.m.sok/c2/pdf.
4. Pavlidou, S. and C.D. Papaspyrides, The effect of hygrothermal history on water sorption and interlaminar shear strength of glass/polyester composites with different interfacial strength. *Composites Part A: Applied Science and Manufacturing (Incorporating Composites and Composites Manufacturing)*, November 2003.
5. Dhingra, S.S., Mixed Gas Transport Study Through Polymeric Membranes: A Novel Technic. Ph.D. Dissertation, Virginia Tech, 1997, http://scholar.lib.vt.edu/theses/public/etd-52497-133245/materials/ETD.PDF.
6. www.matweb.com.
7. Jones, F.R., The effects of aggressive environments on long-term behaviour. In Harris, B. (Ed.), *Fatigue in Composites*. Woodhead Publishing Ltd, Boca Raton, 2003.
8. Karad, S.K., F.R. Jones and D. Attwood, Moisture absorption by cyanate ester modified epoxy resin matrices. Part II. The reverse thermal effect. *Polymer*, Oct 2002.
9. Verghese, K.N.E., M.D. Hayes, K. Garcia, C. Carrier, J. Wood and J.J. Lesko, Effects of matrix chemistry on short-term hygrothermal aging of vinyl ester matrix composites. *Journal of Composite Materials*, 1999, 33(20).
10. Hough, J.A., F.R. Jones and Z.D. Xiang, The effect of thermal spiking and resultant enhanced moisture absorption on the mechanical and viscoelastic properties of carbon fiber reinforced epoxy laminates. *Key Engineering Materials*, 1998, 144, 27–42.
11. rredc.nrel.gov/solar/glossary/gloss_r.html.
12. Springer, G.S. (Ed.), *Environmental Effects on Composite Materials*. 3 Volumes, Technomic Publishing Company, Westport, 1981.
13. Jost, W., *Diffusion in Solids, Liquids, Gases*. Academic Press, 1960.
14. Springer, G.S. and S.W. Tsai, Thermal Conductivities of Unidirectional Materials. *Journal of Composite Materials*, 1967, 1, 166.
15. Koros, W.J., Barrier Polymers and Structures, W.J. Koros (Ed.) *ACS Symposium Ser.*, American Chemical Society, Washington DC, 1990, 1–21.

16. Aminabhavi, T.M. and U.S. Aithal, *J. Macromol. Sci-Rev. Macromol. Chem. Phys.*, 1991. C31(2&3), 117.
17. Dunham, M.L., D.l. Bailey and R.Y. Mixer, New curing system for silicon rubber. *Industrial Engineering and Chemistry*, 1957, 49(9), 1373–1376.
18. Kawai, K., T. Nohmi and K. Kamide, *Polym. Prepr., American Chemical Society, Division of Polymer Chemistry*, 1979 20, 309.
19. Wallvik, J. and O. Akerblom, The platelet storage capability of different plastic containers. *Vox Sang*, 1990, 58(1), 40.
20. Sperling, L.H., *Introduction to Physical Polymer Science*, 2nd ed. John Wiley & Sons, Inc., USA, 1992.
21. Pauly, S., in *Polymer Handbook*, 3rd ed., J. Brandrup and E.H. immergut (Eds), Wiley-Interscience, New York, Sec. VI, 1989, p. 435.
22. http://www.psrc.usm.edu/mauritz/diffuse.html
23. Ku, B.C., Barrier Properties of Ordered Multilayer Polymer Nanocomposites. In *Dekker Encyclopedia of Nanoscience and Nanotechnology*, 2004, 213–224.
24. Rosato, D.V., D.P. DiMattia and D.V. Rosato, *Designing with Plastics and Composites – A Handbook*. Chapman & Hall, USA, 1991.
25. Le Carbone – Lorraine Product Bulletin Proc. Eng., 1994, Jan., 30-031.
26. Energy Information Administration. www.eia.doe.gov/iea/.
27. Ritter, J.A., A.D. Ebner, J. Wang and R. Zidan, Implementing a hydrogen economy. *Materials Today*, September 2003.
28. Brosius, D., Composites energize fuel cells. *Composites Technology*, November/ December 2001.
29. Zuettle, A., Materials for hydrogen storage. *Materials Today*, September 2003.
30. Schutte, C., Environmental durability of glass-fiber composites. *Materials Science and Engineering*, R13, 1994, 265–324.
31. Morgan, R.J., Thermal characterization of composites. In E.A. Turi, (Ed.), *Thermal Characterization of Polymeric Materials*, 2 Volumes, 2nd ed. Academic Press, New York, 1997.
32. Bair, H.E., in *Thermal Characterization of Polymeric Materials*, E.A. Turi (Ed.), Academic Press, Orlando, FL, 1981, 408–433, 845–909.
33. Wright, W.W., The effect of diffusion of water into epoxy resins and their carbon-fibre reinforced composites. *Composites*, 1981, 12(3), 201–205.
34. Kelley, F.N. and F. Bueche, *Journal of Polymer Science*, 1961, 50, 549–556.
35. Giovambattista, N., C.A. Angell, F. Sciortino and H.E. Stanley, Glass transition temperature of water: A simulation study. *Physical Review Letters*, July 04, 93(4).
36. Nielsen, L.E., *Mechanical Properties of Polymers*, Reinhold, New York, 1962.
37. Morgan, R.J., J.E. O'Neal and D.B. Miller, *Journal of Materials Science*, 1979, 14, 109–124.
38. Starkweather, H.W., Jr., Water in Nylon. *ACS Symposium Series* 127, 1980, 433–440.
39. Matthews, F.L. and R.D. Rawlings, *Composite Materials: Engineering and Science*. Chapman & Hall, Oxford, 1994.
40. Bruce Prime, R., Thermosets, in E.A. Turi, (Ed.), *Thermal Characterization of Polymeric Materials*, 2 Volumes, 2nd ed. Academic Press, New York, 1997.
41. Harper, B.D. and Y. Weitsman, On the effects of environmental conditioning on residual stresses in composite laminates. *International Journal of Solids and Structures*, 1985, 21(8), 907–926.
42. Carter, H.G. and K.G. Kibler, *Journal of Composite Materials*, 1978, 12, 118–131.

43. Callister, W.D. (Jr), *Materials Science and Engineering – An Introduction*, 4th ed. John Wiley & Sons, USA, 1997.
44. Krisyuk, V.E. and K.L. Smirnov, Kinetics of the mechanically activated hydrolysis of oriented polyamide-6. *Polymer Science*, U.S.S.R., 1989, 31(2), 360–366.
45. Boron-free glass fibres- the trend for the future? *Reinforced Plastics*, June 2003, 36–40.
46. Verghese, K.N.E., N.S. Broyles, J.J. Lesko, R.M. Davis and J.S. Riffle, Pultruded carbon fiber/vinyl ester composites processed with different fiber sizing agents. Part I: Processing and static mechanical performance. *Journal of Materials in Civil Engineering*, 2005, 17(3), 320–333.
47. Verghese, K.N.E., N.S. Broyles, J.J. Lesko, R.M. Davis and J.S. Riffle, Pultruded carbon fiber/vinyl ester composites processed with different fiber sizing agents. Part II: Enviro-mechanical durability. *Journal of Materials in Civil Engineering*, 2005, 17(3), 334–342.
48. Verghese, K.N.E., N.S. Broyles, J.J. Lesko, R.M. Davis, J.S. Riffle, Pultruded carbon fiber/vinyl ester composites processed with different fiber sizing agents. Part III: Theoretical aspects. *Journal of Materials in Civil Engineering*, 2005, 17(3), 343–352.
49. Marom, G. and L.J. Broutman, Moisture penetration into composites under external stress, *Polymer Composites*, 1981, 2(3), 132–136.
50. Marom, G., Swelling and hygroelasticity of polymeric composites. *Polymer Engineering and Science*, 1977, 17(11), 799–802.
51. Neumann, S. and G. Marom, Stress dependence of the coefficient of moisture diffusion in composite materials. *Polymer Composites*, 1985, 6(1), 9–12.
52. Mahieux, C.A., D. Lehmann and A. desLigneris, Experimental determination of the effects of industrial oil on polymer-based composites. *Polymer Testing*, June 2002, 21(7), 751–756.
53. Williams, J.G., *Journal of Materials Science*, 1982, 17, 1427–1433.
54. Williams, J.G., The beta relaxation in epoxy-based networks. *Journal of Applied Polymer Science*, 1979, 23(12), 3433–3444.
55. Schrager, M., Fatigue as monitored by the torsion pendulum. *Journal of Polymer Science, Part A-2: Polymer Physics*, 1970, 8(11), 1999–2014.
56. Patterson-Jones, J.C. and D.A. Smith, The thermal degradation of an amine-cured epoxide resin at temperatures between 200°C and 310°C. *Journal of Applied Polymer Science*, 1968, 12(7), 1601–1320.
57. Pogany, G.A., Gamma relaxation in epoxy resins and related polymers. *Polymer*, 1970, 11(2), 66–78.
58. Ebdon, M.P., O. Delatycki and J.G. Williams, Dynamic mechanical properties of glass-filled epoxy resin, *Journal of Polymer Science: Polymer Physics Edition*, 1974, 12(8), 1555–1564.
59. Mandell, J.F., Origin of moisture effects on crack propadation in composites. *Polymer Engineering and Science*, 1979, 19(5), 353–358.
60. Jones, F.R., et al. Reinforced Plastics Congress, Brighton, UK, 1982, British plastic federation, London, UK, Paper 32.
61. Aveston, J., A. Kelly and J.M. Sillwood, Long-term strength of glass reinforced plastics in wet environments. In *Advances in Composite Materials*. Vol. 1, Bunsel et al. (Eds), Paris, Pergamon, 556–568.
62. Howard, A. and N.J. Parrat, Life prediction for aromatic polyamide reinforcements. In ICCM5, W. Harrington et al. (Ed.) The Metallurgical Society, Warrendale, PA, USA, 1985, 277–292.
63. Morgan, R.J., et al., *Proc. Int. SAMPE. Tech. Conf.*, 22nd, Boston, 1990, 145–156.

64. Ely, T., D. Armentrout and M. Kumosa, Evaluation of stress corrosion properties of pultruded glass fiber/polymer composite materials. *Journal of Composite Materials*, 2001, 35(9), 751–773.
65. Rose, D.H., Reducing Acquisition Risk by "Designing in" Corrosion Resistance, *AMP-TIAC Quaterly*, 2004, 8(2).
66. Black, S., Vinyl esters make tough parts for highly corrosive applications. *Composites Technology*, August 2003.
67. Jacob, A., Vinyl esters lead the corrosion challenge. *Reinforced Plastics*, June 2003, 32–35.
68. Composites renovate deteriorating sewers. *Reinforced Plastics*, June 2004, 20–24.
69. www.corrosionresins.com
70. Hollaway, L.C., The evolution of and the way forward for advanced polymer composites in the civil infrastructure. *Construction and Building Materials*, September–October 2003, 17(6–7), 365–378.
71. Aref, A.J. and I.D. Parsons, Design and performance of a modular fiber reinforced plastic bridge. *Composites Part B: Engineering*, October 2000, 31(6–7), 619–628.
72. Kaufmann, J.P., Experimental identification of ice formation in small concrete pores. *Cement and Concrete Research*, August 2004, 34(8), 1421–1427.
73. Penttala, V. and F. Al-Neshawy, Stress and strain state of concrete during freezing and thawing cycles. *Cement and Concrete Research*, September 2002, 32(9), 1407–1420.
74. Chatterji, S., Freezing of air-entrained cement-based materials and specific actions of air-entraining agents. *Cement and Concrete Composites*, October 2003, 25(7), 759–765.
75. Bazant, Z.P. and L.J. Najjar, Drying of concrete as a nonlinear diffusion problem. *Cement and Concrete Research*, September 1971, 1(5), 461–473.
76. Litvan, G.G., Adsorption systems at temperatures below the freezing point of the adsorptive. *Advances in Colloid and Interface Science*, June 1978, 9(4), 253–302.
77. Verghese, K.N.E., J. Haramis, M.R. Morrell, M.R. Horne and J.J. Lesko, Freeze-thaw durability of polymer matrix composites in infrastructure. *Proceedings of the 4th International Conference on Durability Analysis of Composite Systems (DURACOSYS)*, Brussels, Belgium, Published by A.A. Balkema, 2000, 457–463.
78. Lopez-Anido, R., I. Harik, P. Dutta and B. Shahrooz, Field Performance Evaluation of Multiple Fiber-Reinforced Polymer Bridge Deck Systems Over Existing Girders – Phase I, Final Report, June 7, 2001. Prepared in cooperation with the Ohio Department of Transportation and the U.S. Department of Transportation, Federal Highway Administration.
79. Kallas, D.H. and J. Lichtman, Cavitation erosion. In D.V. Rosato and R.T. Schwartz (Eds), *Environmental Effects on Polymeric Materials*. Interscience Publishers, 2 vols, New York, 1968.

4

EFFECTS OF ELECTRICAL FIELDS AND RADIATIONS ON POLYMER MATRIX COMPOSITES

4.1 INTRODUCTION

Polymer-based materials are extensively used as insulators for a broad range of voltage applications. Electrical fields have specific effects on polymer matrix composites including polarization and electrical losses. The present chapter provides the elements necessary for the development of an in-depth understanding of the materials electrical and durability responses. Different types of electrical applications such as high voltage cables and motor stator windings are introduced. Principal electrical quantities are reviewed and a special emphasis is placed on understanding short-term and long-term phenomena contributing to electrical losses. Composite specificities, breakdown mechanisms and practical implications are summarized and illustrated by the industrial example of a hydrogenerator stator winding.

Insulation is usually the domain of electrical engineers. However, beyond the polarization phenomena, all other environmental influences (temperature, moisture etc.) should be accounted for when designing the insulation. Indeed, we have seen in Chapter 2 that viscoelasticity leads to time-dependent dielectric responses as well. Electricity also often translates into elevated temperatures that were shown to alter the composites mechanical and electrical responses (Chapter 2) significantly. Therefore, composite insulation materials should be seen as more than dielectrics and holistic durability approaches (Chapter 6) should always be applied. The utility pole case study developed in the present chapter illustrates this concept.

Other types of radiation environments are introduced in a second step. Much alike electrical fields, radiation-induced materials changes and degradation introduced in Section 4.3 can directly be included in the durability approach proposed in Chapter 6.

4.2 EFFECTS OF ELECTRICAL FIELD ON POLYMER MATRIX COMPOSITES

4.2.1 Introduction to Insulation Materials

4.2.1.1 Type of applications

Polymer composites are used for a large range of electrical field strengths. Indeed, the range of applications of polymer composites for electrical insulation is broad, ranging from wire insulation and electronic packaging to high voltage cables and conductors. Traditionally, low, medium, high, extra high and ultra high voltages designate voltages up to 1 kV, 70 kV, 230 kV, 800 kV and 1000 kV respectively. Conduction and polarization principles are similar for all field strengths with the exception that higher fields can induce additional mechanisms in the insulator such as cavity discharges and space charge accumulation. For completeness reasons, we will from now on mostly focus our attention on higher voltages. Main high voltage applications include cables, stator windings and stator laminations for rotating electrical machinery (motor or generator).

Cables usually comprise copper conductors covered by an insulation layer (PVC for low voltage, modified PE for medium and high voltages), diverse shields and mechanically resistant armors [1] (Figure 4.1). They are meant to transport electrical

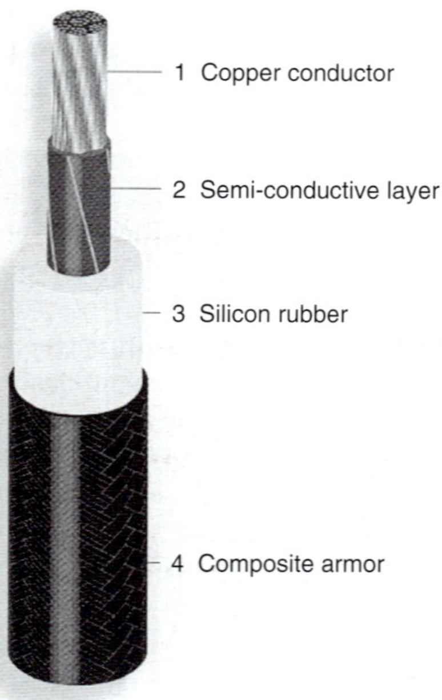

Figure 4.1. Omerin single core cable [2]. 13.8 kV SILICOUL Cable. (Picture courtesy of OMERIN SAS Ambert.)

power over large distances with minimum losses (see Section 4.2.2.2). Cables are generally underground (occasionally overhead) and should therefore be extremely resistant to weathering. To date extreme application electrical cables can resist temperatures from −190 to 1400°C [2]. However, typical operation temperatures generally range from −50 to +200°C.

Polymer-based materials are also used extensively as insulators in large electrical machines such as transformers, motors and generators. Generator winding stator bars are perfect examples of composite materials under high voltages. At the time of writing, state of the art for the main insulation relied upon the use of glass-fiber/mica epoxy composite tapes. Typical stator winding layers and cross section are shown in Figures 4.2 and 4.3 and illustrate the complexity of the bar geometry. The mechanical and environmental loads on the insulation layers are also complex with voltages up to 30 kV in the insulation and temperatures ranging from −20 to 180°C for turbogenerators. For such machines the large winding length (up to 8 m) can additionally create important bending moments during the bar handling and manufacturing.

Polymers and reinforced composites are used as insulation materials in tape, varnish, fleece or laminated forms. But beyond the nature of the materials, entire parts such as cables or electrical windings can be considered as copper reinforced polymer multilayered composite structures and should be analyzed as such. According to this holistic view, the calculation methods in Chapters 5 and 6 apply not only to the polymer layers but also to the entire cable or bar taken as a global composite.

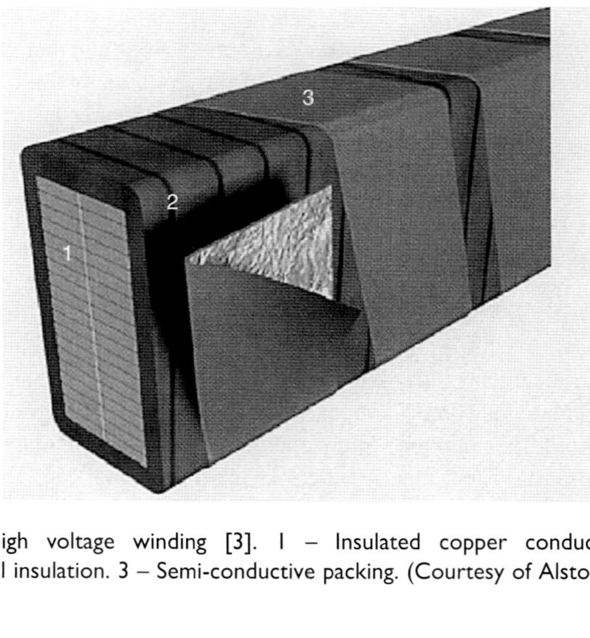

Figure 4.2. High voltage winding [3]. 1 – Insulated copper conductors (strands). 2 – Groundwall insulation. 3 – Semi-conductive packing. (Courtesy of Alstom.)

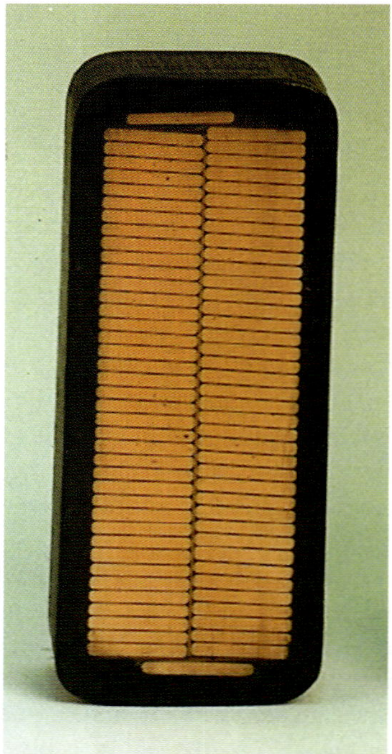

Figure 4.3. Winding cross-section. (Courtesy of Alstom.)

4.2.1.2 Most common materials

Pure, particle reinforced and fiber reinforced thermoplastics and thermosets are used as insulation materials. Detailed descriptions of those materials can be found in the literature [4]. We will list here only the most common polymer-based materials for high voltage applications.

(a) *Neat polymers and particulate reinforced polymers*: Thanks to a low dielectric constant, low loss factor and low cost PE is broadly used as insulation material. Low density polyethylene (LDPE) can be operated from −50 to 75°C. The upper temperature limit can be extended to 125°C by modifying the molecular structure and promoting three-dimensional network arrangements thus obtaining highly crosslinked polyethylene (XLPE).

Particulate reinforced composites such as filled PVC, alumina filled elastomers and filled ethylene propylene rubber are common materials for cable applications. When superior thermal, environmental and chemical resistances are required, silicone-based materials might be attractive candidates. Despite a high raw material cost, silicones allow for continuous operation over a broad −50 to 200°C temperature span.

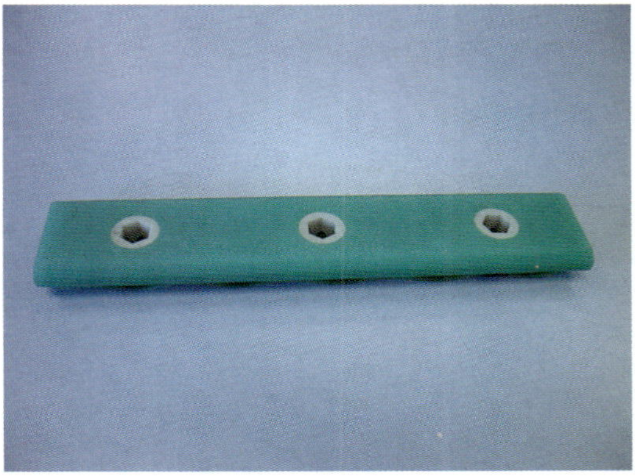

Figure 4.4. Glass-fiber epoxy reinforced wedging system.

Figure 4.5. Carbon-fiber reinforced epoxy ripple spring.

(b) Laminates: Glass-fiber reinforced epoxies and polyesters are used as laminates for various functions in electrical machines such as wedges, separation blocks or stator insulation caps (Figures 4.4, 4.5 and 4.6). These components, though not as critical as the main insulation, have strong mechanical constraints. Moderate electrical stresses experienced by those parts allow for larger void contents in the material than in the case of the main insulation. The use of long-fiber composites, usually containing more voids than homogeneous or particulate reinforced composites is therefore permitted.

(c) Mica tapes: High voltage winding insulation for rotating electrical machines relies upon the use of mica-reinforced polymers. Mica has excellent

Figure 4.6. Glass fiber – Polyester insulating cap.

mechanical and insulation properties. Historically, polyester resins were first used to impregnate the mica particles [4]. Epoxy/mica tapes were introduced at a later date to improve the thermo-mechanical resistance of the insulation. In addition to the Epoxy/mica layer, the main insulation comprises an E-glass-fiber woving conferring a high degree of thermo-mechanical stability to the insulation. Glass fibers also enhance the durability of the electric components thanks to an excellent resistance to partial discharge erosion (Section 4.2.3).

For rotating machinery such as motors and generators, two main manufacturing processes related to insulation systems dominate the market: resin rich (RR) and vacuum pressure impregnation (VPI). The RR system uses glass fiber-mica tapes, which are pre-impregnated with a partially cured epoxy (B-stage). The tapes are wound onto the bars then cured during a subsequent heating process. On the other hand, the VPI system employs dry glass fiber-mica tapes (Table 4.1) wrapped around the conductors. The bars are pressure impregnated with Epoxy after vacuum application for a specific time in the autoclave.

4.2.2 Definition of Electrical Quantities and Properties

High voltage cables and windings are made of a series of conductive, semi-conductive and insulating materials resulting in composites with complex

Table 4.1. Typical properties of Muscovite Mica and standard VPI tape

Properties	Mica (Muscovite) [5]	Typical VPI tape (60% mica content)
Density	2.6–3.2	1.8–2.0
Tensile strength	170 MPa	190 MPa (20°C) 95 MPa (135°C)
Maximum operating temperature	500–600°C	180°C
Breakdown voltage	120–200 kV/mm	>60 kV/mm
Loss tangent	0.0001–0.0004 (at 15°C)	<0.01 (at 23°C) <0.001 (at 120°C)

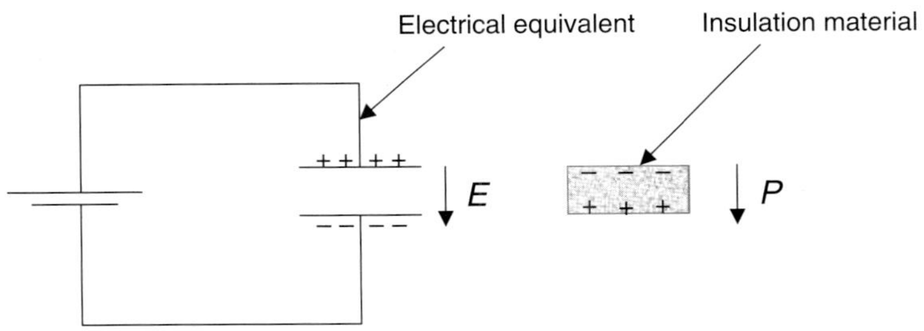

Figure 4.7. Insulation polarization and capacitor model.

mechanical and electrical responses (Figure 4.2). Environmental and mechanical considerations presented in Chapters 2–3 and 5–6 fully apply to such systems. We will now focus our attention on the additional tools necessary to understand the specificities of the electrical field exposure.

4.2.2.1 Capacitance, resistivity, conductivity, polarization

Insulating polymers and polymer matrix composites generally belong to the class of dielectrics (low conductivity). In a first approximation, the insulating material in an electric field can be thought of as a capacitor, with two electrodes separated by the material (Figure 4.7). The storage ability of the capacitor is measured by the capacitance C (in Farads) and is the ratio of the charge stored over the voltage:

$$C = Q/V \tag{4.1}$$

Let us now consider a dielectric of thickness "t". The *capacitance* of the material over the distance "t" can be calculated:

$$C = \varepsilon A/t \tag{4.2}$$

where A is the area of the (fictive) plates of the capacitor. If the material is removed, the capacitance of the vacuum-filled capacitor becomes:

$$C_o = \varepsilon_o A/t \tag{4.3}$$

with ε_o the permittivity of vacuum ($\varepsilon_o = 8.85 \times 10^{-12}\,\mathrm{F/m}$).

By convention the *relative permittivity* or *dielectric constant* (ε_r) is the ratio of the permittivity of the insulating material over the permittivity of vacuum:

$$\varepsilon_r = \varepsilon/\varepsilon_o = C/C_o \tag{4.4}$$

Solid materials are also characterized by a *bulk electrical resistivity* ρ (in $\Omega^{-1}\,\mathrm{m}^{-1}$). The *direct current* (DC) *conductivity* σ is defined as the invert of the resistivity. It is proportional to the current density and the applied electrical field:

$$\sigma = 1/\rho = J/E \tag{4.5}$$

where J is the current density ($\mathrm{A/m^2}$) and E is the electric stress (V/m). The electrical conductivity depends on the materials nature (especially the presence of impurity) and the environment. More specifically, the temperature dependence of the conductivity generally follows an Arrhenius law:

$$\sigma(T) = A\exp(-E/kT) \tag{4.6}$$

where E is the electric field, A a constant and k the Boltzmann constant ($k = 1.380\,662 \times 10^{-23}\,\mathrm{J/K}$).

Under the effects of the electric field, electrons and polymer molecules tend to re-arrange. This re-arrangement is referred to as polarization. Conduction might also take place depending on the nature of the composite. Generally speaking, the *polarization* P is related to the electric field E and the permittivity by:

$$P = \varepsilon_o(\varepsilon_r - 1)E \tag{4.7}$$

Under an *alternating current* (AC), the material experiences losses due to conduction and polarization. For a perfect dielectric under a sinusoidal alternating current ($\sin \omega t$), the phase difference between the tension and the current is 90°. For the real composite, the phase angle is actually smaller. The complementary angle (δ) represents the loss angle (Figure 4.8). For industrial applications, the *loss factor* ($\tan \delta$) is the quantity used for quantifying the electrical losses in the material.

$$\tan \delta = \frac{\sigma}{\omega \varepsilon_o \varepsilon_r} \tag{4.8}$$

The loss factor is strongly influenced by the nature of the material and the environment.

With the electrical strength, the loss factor is probably the most important characteristic of the material for industrial applications. Electrical losses mechanisms are therefore studied in greater detail thereafter.

4.2.2.2 Losses

Figure 4.8 shows the simultaneous ability of the material to store and disperse electrical energy. The similitude with the polymer response to cycling load in Chapter 2 is striking. To model the materials dielectric response, we therefore introduce analogs to the Voigt and Maxwell models of Chapter 2 (Table 4.2). Storing energy corresponds to the use of a spring element for mechanical considerations and to a capacitor for electrical considerations. Losses are represented by dashpots for mechanical models and resistances for electrical models. Resistance and capacitors can be placed in series or parallel to model the electrical response of the material (Figure 4.9).

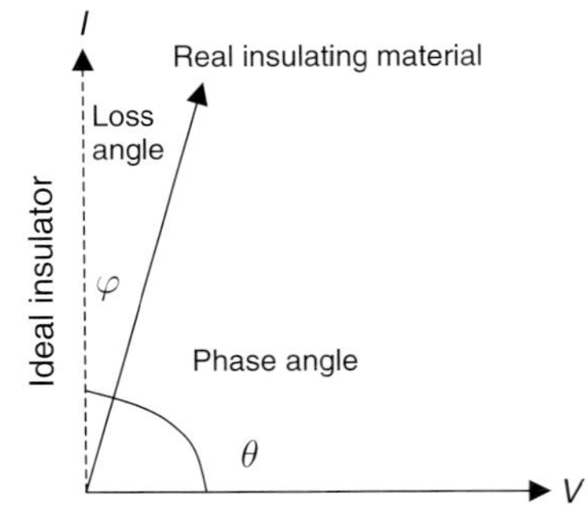

Figure 4.8. Loss and phase angles. (Note that the loss angle is proportional to the energy lost per cycle and is kept as small as possible for most high voltage insulators, typically 0.001).

Table 4.2. Mechanical and electrical analogs

Mechanical quantity	Electrical analog
Stress σ	Voltage V
Strain ε	Charge Q
Compliance D	Capacitance C
Voigt model	R–C circuit
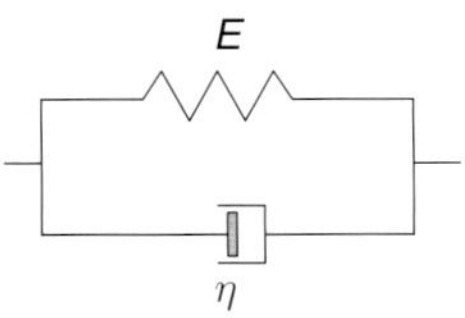	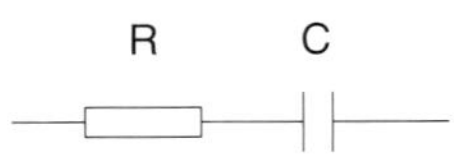
Differential equation for Voigt model	Differential equation for R–C model
$\sigma = \eta \frac{d\varepsilon}{dt} + E\varepsilon$	$V = R\frac{dQ}{dt} + \frac{Q}{C}$ (4.12)
Solution example for creep (2.6)	Solution for constant voltage U_o at $t = 0$ (4.13)
$D(t) = D_o(1 - \exp(-t/\tau))$	$Q(t) = CV_o(1 - \exp(-t/\tau))$

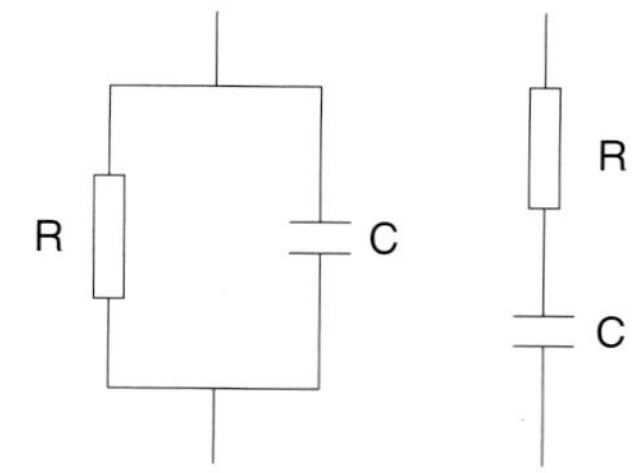

Figure 4.9. Insulation electrical parallel and series analog models.

A major distinction between mechanical and electrical models is, however, that stresses add when in parallel when voltages add when in series. Therefore, a capacitor and a resistance in series lead to a *Voigt* type equation (resulting from mechanical elements in parallel). Inversely, a capacitor and a resistance in parallel lead to a *Maxwell* type equation (resulting from mechanical elements in series).

Let us consider a capacitor and a resistance in series (Figure 4.9):

$$V = V_c + V_R \tag{4.9}$$

Using the well-known electrical relations:

$$V = RI \tag{4.10}$$

and

$$I = \frac{dQ}{dt} \tag{4.11}$$

we obtain:

$$V = R\frac{dQ}{dt} + \frac{Q}{C} \tag{4.12}$$

which can be solved using the methods of Chapter 2 leading to:

$$Q(t) = CV_o(1 - \exp(-t/\tau)) \tag{4.13}$$

For a capacitor and a resistance in parallel, we now write:

$$I = I_c + I_R \tag{4.14}$$

or

$$\frac{dQ}{dt} = C\frac{dU}{dt} + \frac{U}{R} \tag{4.15}$$

which can also be solved for given initial and boundary conditions using the methods of Chapter 2.

If we now introduce a complex dielectric constant for the composite:

$$\varepsilon_r^* = \varepsilon_r' - i\varepsilon_r'' \tag{4.16}$$

Equation 4.14 becomes:

$$I = CVi\omega\,(\varepsilon_R' - i\varepsilon_R'') + \frac{V}{R} = \omega CV\varepsilon_r' i + V\left(C\omega\varepsilon_r'' + \frac{1}{R}\right) \tag{4.17}$$

where $i^2 = -1$.

Beyer et al. [6] propose to quantify the imaginary and real part of the current:

$$I = I_{im} i + I_{re} \tag{4.18}$$

where

$$I_{im} = \omega CV\varepsilon_r' \tag{4.19}$$

and

$$I_{re} = V\left(C\omega\varepsilon_r'' + \frac{1}{R}\right) \tag{4.20}$$

The form of Equations (4.19) and (4.20) allows to distinguish between short-term and long-term contributions of the different phenomena to the losses. Indeed, knowing that the loss factor is also the ratio of the real and imaginary components enables us to write:

$$\tan\delta = \frac{I_{re}}{I_{im}} = \frac{1}{RC\omega\varepsilon_r'} + \frac{\varepsilon_r''}{\varepsilon_r'} = \tan\delta_c + \tan\delta_P \tag{4.21}$$

Based on Equation (4.21), Beyer et al. [6] distinguish between two loss components in liquid insulations. The first component $\tan\delta_c$ being related to the conduction phenomenon while the second component $\tan\delta_p$ is linked to the polarization mechanisms.

Aklonis and McKnight [7] on the other hand prefer to talk in terms of instantaneous and time-dependent polarization. We will attempt to unite those theories by analyzing the phenomena responsible for losses and by grouping them into two groups: (1) Short-term loss contributors and (2) Long-term loss contributors.

(1) Short-term losses
  (a) Conduction (free carriers) and interfacial polarization: Conduction by free carriers (ions or electrons) is responsible for the first part of losses expressed in Equation (4.21). In good solid insulators such as polymers, the charge migration within the material is often negligible. However, the presence of reinforcement or defects can increase the conduction loss contributions,

especially under high voltages. Those defects include ionic impurities, trapped space charges and voids.

Conduction can also occur in fibers. For example, E-glass mentioned in Section 4.2.1.2 is often used in combination with epoxy as an insulating material. E-glass composites generally allow for ion migration (such as sodium) leading to significant conduction losses [1].

Conduction losses are temperature and frequency dependent. Additionally, close to the breakdown voltage, Schottky or tunneling effect might lead to electron emission originating from the electrode, resulting in increased electrical losses [1].

Migrating charges might accumulate at an interface or *defect* region. This mechanism is referred to as boundary polarization (Figure 4.10) and tends to further increase losses. Boundary polarization can occur in inhomogeneous (such as semi-crystalline) polymers and composites. Interfacial polarization is much slower than conduction and generally requires several minutes. Figure 4.11 illustrates the differences in the dielectric constant for pure and filled XLPE. The composite exhibits a behavior strongly influenced by interfacial polarizability characterized by a higher dielectric constant and higher losses [1].

(b) Induced dipole moments (also called *electronic*, *optic*, *vibration* or *deformation polarization*): Vibration polarization occurs in all dielectrics including non-polar composites. The electronic polarization occurs when the field is applied: the electrons tend to displace to accommodate the electric field. This displacement combined with the non-symmetry of the original molecular structure results in displacements of the nuclei of the atoms. Dipoles are therefore induced from the electrical field generating losses in the material. These dipoles disappear upon removal of the load. The induced dipole moments contribution to the total losses can be predominant in the case of non-polar polymers and composites. Such polarization occurs extremely quickly and takes place at high frequencies beyond visible light.

(2) Long-term losses

(a) Orientation polarization: Unlike short-term conduction and polarization losses, long-term losses occur only in polar materials. Indeed, polar

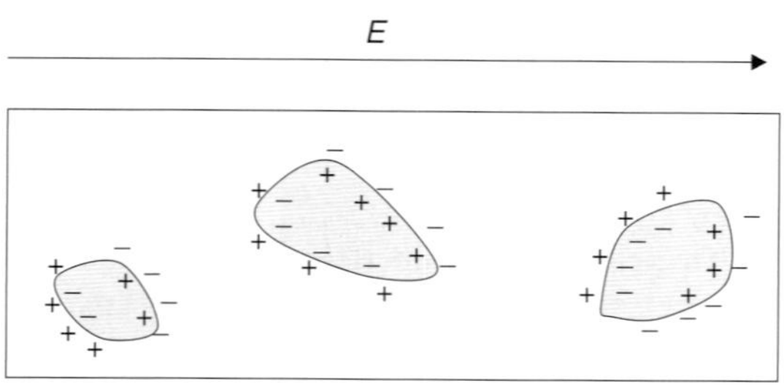

Figure 4.10. Interfacial polarization in a particulate composite.

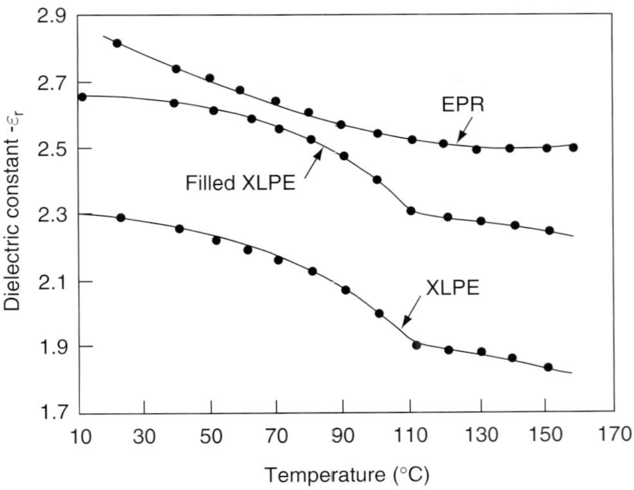

Figure 4.11. Effect of fillers on the dielectric constant. (Copyright 1998, from *Electrical Insulation in Power Systems*, by N.H. Malik [1], reproduced by permission of Routledge/Taylor and Francis Group, LLC.)

polymers present an asymmetric distribution of positive and negative charges that form permanent dipoles. This distribution is *spontaneous*, i.e. it exists independently of the presence of an electrical field. If no current is applied, the material is neutral at the macroscopic scale due to the random distribution of the dipoles. In the presence of an electric field, the electrical moments tend to align creating an orientation polarization (Figure 4.12). This polarization can lead to the actual displacement of entire molecules motivated by changes in their electrical moments. Re-orientation is slowed down by concomitant large-scale molecular motion and molecular collision. Orientation polarization is therefore a much slower process than conduction and vibration mechanisms. Indeed, the polarization will be maximal only

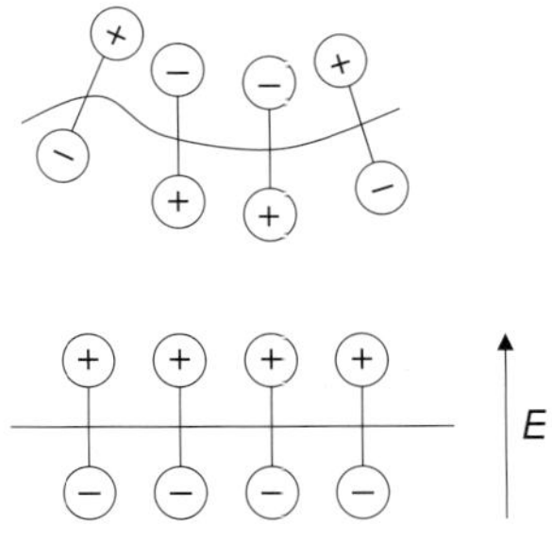

Figure 4.12. Molecule permanent dipole orientation random (no field) and under field.

after a relaxation time τ_{op}. Under a cycling electrical load, the dipoles try to re-orient themselves for every half cycle. If the relaxation time is longer than the cycle half-period, the polarization will only be partial. For high frequencies, the orientation polarization does not take place anymore. This process typically occurs in a 1 kHz–1 MHz range [8].

(b) Notes on the dielectric relaxation: The process of dielectric relaxation is complex. Debye, Onsager and Kirkwood-Frölich [9]–[11] propose equations attempting to include the physics of the molecular relaxation. Unfortunately, these equations are difficult to apply to complex composites systems in an industrial context.

It is however worth mentioning the use of the Cole-Cole plot which is quite common when studying dielectric relaxation [12]: let us consider ε_s and ε_l, relative permittivities of the material at short (instantaneous) and long times. The complex dielectric constant can then be written in terms of ε_s and ε_l:

$$\varepsilon^*(\omega) = \varepsilon' - i\varepsilon'' = \varepsilon_u + \frac{\varepsilon_l - \varepsilon_s}{1+\omega^2\tau^2} - i\frac{(\varepsilon_l - \varepsilon_s)\omega\tau}{1+\omega^2\tau^2} \tag{4.22}$$

Remembering that the loss factor is the ratio of the imaginary and real components:

$$\tan\delta = \varepsilon''/\varepsilon' \tag{4.23}$$

it is possible to plot the various quantities as illustrated in Figure 4.13. Figure 4.13 is commonly referred to as a Debye plot.

Another form of Equation (4.22) is:

$$\left(\varepsilon' - \frac{\varepsilon_l + \varepsilon_s}{2}\right)^2 + (\varepsilon'')^2 = \left(\frac{\varepsilon_l - \varepsilon_s}{2}\right)^2 \tag{4.24}$$

This equation leads to a Cole-Cole plot [12] where ε'' is plotted against ε'. The curve is a half-circle in the case of single relaxation times but loses its symmetry in the case of multiple relaxation times (Figure 4.14). Actual composites generally lead to dissymmetrical curves.

4.2.2.3 *Specificity of composites*

The composites dielectric properties can be estimated based on the properties of the constituents. Indeed, if we consider two materials in perfect contact, a rule of mixtures can be used to express the composite dielectric constant and the resulting tan δ as a function of the matrix and fiber dielectric constants (ε_m, ε_f) and the

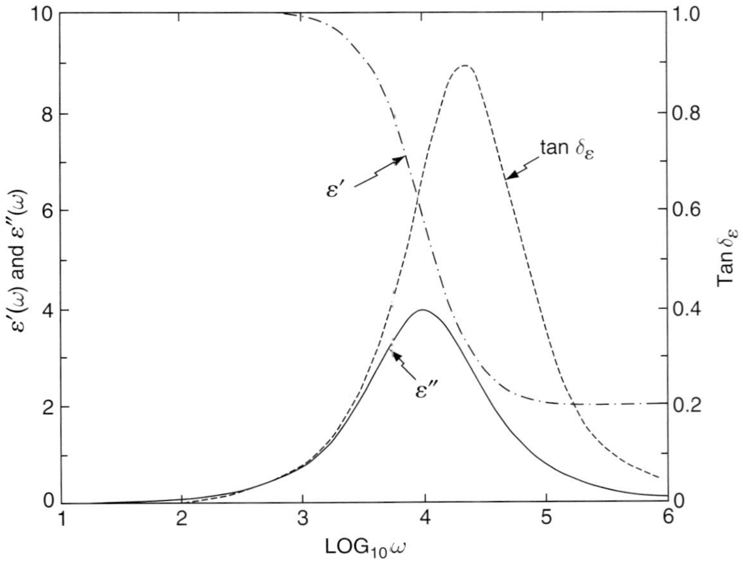

Figure 4.13. Debye Plot. (Copyright 1983, from *Introduction to Polymer Viscoelasticity* by Aklonis and McKnight [7], reproduced by permission of John Wiley & Sons, Inc.)

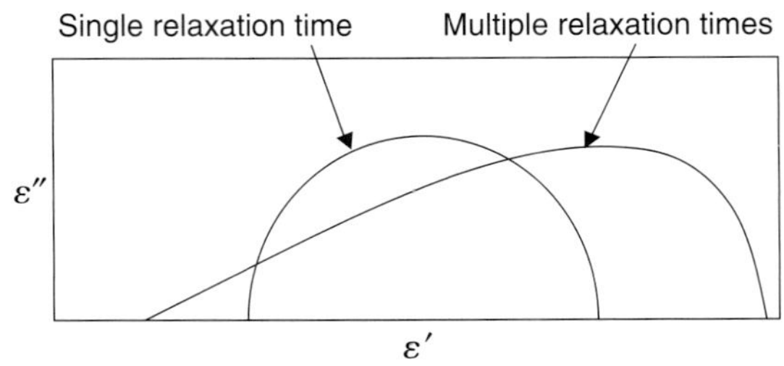

Figure 4.14. Cole-Cole plot for a pure linear polymer with single relaxation time. Cole-Cole plot for a typical multiple relaxation time insulation polymer composite.

matrix and fiber volume fractions (V_m, V_f). Malik [1] proposes for a perfect bi-layer structure:

$$\varepsilon_c = \frac{\varepsilon_m \varepsilon_f}{\varepsilon_m V_m + \varepsilon_f V_f} \tag{4.25}$$

and

$$\tan \delta_c = \frac{\tan \delta_m + (\varepsilon_m V_f / \varepsilon_f V_m) \tan \delta_f}{1 + (\varepsilon_m V_f / \varepsilon_f V_m)} \tag{4.26}$$

If the resulting values can be used as a first approximation to evaluate the impact of the second material, it is however recommended to use experimental macroscopic data of the complete composite. Indeed, interfacial polarization is not accounted for in Equations (4.25) and (4.26). Due to a large number of defects and interfaces in composites, ignoring interfacial polarization might lead to erroneous data.

4.2.2.4 Practical consequences

Considering the tight link between polarizations and relaxation phenomena in polymers, it is no surprise that the loss factor measurements are strongly influenced by environmental parameters such as temperature and voltage. Figure 4.15 shows the different shapes of the tan δ versus temperature curve as a function of temperature depending on the polarization and conduction processes.

The resulting loss factor is the sum of the contributing loss tangents. Indeed, Robert [8] mentions that polarization effects add linearly and do not usually interact. The polarization tangent might exhibit the presence of multiple peaks depending on the activated polarization process.

Tan δ measurements are sensitive to the applied voltage as well. The degree of sensitivity depends upon the polarization processes activated in the system (Figure 4.16).

The quality of the insulation is generally judged by the absolute value of the tan δ at two times rated voltage ($2U_n$). According to the norm DIN EN50209 for example [13], this value should be no greater than 30×10^{-3}.

The absolute value of the loss factor globally reflects the different loss processes in the material. If we reproduce the tan δ measurements within a few hours for very high quality insulation materials, a monotonic decrease of the values can be observed (Figure 4.17). If the material is left to rest for a few days, a recovery can be observed. This example illustrates the contributions of short-term and long-term components to the losses. During the first measurement, all conduction and polarization effects occur. During the subsequent measurements, short-term effects (conduction and vibration polarization) still take place. However, the magnitude

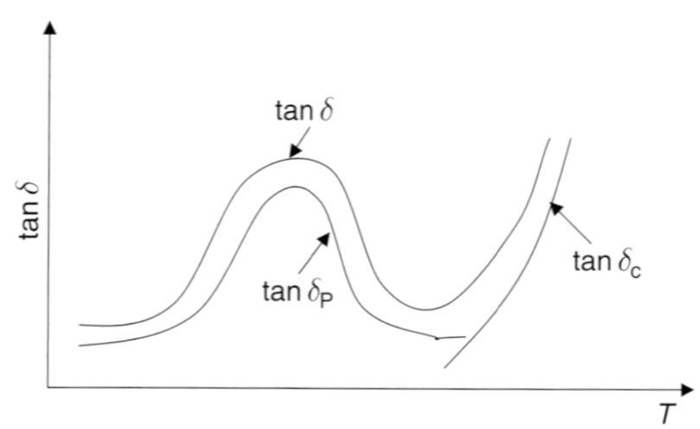

Figure 4.15. Loss tangent versus temperature.

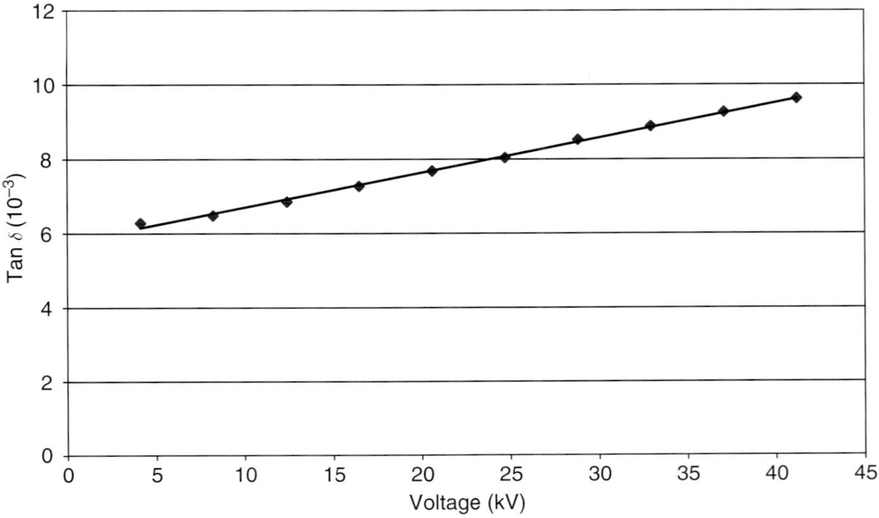

Figure 4.16. Tan δ versus voltage for typical generator stator winding bars.

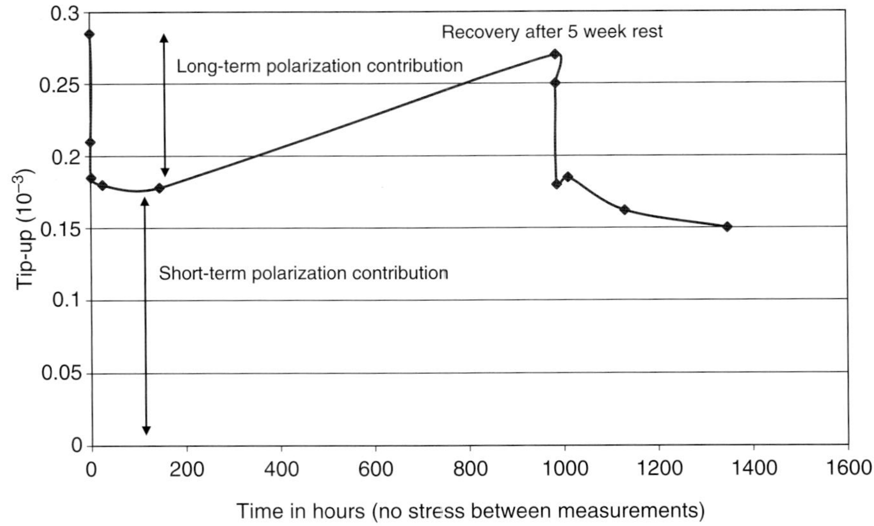

Figure 4.17. Tip-up or tan δ reproducibility curves.

of orientation polarization (long term) is reduced. Consequently, the more the measurement replicates are performed, the lower the orientation losses and therefore the lower the global losses in the bar.

These experimental observations have great consequences: they evidence the fact that results of tan delta measurements can vary depending upon the *resting* time of a bar (resting time after operation or after testing). The data is therefore more relative than absolute: observing lower tan δ values after multiple testing does not mean that the material is really getting *better*. It only means that enough time (related to the materials relaxation times) was not provided for the material to relax to its original state. To enable a one-to-one comparison between two bars or one bar in different states, it is necessary to measure the tan δ allowing identical relaxation time.

Another important consequence is that it is to a certain extend possible to separate and measure the contribution of short- and long-term loss components: after repeated testing, short term (conduction and vibration polarization) become predominant. The difference between original value (after relaxation) and final value shows the contribution of orientation polarization.

The second indicator for the quality of the insulation is the slope of the tan δ versus rated voltage measurement. The norm DIN EN 50209 specifies that the increase in tan δ should not be more than 6×10^{-3} over a "$0.2 \times U_n$" measurement step. The relationship is generally linear and reflects the linearity of the polarization as a function of the applied electrical field. To measure the slope of tan δ an industrial quantity called *tip-up* is introduced. Unfortunately, definitions of the tip-up vary from one industry and one continent to the other. IEEE Std 1310-1996 [14] indicates the following tip-up definition:

$$\text{tip-up} = \frac{\tan \delta_{0.6U_n} - \tan \delta_{0.2U_n}}{2} \tag{4.27}$$

Tan δ measurement is a great non-destructive test to track damage in the material. Indeed we have seen that tan δ is very sensitive to the presence of impurities: conductive impurities lead to a higher conductivity, which increases the losses. If air is present in the material, two situations can occur. For low electrical fields, the air remains insulating and little difference will be observed on the tan δ measurement of the dielectric. However, in high electrical fields (electrical field exceeding the air dielectric strength usually equal to 3 kV/mm), breakdown occurs and the air becomes conductive. This breakdown translates in partial discharges (Section 4.2.3) and in an instantaneous increase of the losses via the conduction short-term component. The slope of the tan δ versus rated voltage, which is dominated by short-term component, increases due to increased conduction in the cavity. Long-term orientation polarization and short-term vibration polarization remain constant (not affected by the void) but their contribution to the overall losses (which have increased) is reduced. The global magnitude of loss increase depends upon the size of the cavity.

Comparing tip-up measurements in a *new* and *after* stages is therefore an excellent non-destructive evaluation means to provide information on the level of degradation (via debonding or delamination) in the material provided that the tan δ measurements are always measured following an identical relaxation time sequence.

4.2.3 Breakdown and Failure

4.2.3.1 Electrical breakdown

An insulating material can become conductive under very high electric fields, with a conductivity strongly dependent upon the magnitude of the electrical field. This occurs when the energy provided to the atoms is enough for a large number of electrons to move to the conduction band. The electrical properties of the material are then completely modified and irreversible damage of the composite such as localized melting of the crystals (in case of semi-crystalline materials), flow or chain scission might occur.

Electrical breakdown in polymers and composites is a complex field. Ieda et al. [15] provide a good summary of the different theories concerning electrical breakdown in polymeric materials and major highlights are reported thereafter.

Like all materials, insulating composites are characterized by a dielectric strength (E_b), which is a measure of the electrical field necessary to induce an electrical breakdown. The value of the dielectric strength directly defines the necessary thickness of the insulation and is therefore a fundamental parameter used in electrical design. Examples of dielectric strength values are given in Table 4.3 for different polymers and composites.

Table 4.3 shows very wide variations. Maximum working fields for typical applications are indicated in Table 4.4.

Table 4.3. Dielectric strength for different materials [16]

Material	Electrical strength (E_b) in kV/mm
40% Mica/polypropylene	20
Polycarbonate	15–30
30% Glass fiber/polysulfone	16.9–40
40% Glass fiber/nylon 6	19.7–41
Glass fiber/polyketone	24–41
40% Glass fiber/nylon 66	19.7–41
Polytetrafluoroethylene	18–105
Nylon 66/6	18–120
HDPE	19–150
Polyimide, thermoset film	154–339

Table 4.4. Working fields for typical applications [17]

Material	Maximum working field (kV/mm)
XLPE in power cables	12–25
Filled-Epoxy in switchgear	2
Polypropylene (PP) in DC thin film capacitors	1000

Like most phenomena in polymer-based materials, the breakdown mechanisms and therefore the breakdown strengths vary with the application time of the electrical field. The dielectric strengths also depend upon temperature and frequency. Indeed, short times (smaller than the micro-second) are characterized by the intrinsic breakdown phenomenon. Physical models and explanations of the intrinsic breakdown for amorphous and inhomogeneous materials revolve around the electronic avalanche concept where an electron becomes free and excites further electrons that move in their turn into the conduction band (cascading process). The details of intrinsic breakdown mechanisms are complex and were under heavy discussion at the time of writing [18].

Still for short times, electromechanical breakdown can occur if the electrostatic compressive forces (also referred to as electrostriction) exceed the mechanical strength of the material.

For times comprised between the micro and several hundred seconds, the dielectric strength might be dominated by the thermal breakdown process (Figure 4.18). Thermal breakdown is a well-understood mechanism and is directly related to the notions of losses detailed in Section 4.2.2.2. Conduction and polarizations contribute to the apparition of losses. The losses create an elevation in temperature. If not countered by proper cooling, this temperature elevation can lead to increased conductivity (conductivity being a function of temperature, see Equation (4.6), and temperature-induced degradation of the composite such as melting, flow or chain scission (Chapter 2). Losses are generally much higher in an alternating field than in a direct current field, resulting in different thermal breakdown threshold values depending upon the conditions.

For longer times, a more progressive degradation of the dielectric strength resulting from accumulating damage mechanisms can be observed. This degradation

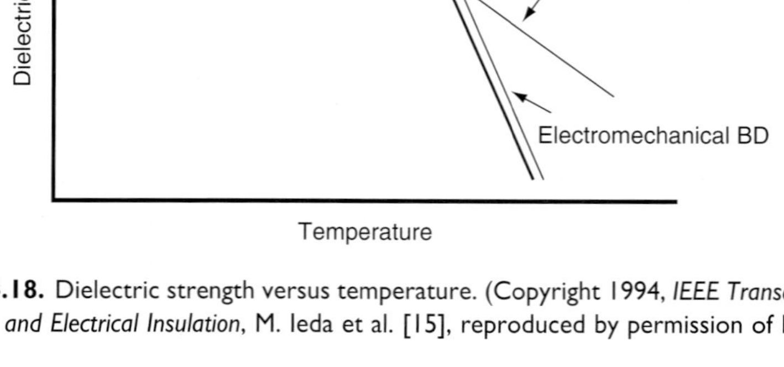

Figure 4.18. Dielectric strength versus temperature. (Copyright 1994, *IEEE Transactions on Dielectrics and Electrical Insulation*, M. Ieda et al. [15], reproduced by permission of IEEE.)

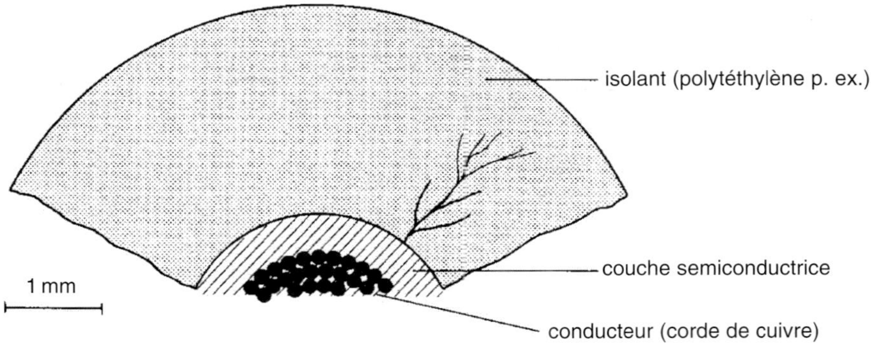

Figure 4.19. Treeing in cables. (Copyright 1989, from *Traité d'Electricité*, Vol. 2, by P. Robert [8], reproduced by permission of Presses Polytechniques et Universitaires Romandes.)

is generally the result of multiple factors including all environmental degradation mechanisms reviewed in this book such as moisture absorption or UV radiation and additional electric phenomena such as partial discharges in which the repeated arching in the cavity can start an erosion process in the material. A common damage phenomenon in high voltage cables is, for example, the formation of *trees* (Figure 4.19). Such defects usually originate on an impurity and propagate with time. This growth is pursued until it reaches a breakdown critical size which might lead to cable failure.

4.2.3.2 *Physical degradation and failure (e.g. cycling)*

The physical degradation of composite insulators occurs via numerous and sometimes competing mechanisms. The utility pole case study thereafter shows the diversity of environmental loads responsible for degradation of high voltage line components.

Case study: Utility poles

Thanks to corrosion resistance and excellent dielectric properties glass fiber reinforced composites are being introduced as poles supporting transmission lines. This application is a large market for composite materials. Indeed, in most developed countries, transmission lines were installed around 50 years ago and need replacement within the next 10 years. The North American utility pole refurbishment market alone is estimated at one billion dollars [21].

Utility poles are constantly exposed to mixed environmental loads such as moisture, animal and insect damage (e.g. repeated impact from woodpeckers), wind, ice, ultra violet radiation or guy and brace forces. Wood, concrete and steel are traditional materials for power lines but replacement by composite materials provides many benefits ranging from weight savings to higher energy absorption

Figure 4.20. Composite pole installation. (Courtesy of Strongwell.)

upon vehicle impact. Indeed, the composite poles are expected to have lifetimes three times longer than wooden ones. Typical weight savings obtained by using polymer composites are around 50% for wood and 30% for steel poles replacement. Such drastic weight differences translate in significant cost savings for installation and transportation. For example, larger quantities of poles can be carried on a truck (Figure 4.20) or composite pole installations in remote areas may necessitate smaller investments, such as the use of smaller helicopters (Figure 4.21).

Glass-fiber reinforced polymers can be tailored to provide excellent insulation characteristics, thereby enhancing safety during repair and maintenance. Lightning strikes on concrete poles were seen to induce flashovers resulting in surges large enough to damage household appliances [21]. The use of glass-fiber reinforced polymers can help prevent such occurrences.

Carbon-fiber reinforced poles can also exhibit an excellent resistance to fire: unlike 3000 wooden poles that were fully destroyed, a Powertrusion composite pole installed in San Diego for demonstration survived the 2003 California fire which destroyed over 2000 km^2 of land [19].

Stricter environmental regulations are further motivating the use of composite materials. Indeed, wood poles are usually treated with preservatives (creosote, copper chromium arsenate or pentachlorophenol) which are detrimental to the environment and in the process of getting banned [20].

Figure 4.21. Helicopter installation of composite pole. (Courtesy of Strongwell.)

Despite those advantages, the introduction of composites as utility poles and cross-arms that started around 40 years ago is still very slow. Grid providers and utility companies constitute a group of very traditional industries and innovations are accepted under the condition of the demonstration of multiple operating references. Despite original difficulties in the early introduction, composite cross-arms are generally better accepted than poles by the transmission community, probably due to the fact that the product can be cost-competitive even on a *as sold* basis. In the 1960s and 1970s, composite cross-arms degraded strongly with UV radiation and exhibited fiber blooming. Erosion of the matrix would lead to fiber exposure on the surface [21]. This problem was solved by applying more elaborate UV protections such as inhibiting polyurethane-based paints. Today, necessary ultra-violet protection and required esthetically pleasant surface finish can also be achieved in one step by using a polymer film such as a polyester veil.

Poles are traditionally being manufactured using filament winding or pultrusion process. The fiber reinforced profiles are then generally filled with foam to eliminate nesting pest problems. Pultrusion manufacturing allows precise fiber orientation.

The use of multiaxial fabrics and continuous strand mat results in excellent axial and bending strength even for large poles. Obtaining the proper axial strength by filament winding however can be challenging due to the natural difficulty to orient fibers axially. Despite this, 21 m Class I filament wound poles were successfully manufactured and 26 m poles achieved for lower classes [35].

In addition to the environmental factors detailed in the various chapters of this book, two mechanisms tightly related to the electrical field environments accelerate the materials degradation, namely partial discharge erosion and tracking and surface erosion.

By definition, composite materials possess large intern interface surfaces. These interfaces are preferred locations for defects such as contamination and voids. Voids can, for example, facilitate moisture absorption. Moisture usually decreases resistivity and dielectric strength while increasing the permittivity of the composite [1]. Voids might also favor internal erosion. Indeed, cavities in the composite are generally filled with a gas of lower dielectric strength. Therefore, breakdowns within the cavity, also called *partial discharges*, can occur. Discharges occurring in air are a special case of partial discharges and are generally referred to as corona emissions.

Partial discharges occur only in cavities and not in the material. However, the electrons traveling in the cavity hit the materials surface and might induce chain scission and irreversible material degradation. This erosion phenomenon increases the size of the cavity, which increases in turn the partial discharge process.

Partial discharge erosion is a major concern for many electrical applications. The size and the number of voids in the composite need to be precisely controlled for medium and high voltage applications. In power cables, no voids that could be seen (i.e. bigger than about one micrometer) would be tolerated [17]. Partial discharge erosion is often a limiting factor for the use of long-fiber composite materials in which the presence of microscopic voids is almost inevitable.

To make matter worse, ozone (O_3) is created as a by-product of air cavity discharges. Therefore, an excellent resistance to ozone exposure is required for composite insulators. Indeed, partial discharges are inevitable in the composite and generation and diffusion of ozone will occur (see Chapter 3 for gaseous diffusion). Therefore, experimental methods of Chapter 3 should be used to insure that ozone does not create irreversible damage in the material.

The loss tangent was shown to be a powerful tool in measuring the degree of damage in the composite. Partial discharges can also be used to assess the degree of void content in the material. However, loss factor increase and partial discharge measurements usually do not coincide perfectly. Indeed, loss tangent ($\tan \delta$) measurements are influenced by the defects in the material as well as by polarization mechanisms (which do not mean degradation) when partial discharges are affected only by the voids: unlike partial discharges, loss measurements are, for example, strongly influenced by relaxation mechanisms and the viscoelastic nature of the composite. Furthermore, partial discharges measurement methods heavily rely upon statistical analysis and empirical rules, and do not always provide a clear picture of the state of damage in the material. Indeed, under the presence of the electric field, a given void might experience a partial discharge which can be measured. However, repeated discharges in the cavity may lead to carbonization in which a thin carbon layer forms on the surface of the cavity. The conductivity of this thin layer might suffice to equalize the potentials and new measurements show a recession in the partial discharge activity even though the void is still present. The thin carbonized layer does not, however, significantly influence the tan δ measurements due to the small dimensions of the layer. Therefore, a recession in the partial discharge activity (indicating that the bar would actually become better) should be clearly distinguished from the loss relaxation phenomenon discussed in Section 4.2.2.4 (that also showed a *bettering* of the insulation).

Based on these observations, partial discharge measurements should be interpreted with care and it is commonly agreed within the industrial community that such measurements are more an art than a science.

Discharges can also occur on the surface of the material. At the surface of insulators, an erosion phenomenon similar to the cavity process can take place. This erosion can be accompanied by an accumulation of carbon on the surface creating a conductive area (tracking). Moisture accumulation in this conductive layer results in low surface resistance and losses. The losses produce heat which tends to dry off some moisture from given areas. Discharges can then occur between dry and wet areas leading to further damage and eventually to carbonization. Carbonization on the surface can be such that the insulation cannot withstand the operation voltage anymore. Surface discharges in air can also combine with ambient molecules to form corrosive gases. This is the case of PE where surface discharges result in nitric acid when combined with ambient moisture. The nitric acid will in turn attack the insulation resulting in irreversible damage [1].

4.2.4 Special Focus: Thermal Cycling of Generator Bars

For rotating machinery, the passage of current in the insulated conductors can create high thermal loads onto the insulation (Figure 4.22). Switching the machine on and off imposes thermal cycles on the material. This issue is becoming increasingly important for hydrogenerators with the expansion of pump-storage turbine applications necessitating multiple starts and stops.

Accelerate testing simulating a large number of thermal cycles can be performed in the laboratory. IEEE recently emitted a guideline for the thermal cycling of stator winding bars (IEEE Std 1310-1996 [14]). A typical test arrangement for thermal cycling is shown in Figure 4.23. Recorded thermal cycles are reported in Figure 4.24. The IEEE guideline recommends subjecting the bars to partial discharges, loss tangent and tip-up measurements at different stage of the experiment. The tests end after 500 cycles and bars can be dissected in order to inspect potential degradation.

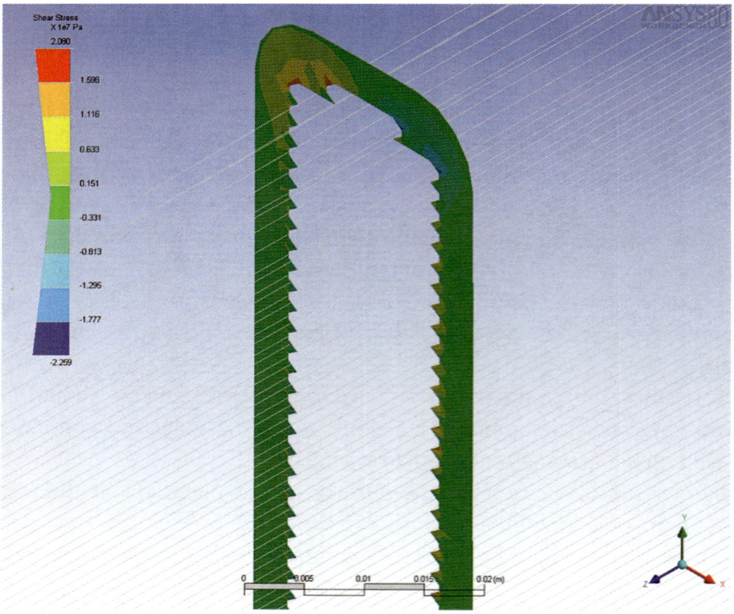

Figure 4.22. Mechanical shear stresses due to current flow in a generator bar. (Courtesy of Alstom.)

As the number of thermal cycles increases, the bars can undergo small or large changes in the tip-up values depending upon the quality of the insulation. Examples of loss factor measurements before and after thermal cycling are presented in Figure 4.25 for mica glass-fiber reinforced epoxy insulated stator bars. One specimen shows an anomalous increase in the loss tangent with increased voltage. This high change in tip-up value can be correlated to the presence of extensive delaminations within the insulation observable under the optical microscope (Figure 4.26). The other bars having very acceptable loss tangent after cycling show no sign of damage (Figures 4.25 and 4.27).

The thermal cycling induces a typical mechanical fatigue phenomenon. Macroscopic static finite element calculations involving temperature gradients encountered during thermal cycling predict a delamination at the copper/insulation interface initiating at corners (Figure 4.22). Crack growth and damage accumulation then follows according to the laws described in Chapter 6. Observed changes in tip-up value (Figure 4.28) follow the typical damage accumulation curves for composite materials (region A:initiation; region B:plateau; region C:instable growth and failure, see Chapter 6) corroborant the assumption that changes in the tip-up values represent damage accumulation when measured under careful conditions (i.e. eliminating the effects of degradation unrelated long-term relaxation).

Figure 4.23. Thermal cycling apparatus. (Courtesy of Alstom.)

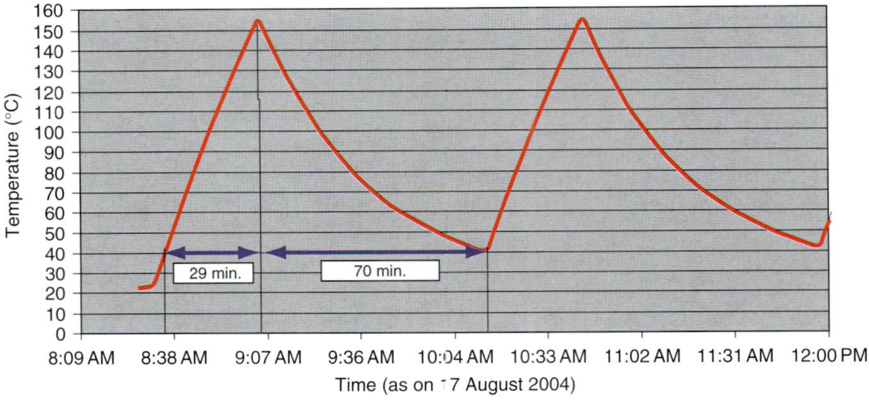

Figure 4.24. Typical recorded thermal cycles.

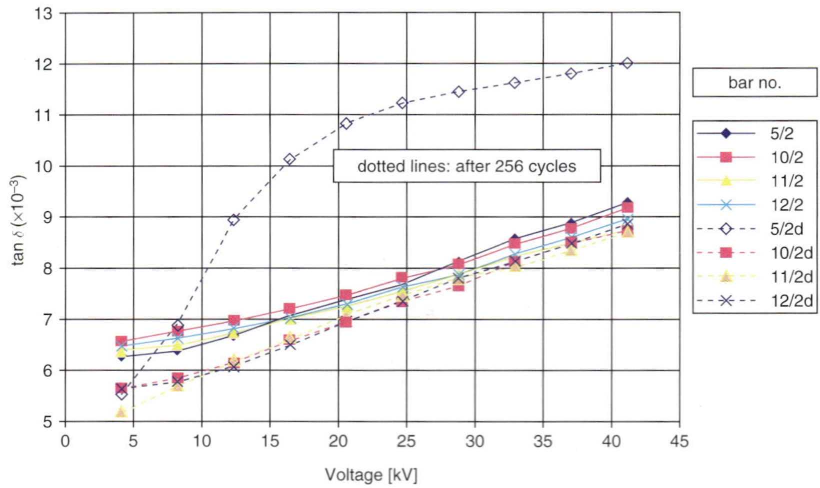

Figure 4.25. Tan δ measurements on five mica glass-fiber reinforced epoxy insulated stator bars before (solid lines) and after (dotted lines) thermal cycling. Bar 5/2 shows an anomalous increase in tan δ with voltage.

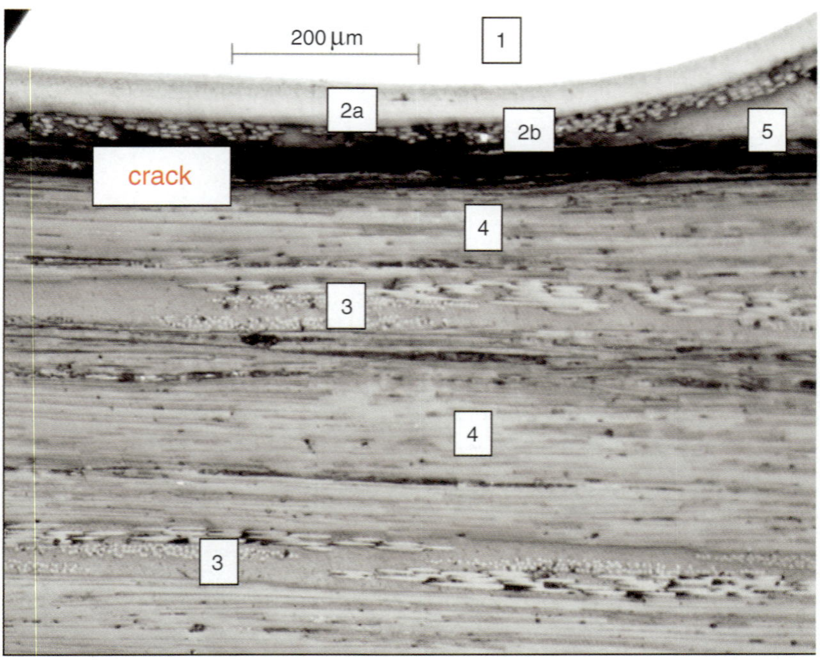

Figure 4.26. Picture of a bar cross-section showing an anomalous increase in tan δ with voltage. Debonding of insulation visible at optical microscope. 1 – Copper conductor, 2a–b – Adhesive and intermediate layers, 3 – Glass, 4 – Mica, 5 – Neat Epoxy.

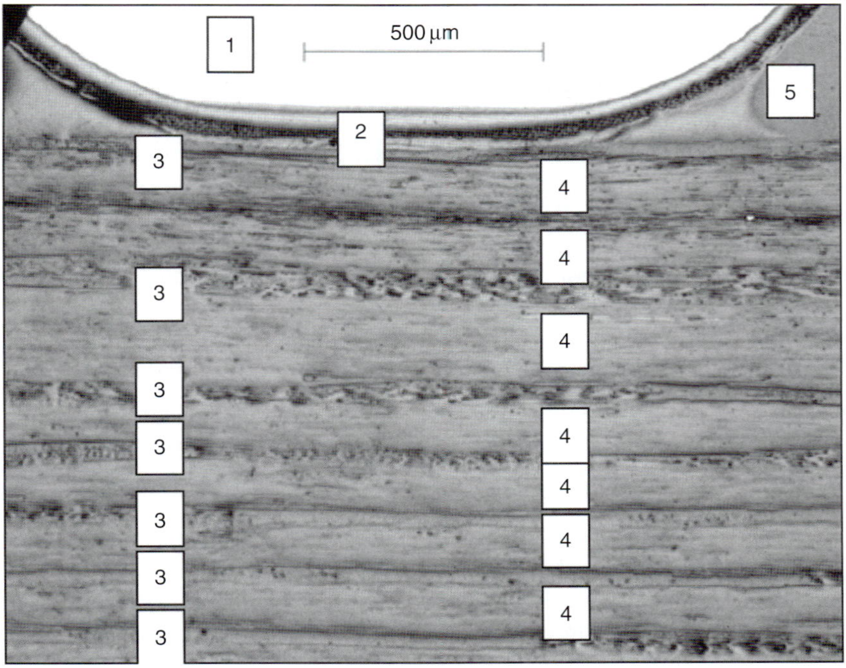

Figure 4.27. Picture of a bar with normal tan δ value. No visible damage at optical microscope. 1 – Copper conductor, 2 – Adhesive and intermediate layers, 3 – Glass, 4 – Mica, 5 – Neat Epoxy.

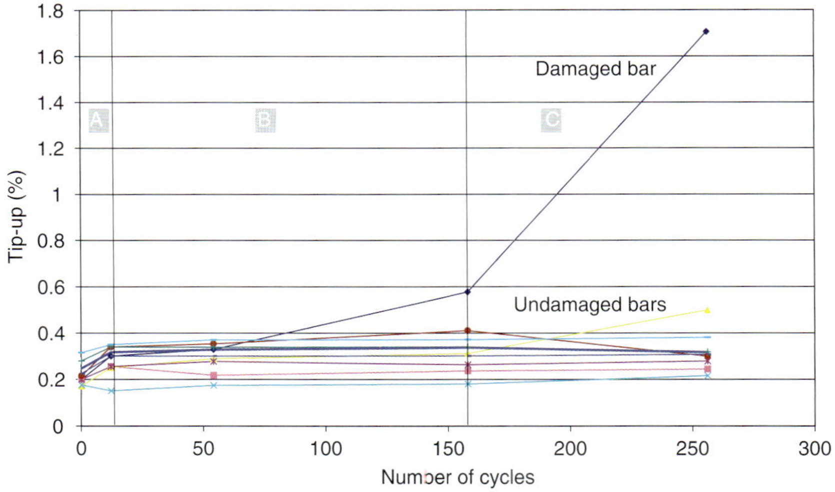

Figure 4.28. Tip-up values versus number of cycles. A – damage initiation region, B – plateau, C – rapid damage growth.

4.3 RADIATIONS

Ultraviolet (UV) radiation exposure is common to most composite materials for limited periods of time during storage, transportation or operation. However, prolonged UV radiations or exposure to radiations of a different nature, such as nuclear radiations, can be very damaging for the material. Radiation exposure is not the most common topic for composite materials. It is however an intrinsic part of environmental degradation. We therefore propose to introduce thereafter the main radiation types and general consequences on the material without going into detail which is beyond the scope of this general book on composite degradation.

4.3.1 The Different Types of Radiations and General Effects

The large variety of radiations is reviewed by D.V. Rosato in [23] and is summarized in Table 4.5. Generally speaking, the level of damage resulting from radiation exposure depends on the strength of radiation flux, distance from source, time of exposure and temperature [24].

Amorphous structures are more strongly affected by radiations than crystalline structures (weakness of bonds). Therefore, glass fibers and polymers are affected in greater extend than metals. Radiations can have beneficial or detrimental effects on composites. Indeed, main effects of radiation on composites include curing as well as degradation, embrittlement and gas emission.

Table 4.5. Main radiation types

Radiation type	Comments
High energy radiations	• Cosmic radiation (low total energy) • γ radiations (from nuclear reactors)
Soft X-radiations	10^{-4} to $10^{-3}\,J/m^2$; 0.1–10 nm
Ultra-violet radiations	• Far UV: $10^{-2}\,J/m^2$; 10–120 nm • Vacuum UV: $2-3 \times 10^{-2}\,J/m^2$; 120–200 nm, 0.2% of solar radiations • Near UV: $10^{-2}\,J/m^2$; 200–400 nm, 7.5% of total solar radiations
Visible radiations	400–700 nm, 41% of total solar radiations
Infrared radiations	• short IR: 22% portion of total • medium IR: 23% portion of total • long IR: 6% portion of total
Radiowaves	Particulate radiations
Cosmic radiations	Low total energy
Van Allen radiations	• Protons: Low total energy • Electrons: Maximum about $3\,J/m^2$-sec
Solar wind	Proton component about $4 \times 10^{-4}\,J/m^2$-sec
Neutrons	Associated with nuclear reactors
α particles	Low penetration

4.3.2 Ultra-Violet (UV) Radiations

Ultra-violet radiation wavelengths vary between 100 and 4000 Å (100×10^{-10} m to 4000×10^{-10} m), leading to a very broad energy spectrum. Only short wavelength UV radiations can provide the necessary minimum activation energies required to initiate reactions in the composites. The amount of such radiations is estimated at only 4% of the total radiations received at the earth surface [25].

Ultra-violet exposure can lead to the formation of three-dimensional networks in the polymer (cross-linking). This method can be used effectively during manufacturing to promote curing. If the material is only partially cross-linked at the end of the manufacturing process, care should be taken that exposure to UV radiation might create property changes due to the additional cross-links. Those property changes can be beneficial (increased strength) or adverse (increased brittleness) depending on the application. Therefore, properties of the partially as well as fully cured components should be used in durability studies (Chapter 6).

Low intensity radiations, representing 96% of sunlight radiation at the earth surface [26], have no detrimental effects on the composite. Indeed, the energy provided into the system is much below the level required to break molecular bonds. Degradation is reported to occur for higher energy levels (wavelengths ranging from 280 to 40 nm). In the absence of oxygen, photodegradation rates are low [27]. Ultra-violet radiations combined with vacuum effect in space (10^{-15} atm beyond earth vicinity) mainly result in out-gasing of polymers.

In general, photodegradation is combined with oxidation leading to the reaction between free radicals and oxygen. Natural PE undergoes a time-dependent photo-oxidation process, in which the relationship between PE bond breakage and exposure time was found to vary linearly [26].

Ultra-violet degradation resistance can be improved by using photostabilizers. Photostabilizers are usually classified as a function of their mode of actions:

- Light screeners (such as painting or pigment addition) – most common method
- UV light absorbers
- Excited state quenchers
- Antioxidants additives.

4.3.3 Electron-beam Radiations

Electron beam radiations are important not only for electrical devices but also for space applications. Indeed, Ushakov [26] mentions that space vehicles are commonly subjected to electrons and ions radiations in the 10–20 keV range. Electron beams were already discussed: they may promote curing but can also induce chain-scission (see Section 4.2.3).

4.3.4 Nuclear Radiations

Similitudes between high temperature and nuclear radiations exposures can be drawn; however, with one exception, nuclear radiation effects are stronger on the surface than inside the material. The literature in this field is rather sparse. A good

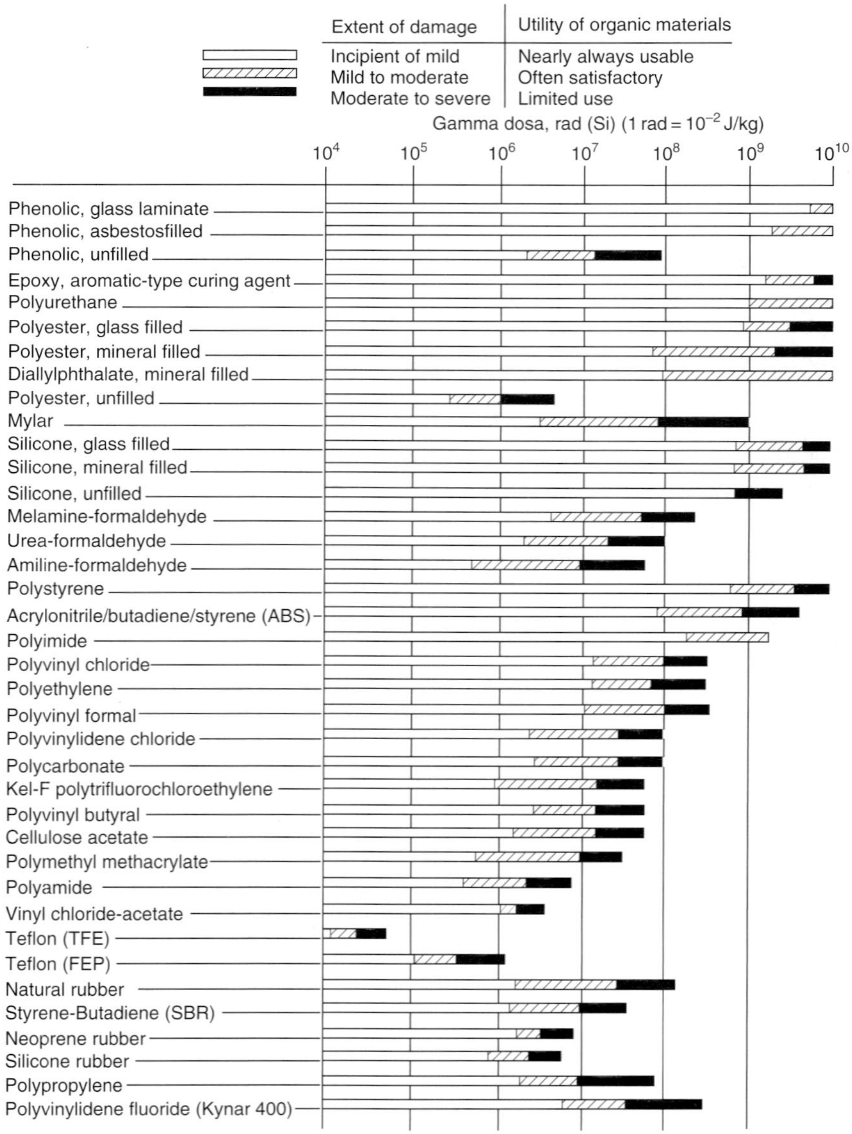

Figure 4.29. Resistance of the indicated materials to γ radiation and their suitability for insulation under different doses. (Copyright 2004, *Insulation of High-Voltage Equipment*, by Ushakov [26], reproduced by permission of John Wiley & Sons.)

summary of radioactive radiation effects on polymers by Bowen and Rosato can, however, be found in the literature [23].

Much alike UV radiations, ionizing radiations can promote the polymer networking and/or lead to chain scission. The two mechanisms are competing and depend upon the nature and rate of the radiation. Degradation in composites under radioactive exposure mainly results from α rays and fast neutrons. Figure 4.29 provides indications about the resistance of common polymer-based materials against different doses of γ radiations.

4.4 TESTING

Radiation norms are not specific to composite materials and are therefore not listed here. A few notes are made on selected electrical experiments, the interpretation of which is directly related to the concepts introduced in the chapter. The summary of standard electrical tests is provided at the end of this section.

4.4.1 High Voltage Test

In a high voltage test, the insulation is subjected for a limited time (typically one minute) to a voltage several times higher than during normal operation. High voltage tests (also called *highpot* or *hipot*) are performed as standard procedures for most insulating materials. Indeed, survival at a higher voltage guaranties a dielectric strength in the new state well in excess of the operational requirement. The voltage level is generally $(2 \times U_n + 1)$ for an AC test and $(2 \times U_n + 1) \times 1.7$ for a DC test. This last test voltage (DC) is however under debate and many manufacturers prefer lower values for qualification tests [17].

4.4.2 Life Endurance Test

In a life endurance test, the material is submitted to a higher voltage (typically $2U_n$) for 500 continuous hours. This test is widely used to qualify generator stator winding bars. It provides indications concerning the quality of the bar based on a minimum remaining life principle. However, this test does not detect damages (such as debonding) that do not alter the functioning of the part.

4.4.3 Loss Tangent (tan δ) Measurement

Tan δ measurements can be very powerful in determining the performance and level of degradation of the insulation (see Section 4.2.2.2). The most common apparatus to measure the electrical losses is the Schering bridge.

4.4.4 Partial Discharge Test

It was mentioned in Section 4.2.3 that absolute values quantifying admissible levels of partial discharges did not exist to date, except for very specific applications such as XLPE cables, in which partial discharges greater than 10 pC would cause them to be ejected [17]. Current commercial apparatus rather focus on detecting changes in partial discharge activities (Figure 4.30).

4.4.5 Related ASTM Norms

Norms are quite abundant on the testing of materials for electrical applications. Only most relevant norms are quoted here (Table 4.6). Table 4.6 is in no way exhaustive. ASTM test standards relevant to rotating insulation materials are further reviewed by Stone et al. [4].

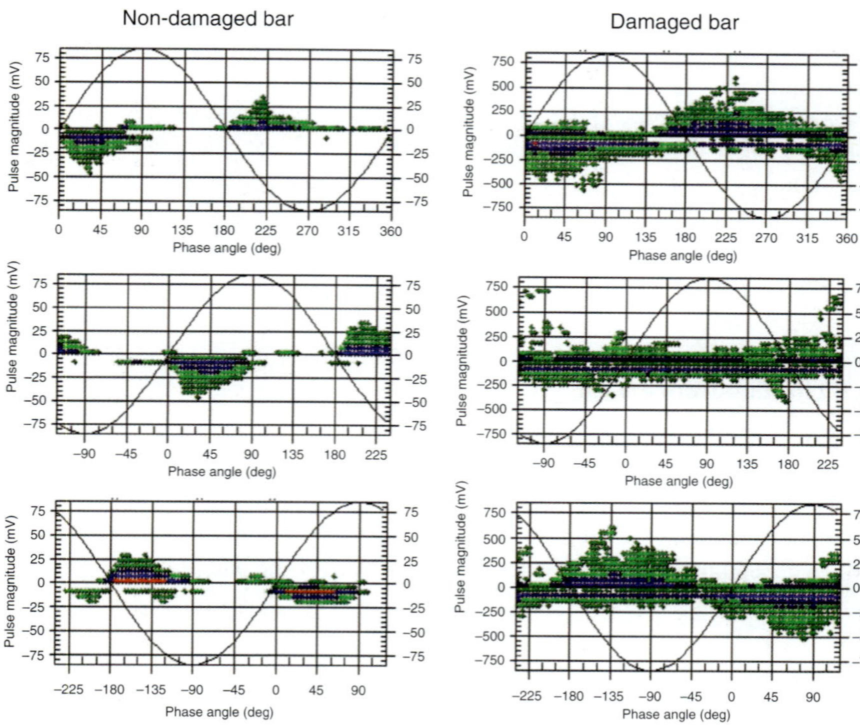

Figure 4.30. Comparison of PD activity in two stators. The stator with higher PD activity (right) is most deteriorated. (Data courtesy of Iris Power.)

Table 4.6. Selected methods for insulation electrical testing

Standard	Designation	Title
IEEE	Std 930-2004	IEEE Guide for the Statistical Analysis of Electrical Insulation Breakdown Data
ASTM	D150-81	Test Methods for AC Loss Characteristics and Permittivity (Dielectric Constant) of Solid Electrical Insulating Materials
ASTM	D1868-81	Method for Detection and Measurement of Partial Discharge (Corona) Pulses in Evaluation of Insulation Systems
ASTM	D2303-85	Test Methods for Liquid-Containment, Inclined Plane Tracking, and Erosion of Insulating Materials
ASTM	D3151-79	Test Methods for Thermal Failure Under Electric Stress of Solid Electrical Insulating Materials
ASTM	D3755-79	Test Method for Dielectric Breakdown Voltage and Dielectric Strength of Solid Electrical Insulating Materials Under Direct-Voltage Stress
ASTM	D4566-98	Standard Test Methods for Electrical Performance Properties of insulations and jackets for telecommunications wire and cable
IEC	270	*Partial Discharge Measurements*
IEC	60	*High Voltage Testing Techniques* (Part 1 and 2).
IEEE	4134-2000	IEEE Trial Use Guide to Measurement of Partial Discharges in Rotating Machinery
IEEE	1310-1996	Recommended Practice for Thermal Cycle Testing of Form-Wound Stator Bars and Coils for Large Generators
EN	50209	Test of Insulation of Bars and Coils of High-Voltage Machines
IECA	E-24-380-2	*Guide for Partial Discharge Measurements*

4.5 TOOL KIT

Topic	Equation	Assumptions	Importance
Basics	Capacitance $C = Q/V$ or $C = \varepsilon A/t$ Permittivity $\varepsilon_r = \varepsilon/\varepsilon_o = C/C_o$ Conductivity $\sigma = 1/\rho = J/E$ and $\sigma(T) = A\exp(-E/\mathrm{kT})$ Polarization $P = \varepsilon_o(\varepsilon_r - 1)E$	None	Electrical definitions
Loss factor	$\tan\delta = \dfrac{\sigma}{\omega\varepsilon_o\varepsilon_r}$ $\tan\delta = \varepsilon''/\varepsilon'$ $\tan\delta = \tan\delta_c + \tan\delta_P$	None	Insulation performance and damage metric
R–C model	$V = R\dfrac{dQ}{dt} + \dfrac{Q}{C}$	R–C series model	Model for curve-fit

(*Continued*)

Topic	Equation	Assumptions	Importance
R–C model	$\frac{dQ}{dt} = C\frac{dU}{dt} + \frac{U}{R}$	R–C parallel model	Model for curve-fit
Cole-Cole plot equation	$\left(\varepsilon' - \frac{\varepsilon_l + \varepsilon_s}{2}\right)^2 + (\varepsilon'')^2 = \left(\frac{\varepsilon_l - \varepsilon_s}{2}\right)^2$	None	Indications on single/multiple dielectric relaxation times
Composite permittivity	$\varepsilon_c = \frac{\varepsilon_m \varepsilon_f}{\varepsilon_m V_m + \varepsilon_f V_f}$	Perfect composite (i.e. bonding)	Composite permittivity estimation from constituents
Composite loss factor	$\tan\delta_c = \frac{\tan\delta_m + (\varepsilon_m V_f/\varepsilon_f V_m)\tan\delta_f}{1 + (\varepsilon_m V_f/\varepsilon_f V_m)}$	Perfect composite (i.e. bonding)	Composite loss factor estimation from constituents

REFERENCES

1. Malik, N.H., A.A. Al-Arainy and M.I. Qureshi, *Electrical Insulation in Power Systems*. Marcel Dekker, Inc., New York, 1998.
2. OMERIN, the cables for hazardous conditions, 2000 Edition.
3. ALSTOM, High Performance Insulating Systems for Hydro Generators. Made by ALSTOM, 2002.
4. Stone, G.C., E.A. Boulter, I. Culbert and H. Dhirani, *Electrical Insulation for Rotating Machines: Design, Evaluation, Aging, Testing, and Repair*, IEEE Press series on Power Engineering, Mohamed E. El-Hawary, Series Editor, 2004.
5. Inderchand Rajgarhia & Sons Ltd Brochure, http://www.icrmica.com/icrmica_physical_properties.html.
6. Beyer, M., W. Boeck, K. Möller and W. Zaengl, *Hochspannungstechnik*. Springer Verlag, 1992.
7. Aklonis, J.J. and W.J. MacKnight, *Introduction to Polymer Viscoelasticity*, 2nd ed. John Wiley & Sons, 1983.
8. Robert, P., *Traite d'Electricite*, Vol. 2. Presses polytechniques romandes, Lausanne, 1989.
9. Debye, P., *Polar Molecules*. Dover Publications, New York, 1945.
10. Onsager, L., Electric moments of molecules in liquids. *Journal of the American Chemical Society*, 1936, 58(8), 1486–1493.
11. Frolich, H., *Theory of Dielectrics*, 2nd ed. Oxford University Press, Oxford, 1958.
12. Cole, R.H. and K.S. Cole, Dispersion and absorption in dielectrics I. Alternating current characteristics. *Journal of Chemical Physics*, 1941, 9, 341.
13. DIN EN 50209, Prüfung der Isolierung von Stäben und Spulen von Hochspanungs-maschinen, VDE 0530 Teil 33, 1998.
14. IEEE STD 1310-1996, IEEE Trial Use Recommended Practice for Thermal Cycle Testing of Form-Wound Stator Bars and Coils for Large Generators, published by the IEEE, New York, 1996.

15. Ieda, M., M. Nagao and M. Hikita, High-field conduction and breakdown in insulating polymers. *IEEE Transactions on Dielectrics and Electrical Insulation*, 1994, 1(5), 934–945.
16. www.matweb.com.
17. Forthergill, J.C., Private communication.
18. Dissado, L.A. and J.C. Forthergill, *Electrical Degradation and Breakdown in Polymers*, Peter Peregrinus Ltd. for the IEE, 1992.
19. Composite poles developed to support power cables. *Reinforced Plastics*, September 2004, p. 6.
20. Gangaro, H. and R. Liang, Opening doors for composite infrastructure. *Composites Technology*, April 2004, p. 6.
21. Pultruded poles carry power. *Reinforced Plastics*, January 2003, 20–24.
22. Fisher Mason, K., Composite on the line. *Composites Technology*, August 2004, 29–33.
23. Bowen, J.H. and D.V. Rosato, Radiation. In D.V. Rosato and R.T. Schwartz (Eds), *Environmental Effects on Polymeric Materials*, Interscience Publishers, 2 vols, New York, 1968.
24. Skinner, W. and J.D. Goldhar, An introduction to the plastic industry. In D.V. Rosato and R.T. Schwartz (Eds), *Environmental Effects on Polymeric Materials*, Interscience Publishers, 2 vols, New York, 1968.
25. Rugger, G.R., Weathering. In D.V. Rosato and R.T. Schwartz (Eds), *Environmental Effects on Polymeric Materials*, Interscience Publishers, 2 vols, New York, 1968.
26. Ushakov, V.Y., *Insulation of High-Voltage Equipment*. Springer-Verlag, Berlin, 2004.
27. Schnabel, W., *Polymer Degradation: Principle and Practical Applications*. Hanser International, 1981.

5

ENVIRONMENTAL IMPACT ON MICROMECHANICAL AND MACROMECHANICAL CALCULATIONS

5.1 INTRODUCTION

The use of light materials is essential in the transportation sector, where a few kilograms saved on the vehicle structure can translate into significantly lower fuel costs. Carbon-fiber materials such as carbon-fiber reinforced polymers also convey a certain technological prestige. The commemorative Edition Z06 Corvette, for example, makes optimized use of carbon fibers for the car's hood. The hood is made of a carbon-fiber/glass sheet molding compound covered by a 100% carbon-fiber epoxy prepreg skin. The carbon-fiber epoxy hood is respectively 50 and 30% lighter than its metallic and glass-fiber reinforced equivalents [1]. The material selection for automotive products is further complicated by the necessity of having a reasonable manufacturing time. For the Z06 hood, a production rate of 16 parts per day was targeted. A TORAY P 3831C-190 prepreg combining 24 K carbon-fiber reinforcement and a *quick cure* epoxy resin necessitating an autoclave curing time below 10 minutes at 150°C was found to be the best alternative in response to the stringent thermal and mechanical requirements on the hood, which included dent and hail, hood slam, deflection, crash and torsion. Like most actual industrial products, the hood geometry is complex. A proper design requires a precise determination of the initial material properties as well as strain and stress calculations under a variety of loads, typically performed by finite element analysis (Figure 5.1). Furthermore, time-dependent environmental effects should be included in order to predict the long-term mechanical response of the part. This last step still remains a challenge for composite specialists worldwide.

The present chapter describes the basis for introducing time and environmentally dependent properties into classical micromechanical and macromechanical models. Indeed, the previous chapters presented the effects of individual environmental conditions on the composite constituents. From the knowledge of the constituent

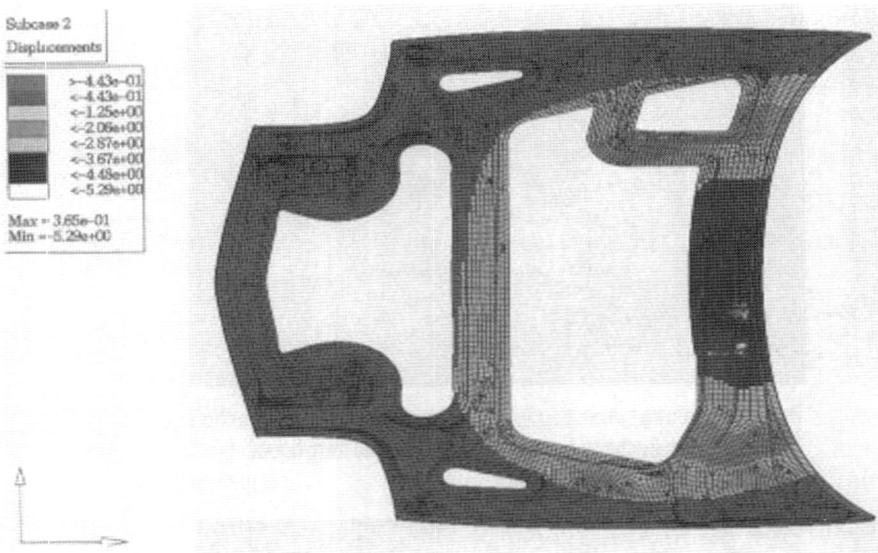

Figure 5.1. Example of a finite element computation result for the corvette hood. Prior to production, load simulations were conducted on the composite hood, including deflection analysis [1]. (Copyright 2004, High Performance Composites, D. Brosius, Ray Publishing.)

properties, it is possible to anticipate (to a certain extent) the properties of the composite.

Let us consider a composite component "m" (e.g. for a matrix) with a property $X_{m\ (t=0)}$ at the time of the initial design ($t = 0$) and a second composite component "f" (for fiber, fiber being used in the broader sense of reinforcement) with a property $X_{f\ (t=0)}$ at the time of the initial design. Let us also assume that the response of the composite is linear elastic (i.e. time dependent but independent of stress level, see Chapter 2, Section 2.6.2). Traditional models of micromechanics enable the estimation (with an accuracy within the limits of the model) of the composite ply properties. However, if the composite is a thin laminate, CLT enables a calculation of the properties and global response (stresses and strains) of the laminate. On the other hand, if the composite is a complex or thick laminate, finite element calculations using $X_{m\ (t=0)}$ and $X_{f\ (t=0)}$ can be performed to model the initial response of the composite part.

We demonstrated in the previous chapters that the properties of the composite constituents might change over time under the influence of environmental conditions. To model the consequences of this evolution, it is proposed to perform micromechanical and macromechanical calculations using the same traditional models, but with altered properties: $X'_{m\ (t>0)}$ and $X'_{f\ (t>0)}$. If the property changes can be expressed analytically as a function of time and environmental condition (thanks to the various models given in the previous chapters), it will be possible to estimate the microscopic and macroscopic response of the composite at all times. This approach is very useful as it can help reduce the number of experiments to be performed and provides a basis for *virtual design* [2]. If the properties are only defined experimentally for some times and environmental conditions, then the microscopic and macroscopic calculations will most likely be performed incrementally.

In the present text, we will restrict our discussion to the most basic and commonly used micromechanical and macromechanical models: references are made to the literature studies presenting the original equations without environmental dependency. The present philosophy and approach can also be adapted in order to integrate time and environmental dependency in the more complex models available in the literature [3,4]. This approach is convenient and pragmatic. However, one should note the following inherent limits:

(a) The results depend heavily on the accuracy of the modeling of the properties evolution of the single constituents with time and environmental conditions. If the initial models do not match experimental data, the final results (stress calculation, lifetime etc.) will be irrelevant.
(b) Microscopic and macroscopic mechanical models have their own limits and degrees of inaccuracy. The limits of the models are clearly summarized and should be considered when analyzing calculations results.
(c) The present chapter mainly deals with linear viscoelastic materials (see Chapter 2, Section 2.6.2 for definition). If the material exhibits a load-dependent creep behavior, the approach of Section 5.3 can lead to erroneous results: the larger the non-linearity of the material, the larger the error in the stress and strain calculations. Indeed, if the non-linearity is small, at any instant, for constant external variables, a solution is valid. It is then possible to construct a series of these solutions and extrapolate between them to get a piecewise linear solution as proposed in Section 5.3. Therefore, if the functions are slowly varying with the parameters, and the non-linearity of dependence is small, the method of Section 5.3 will work only for engineering purposes. However, if the material behavior is not linear (e.g. linear viscoelastic), that solution will be in error, and may be greatly in error [5]. Section 5.3.2 deals more in detail with effects of non-linear viscoelasticity on the composite mechanical response.
(d) The proposed approach only considers the changes in the constituent's interface if the changes in the constituent's interface are analytically modeled and built into the equations. If this step is omitted, one should always keep in mind the probability of a different evolution due to aging of the interface.

This approach only yields *approximate results*. A further discussion on the validity of the approach and a more exact procedure is proposed in Section 5.2.4. As always, it is best to rely on experimental data performed at the macroscopic scale in order to model the global response of a composite part to mechanical and environmental loads. However, as this approach might be too costly and unrealistic in the first step of a project (pre-study, screening), we propose to start with microscopic or individual component properties.

5.2 ENVIRONMENTAL EFFECTS ON SINGLE LAYER COMPOSITES: MICROMECHANICS

5.2.1 Environmental Impact on Micromechanical Calculations of Stiffness

5.2.1.1 Definitions

By definition, composite materials are inhomogeneous at the microscopic scale. The volume fractions are generally used to quantify the degree of reinforcement. The volume fraction of fibers V_f (resp. matrix V_m) is the total fiber volume Vol_f (resp. matrix volume Vol_m) divided by the total composite volume Vol_c:

$$V_f = \frac{Vol_f}{Vol_c} \quad \text{and} \quad V_m = \frac{Vol_m}{Vol_c} \tag{5.1}$$

and by definition:

$$V_f + V_m = 1 \tag{5.2}$$

More rarely, the mass fractions (weight of the constituents W_f and W_m over the total composite weight W_c) are used:

$$M_f = \frac{W_f}{W_c} \quad \text{and} \quad M_m = \frac{W_m}{W_c} \tag{5.3}$$

and here again:

$$M_f + M_m = 1 \tag{5.4}$$

Volume or mass fractions are generally selected at an early stage of the design process. Suppliers typically propose composites with large ranges of reinforcement contents (typically between 15 and 40%) suiting most industrial applications.

It is generally assumed that the volume and weight fractions of the composite remain constant over time. However, if one of the constituents undergoes *significant* dimensional or weight changes with time, these should be accounted for in the calculations. For example, some polymers experience weight losses under high temperature exposure (Chapter 2). In this case, the weight fraction of the polymer decreases with increasing exposure time. However, if the polymer experiences swelling due to combined temperature exposure and moisture absorption (Chapter 3) and if the volume of the fibers remains constant, then the volume fraction of the matrix increases with time. Significant changes should be explicitly considered using analytical models or simple curve fits of experimental data. Most changes can be quantified according to the tools provided in the previous chapters.

As a convention in this book, the dependence upon time and environmental conditions of a given property will be indicated by a t (for time) and e (for environment) and will be placed above the specific property. By this we mean that there is a history of dependence of properties that involve external influences that may alter properties (e.g. humidity leading to hydration). Such changes are slow with time, compared to the stresses and strains due to mechanical loading. In other words, we can always calculate a steady state solution at a given point (slow changes), if we can determine the instantaneous state of the variables and the materials.

In the case of the volume fractions, for example, we will write: $\overbrace{V_\mathrm{f}}^{t,e}$ and $\overbrace{V_\mathrm{m}}^{t,e}$. At all times, Equation (5.2) holds and we therefore can write:

$$\overbrace{V_\mathrm{f}}^{t,e} + \overbrace{V_\mathrm{m}}^{t,e} = 1 \tag{5.5}$$

Homogeneity and isotropy are subjective notions and depend upon the scale considered. Indeed a material is considered isotropic if its properties are the same in all directions in space. Single materials are often considered as isotropic. However, crystallinity in polymers such as modified PE or PP can also develop according to a preferred direction under specific manufacturing conditions [6,7] and lead to microscopic and macroscopic anisotropy. For simplicity, we will neglect such mechanisms at the constituent level and assume here that the matrix and reinforcement are isotropic. For an isotropic material (with time and environmentally dependent properties), the elastic properties are related by

$$\overbrace{G}^{t,e} = \frac{\overbrace{E}^{t,e}}{2(1+\overbrace{\nu}^{t,e})} \tag{5.6}$$

Unfortunately, the situation is more complex for anisotropic materials and such a single equation does not exist. Handling anisotropic materials generally requires the use of two sets of Cartesian coordinates. One set of axes called *materials coordinates* 1, 2, 3 is defined at the ply level. In the case of a polymer matrix reinforced with unidirectional continuous fibers, the 1, 2 and 3 axes are usually taken as the fiber direction, transverse direction and direction perpendicular to the 1-2 plane respectively.

For a short random fiber reinforced composite, properties along the 1 and 2 axis are identical. If the fibers are spread in three dimensions, which is rare, then the resulting ply is isotropic. If the fibers are laid into the 1-2 plane, then the properties along the third axis differ significantly.

To express its directional dependency, a property X can be written in a tensor form $X_{i,j}$ where X is the property of the material along the j-direction in the plane crossed perpendicularly by i (Figure 5.2).

Laminated composites are made of several plies stacked on top of each other. To analyze the laminate, we need to define a second set of axes. By convention, this set is called x, y, z and is referred to as the set of global axes (Figure 5.3).

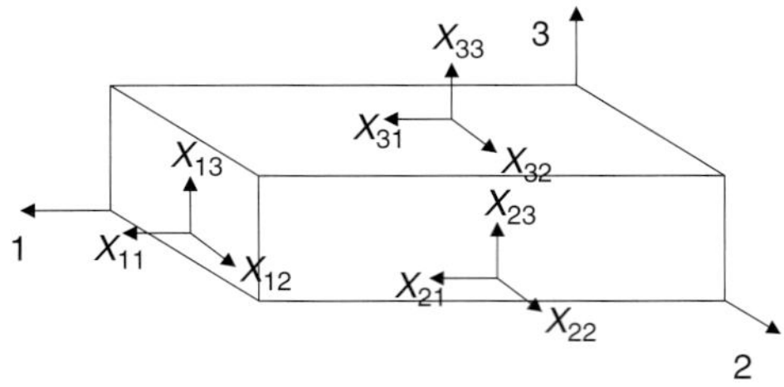

Figure 5.2. Properties and materials axes.

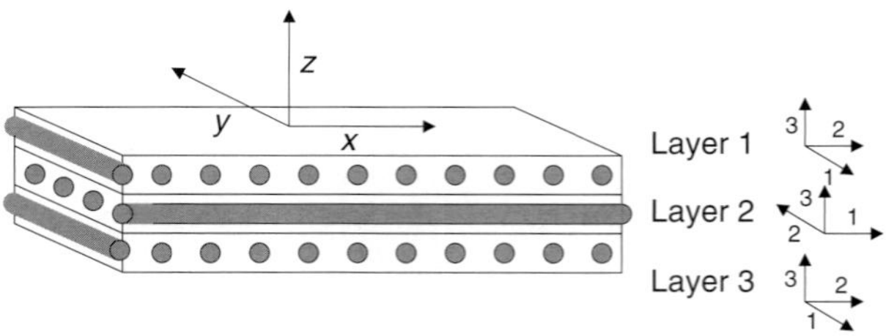

Figure 5.3. Global versus materials coordinates.

Using these conventions, a useful relationship for anisotropic materials can be derived in which the moduli in the (1-2) plane can then be related to the Poisson's ratios:

$$\frac{\overbrace{\nu_{12}}^{t,e}}{\overbrace{E_{11}}^{t,e}} = \frac{\overbrace{\nu_{21}}^{t,e}}{\overbrace{E_{22}}^{t,e}} \tag{5.7}$$

5.2.1.2 Unidirectional composite

For a unidirectional composite, the longitudinal stiffness can quickly be estimated using a rule of mixtures (ROM):

$$\overbrace{E_{11}}^{t,e} = \overbrace{E_{\mathrm{m}}}^{t,e}\overbrace{V_{\mathrm{m}}}^{t,e} + \sum_i \overbrace{E_{\mathrm{f}i}}^{t,e}\overbrace{V_{\mathrm{f}i}}^{t,e} \tag{5.8}$$

where i refers to the number of reinforcement types, m to the matrix and f to the fibers. For a composite made of two constituents (such as carbon fiber/epoxy):

$$\overbrace{E_{11}}^{t,e} = \overbrace{E_{\mathrm{f}}}^{t,e}\overbrace{V_{\mathrm{f}}}^{t,e} + \overbrace{E_{\mathrm{m}}}^{t,e}\overbrace{V_{\mathrm{m}}}^{t,e} = \overbrace{E_{\mathrm{f}}}^{t,e}\overbrace{V_{\mathrm{f}}}^{t,e} + \overbrace{E_{\mathrm{m}}}^{t,e}(1 - \overbrace{V_{\mathrm{f}}}^{t,e}) \tag{5.9}$$

This model assumes perfect bonding between the constituents; in other words, identical strains in the fibers and matrix. On the other hand, assuming identical stresses in the fibers and matrix allows for the use of a rule of mixtures to estimate the transverse modulus of unidirectional composites:

$$\frac{1}{\overbrace{E_{22}}^{t,e}} = \frac{\overbrace{V_{\mathrm{m}}}^{t,e}}{\overbrace{E_{\mathrm{m}}}^{t,e}} + \frac{\overbrace{V_{\mathrm{f}}}^{t,e}}{\overbrace{E_{\mathrm{f}}}^{t,e}} = \frac{1 - \overbrace{V_{\mathrm{f}}}^{t,e}}{\overbrace{E_{\mathrm{m}}}^{t,e}} + \frac{\overbrace{V_{\mathrm{f}}}^{t,e}}{\overbrace{E_{\mathrm{f}}}^{t,e}} \tag{5.10}$$

Equations (5.9) and (5.10) only hold in the case of *ideal* or *perfect* unidirectional composites. However, partial debonding and fiber misalignment are always present in real materials [8] and rules of mixtures only provide estimates for the materials properties.

The more advanced Halpin-Tsai equations [9] are particularly useful to fit real experimental data. The longitudinal modulus (E_{11}) is still calculated according to Equation (5.8), while the transverse properties (E_{22}, G_{12} or ν_{23}) are obtained via Equations (5.11) and (5.12):

$$\overbrace{X}^{t,e} = X_{\mathrm{m}} \left[\frac{1 + \xi \, \eta \overbrace{V_{\mathrm{f}}}^{t,e}}{1 - \eta \overbrace{V_{\mathrm{f}}}^{t,e}} \right] \tag{5.11}$$

and

$$\eta = \frac{\left(\dfrac{\overbrace{X_{\mathrm{f}}}^{t,e}}{\overbrace{X_{\mathrm{m}}}^{t,e}} \right) - 1}{\left(\dfrac{\overbrace{X_{\mathrm{f}}}^{t,e}}{\overbrace{X_{\mathrm{m}}}^{t,e}} \right) + \xi} \tag{5.12}$$

where X is the composite property of interest and the subscripts f and m correspond to the fibers and matrix respectively. ξ is an empirical parameter depending on the geometry of the fiber, the packing geometry, the loading conditions and the state of fiber bonding. It was shown in the previous chapters that exposure of a composite to moisture, for example, can contribute to fiber/matrix debonding. In this case, ξ should vary with water exposure conditions. If the environmental influence on ξ is significant, this aspect should be integrated into the calculations. This integration was not clearly addressed in the literature that focuses more on the influence of fiber geometry. For example, it is known that [4] for perfectly bonded cylindrical fibers $\xi = 2$ and for rectangular fibers $\xi = 2a/b$, with a and b the dimensions of the fibers in the axial and transverse loading directions. We therefore recommend to *absorb* environmental effects into the matrix elastic changes (e.g. $\overbrace{E_{\mathrm{m}}}^{t,e}$) and to keep ξ constant over time.

Example 1: Instantaneous temperature model

Assuming constant fiber properties over the 0–300°C temperature range and a constant reinforcement volume fraction, it is possible to introduce the temperature dependency for the instantaneous response of the material via the model of Equation (2.31):

$$E = \sum_{i=1}^{N} H_i \exp\left(-\left(\frac{T}{T_{\mathrm{ref}_i}}\right)^{\mathrm{m}_i}\right)$$

which gives with the present notation:

$$\overbrace{E_{\mathrm{m}}}^{t,e} = \sum_{i=1}^{N} H_i \exp\left(-\left(\frac{T}{T_{\mathrm{ref}_i}}\right)^{\mathrm{m}_i}\right) \tag{5.13}$$

Equation (5.13) can be combined with Equation (5.9) thus leading to the predictions shown in the following figures for pure crystalline and amorphous PPS (Figure 5.4) and for a carbon-fiber reinforced AS4/PPS composite (Figure 5.5) [10].

Example 2: Use of discrete temperature data

If a model such as Equation 5.13 is not available, it is possible to use discrete data. Let us consider the example of a carbon-fiber reinforced PEEK. We want to anticipate the longitudinal stiffness of the composite in the glassy state (e.g. at room temperature) and in the rubbery stage (above 140°C). Using the mechanical values for the carbon fibers and the PEEK polymer shown in Table 5.1, we can calculate the changes in modulus versus volume fraction by simply using Equations (5.11) and (5.12) for the two temperature regions (Figure 5.6).

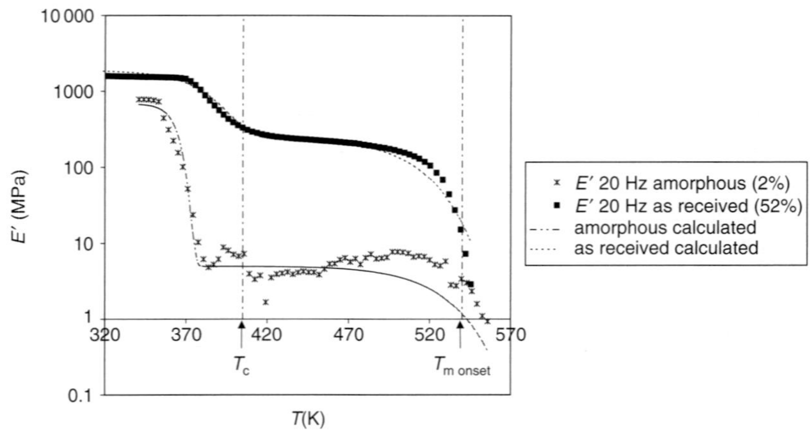

Figure 5.4. Experimental and theoretical variations of the PPS stiffness with temperature as obtained by dynamic mechanical analysis. (Copyright 2001, *Applied Composite Materials*, by C.A. Mahieux et al. [10]. With kind permission of Springer Science and Business Media.)

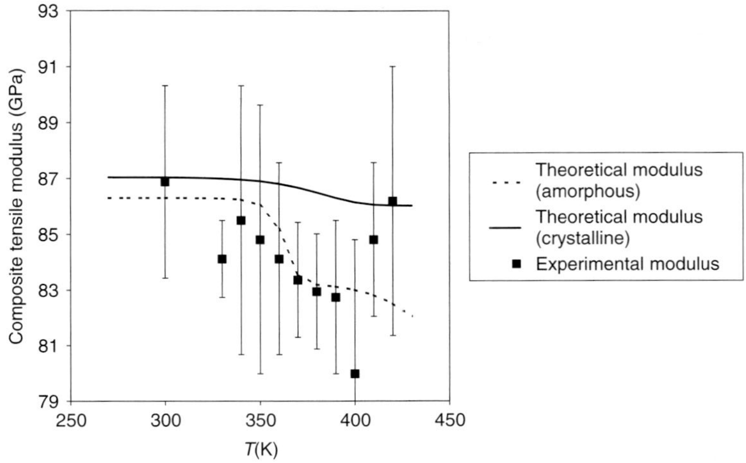

Figure 5.5. Experimental and calculated composite modulus versus temperature for AS4/PPS (from tensile test experiments). (Copyright 2001, *Applied Composite Materials*, by C.A. Mahieux et al. [10]. With kind permission of Springer Science and Business Media.)

Table 5.1. Numerical values for Halpin-Tsai calculations for carbon-fiber PEEK composite

	Glassy state (room temperature)	Rubbery state (160°C)
E_f (GPa)	290 GPa [11]	290 GPa [11]
E_m (GPa)	3 GPa [12]	0.3 GPa
ξ	2	2

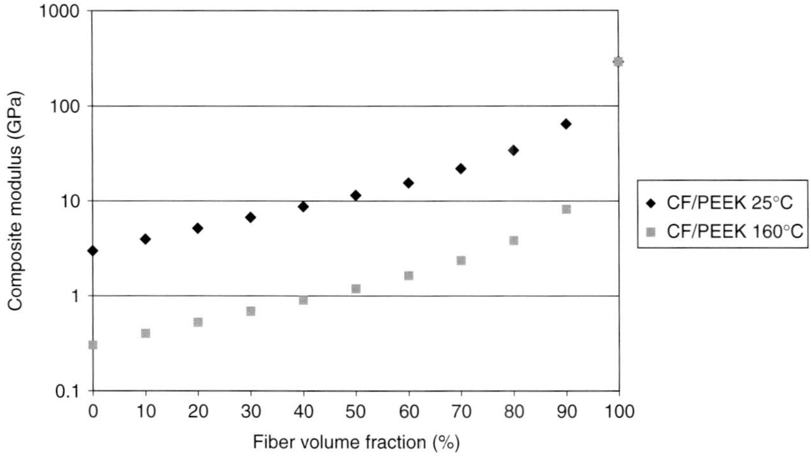

Figure 5.6. Calculated Tensile Modulus E_{11} versus volume fraction at two different temperatures.

The Poisson's ratio (ν_{12}) and in-plane shear modulus G_{12} can also be estimated using simple rules of mixtures, an analog to Equations (5.8) and (5.10):

$$\overbrace{\nu_{12}}^{t,e} = \overbrace{\upsilon_{\mathrm{m}}}^{t,e}\overbrace{V_{\mathrm{m}}}^{t,e} + \sum_{i}\overbrace{\nu_{\mathrm{f}_i}}^{t,e}\overbrace{V_{\mathrm{f}_i}}^{t,e} \tag{5.14}$$

and

$$\frac{1}{\overbrace{G_{12}}^{t,e}} = \frac{\overbrace{V_{\mathrm{m}}}^{t,e}}{\overbrace{G_{\mathrm{m}}}^{t,e}} + \frac{\overbrace{V_{\mathrm{f}}}^{t,e}}{\overbrace{G_{\mathrm{f}}}^{t,e}} = \frac{1-\overbrace{V_{\mathrm{f}}}^{t,e}}{\overbrace{G_{\mathrm{m}}}^{t,e}} + \frac{\overbrace{V_{\mathrm{f}}}^{t,e}}{\overbrace{G_{\mathrm{f}}}^{t,e}}$$

$$= \frac{2\left(1+\overbrace{\nu_{\mathrm{m}}}^{t,e}\right)\left(1-\overbrace{V_{\mathrm{f}}}^{t,e}\right)}{\overbrace{E_{\mathrm{m}}}^{t,e}} + \frac{2\left(1+\overbrace{\nu_{\mathrm{f}}}^{t,e}\right)\overbrace{V_{\mathrm{f}}}^{t,e}}{\overbrace{E_{\mathrm{f}}}^{t,e}} \tag{5.15}$$

More complex and more accurate models are available in the literature [4]. For example, the in-plane shear modulus can be calculated according to the cylindrical assemblage model (CAM) [13]:

$$\overbrace{G_{12}}^{t,e} = \overbrace{G_{\mathrm{m}}}^{t,e}\left[\frac{\left(1+\overbrace{V_{\mathrm{f}}}^{t,e}\right)+\left(1-\overbrace{V_{\mathrm{f}}}^{t,e}\right)\overbrace{G_{\mathrm{m}}}^{t,e}/\overbrace{G_{\mathrm{f}}}^{t,e}}{\left(1-\overbrace{V_{\mathrm{f}}}^{t,e}\right)+\left(1+\overbrace{V_{\mathrm{f}}}^{t,e}\right)\overbrace{G_{\mathrm{m}}}^{t,e}/\overbrace{G_{\mathrm{f}}}^{t,e}}\right] = \left(\frac{\overbrace{E_{\mathrm{m}}}^{t,e}}{2\left(1+\overbrace{V_{\mathrm{m}}}^{t,e}\right)}\right)$$
$$\times\left[\frac{\left(1+\overbrace{V_{\mathrm{f}}}^{t,e}\right)\left(1+\overbrace{\nu_{\mathrm{m}}}^{t,e}\right)\overbrace{E_{\mathrm{f}}}^{t,e}+\left(1-\overbrace{V_{\mathrm{f}}}^{t,e}\right)\left(1+\overbrace{\nu_{\mathrm{f}}}^{t,e}\right)\overbrace{E_{\mathrm{m}}}^{t,e}}{\left(1-\overbrace{V_{\mathrm{f}}}^{t,e}\right)\left(1+\overbrace{\nu_{\mathrm{m}}}^{t,e}\right)\overbrace{E_{\mathrm{f}}}^{t,e}+\left(1+\overbrace{V_{\mathrm{f}}}^{t,e}\right)\left(1+\overbrace{\nu_{\mathrm{f}}}^{t,e}\right)\overbrace{E_{\mathrm{m}}}^{t,e}}\right] \tag{5.16}$$

The in-plane shear modulus can also be evaluated using the periodic microstructure model (PMM) [14]:

$$\overbrace{G_{12}}^{t,e} = \overbrace{G_{\mathrm{m}}}^{t,e}\left[1+\frac{\overbrace{V_{\mathrm{f}}}^{t,e}\left(1-\overbrace{G_{\mathrm{m}}}^{t,e}/\overbrace{G_{\mathrm{f}}}^{t,e}\right)}{\overbrace{G_{\mathrm{m}}}^{t,e}/\overbrace{G_{\mathrm{f}}}^{t,e}+\overbrace{S_3}^{t,e}\left(1-\overbrace{G_{\mathrm{m}}}^{t,e}/\overbrace{G_{\mathrm{f}}}^{t,e}\right)}\right] \tag{5.17}$$

where

$$\overbrace{S_3}^{t,e} = 0.49247 - 0.47603\,\overbrace{V_f}^{t,e} - 0.02748\,\overbrace{V_f^2}^{t,e} \tag{5.18}$$

When required, the interlaminar shear modulus (G_{23}) can be calculated according to SPP (stress partitioning parameter [4,15]) technique:

$$\overbrace{G_{23}}^{t,e} = \overbrace{G_m}^{t,e} \frac{\overbrace{V_f}^{t,e} + \overbrace{\eta_{23}}^{t,e}\left(1 - \overbrace{V_f}^{t,e}\right)}{\overbrace{\eta_{23}}^{t,e}\left(1 - \overbrace{V_f}^{t,e}\right) + \overbrace{V_f}^{t,e}\,\overbrace{G_m}^{t,e} / \overbrace{G_f}^{t,e}} \tag{5.19}$$

with

$$\eta_{23} = \frac{3 - 4\,\overbrace{\nu_m}^{t,e} + \overbrace{G_m}^{t,e} / \overbrace{G_f}^{t,e}}{4\left(1 - \overbrace{\nu_m}^{t,e}\right)} \tag{5.20}$$

However, in most cases [4], it is valid to assume that:

$$\overbrace{G_{13}}^{t,e} = \overbrace{G_{12}}^{t,e} \tag{5.21}$$

which is exact for isotropic materials.

5.2.1.3 Random reinforcement

Due to the great diversity of geometry and fiber types, random reinforcement is very difficult to model in the general case. Nevertheless, it is possible to calculate lower and upper bounds for the elastic properties. For example, the modulus of the composite is always included within the two bounds defined by the axial and transverse rule of mixtures:

$$\frac{\overbrace{E_m}^{t,e}\,\overbrace{E_f}^{t,e}}{\overbrace{V_m}^{t,e}\,\overbrace{E_f}^{t,e} + \overbrace{V_{\nu}}^{t,e}\,\overbrace{E_m}^{t,e}} \le \overbrace{E}^{t,e} \le \overbrace{E_m}^{t,e}\,\overbrace{V_m}^{t,e} + \overbrace{E_f}^{t,e}\,\overbrace{V_f}^{t,e} \tag{5.22}$$

For random reinforced composites, the Halpin-Tsai equations (Equations (5.11) and (5.12)) are applicable as well.

More complex solutions can be derived as proposed by Barbero [4] for a continuous strand mat. In a continuous strand-mat, the continuous fiber rovings are placed randomly in the matrix. The elastic properties of the strand-mat can then be written as:

$$\overbrace{E}^{t,e} = \frac{\overbrace{E_{11}^2}^{t,e} + 4\,\overbrace{E_{11}}^{t,e}\,\overbrace{G_{12}}^{t,e}\,\overbrace{\Delta}^{t,e} + 2\,\overbrace{E_{11}}^{t,e}\,\overbrace{E_{22}}^{t,e} + 8\,\overbrace{\nu_{12}}^{t,e}\,\overbrace{E_{22}}^{t,e}\,\overbrace{G_{12}}^{t,e}\,\overbrace{\Delta}^{t,e} - 4\,\overbrace{\nu_{12}^2}^{t,e}\,\overbrace{E_{22}^2}^{t,e} + 4\,\overbrace{E_{22}}^{t,e}\,\overbrace{G_{12}}^{t,e}\,\overbrace{\Delta}^{t,e} + \overbrace{E_{22}^2}^{t,e}}{\overbrace{\Delta}^{t,e}\left(3\,\overbrace{E_{11}}^{t,e} + 2\,\overbrace{\nu_{12}}^{t,e}\,\overbrace{E_{22}}^{t,e} + 3\,\overbrace{E_{22}}^{t,e} + 4\,\overbrace{G_{12}}^{t,e}\,\overbrace{\Delta}^{t,e}\right)} \tag{5.23}$$

where

$$\overbrace{G}^{t,e} = \frac{\overbrace{E_{11}}^{t,e} - 2\,\overbrace{\nu_{12}}^{t,e}\,\overbrace{E_{22}}^{t,e} + \overbrace{E_{22}}^{t,e} - 4\,\overbrace{G_{12}}^{t,e}\,\overbrace{\Delta}^{t,e}}{8\,\overbrace{\Delta}^{t,e}} \tag{5.24}$$

$$\overbrace{\upsilon}^{t,e} = \frac{\overbrace{E_{11}}^{t,e} + 6\,\overbrace{\nu_{12}}^{t,e}\,\overbrace{E_{22}}^{t,e} + \overbrace{E_{22}}^{t,e} - 4\,\overbrace{G_{12}}^{t,e}\,\overbrace{\Delta}^{t,e}}{3\,\overbrace{E_{11}}^{t,e} + 2\,\overbrace{\nu_{12}}^{t,e}\,\overbrace{E_{22}}^{t,e} + 3\,\overbrace{E_{22}}^{t,e} + 4\,\overbrace{G_{12}}^{t,e}\,\overbrace{\Delta}^{t,e}} \tag{5.25}$$

and

$$\overbrace{\Delta}^{t,e} = 1 - \overbrace{\upsilon_{12}}^{t,e}\,\overbrace{V_{21}}^{t,e} \tag{5.26}$$

with the subscripts 1 and 2 referring to the longitudinal and transverse properties of the equivalent unidirectional composite with same volume fractions.

5.2.2 Environmental Impact on Micromechanical Calculations of Strength

The unidirectional tensile strength in the fiber direction can be very roughly estimated using a rule of mixtures. However, statistical methods such as Bartdorf's model [16–18] are more appropriate to describe the strength of the composite. Modeling the composite transverse strength is even more complex due to a high sensitivity to the quality of bonding and the development of defects in the matrix and the interface. Some empirical models are mentioned in the literature [4,19,20].

However, the literature is rather abundant in models attempting to evaluate the composite compressive strength from the constituent's properties. Undulating

Fiber Models [21], Straight Fiber Models, Microbuckling Models, Kink-band Shear Models [22] and Elastic Fundation models [23] are carefully reviewed by Reifsnider and Case [2].

Introducing time and environmental dependency in the strength models following the approach taken for stiffness is not recommended. Indeed, the calculation of the materials strength depends strongly upon the failure mode of the material. Such failure mechanisms are in turn influenced by the environment. Simply introducing environmental dependence of the constituents into the equations cannot lead to anticipation of failure mode changes. It is therefore recommended to directly measure strength degradation on the composite or at least at the ply level.

5.2.3 Environmental Impact on Micromechanical Calculations of Other Composite Properties

Other materials properties, such as moisture and thermal expansion coefficients, also evolve with environment and can be estimated from the changes in the constituent's properties. Indeed, the coefficients of thermal expansion for a perfectly bonded unidirectional composite can be written as [4,24,25]:

$$\overbrace{\alpha_{11}}^{t,e} = \frac{1}{\overbrace{E_{11}}^{t,e}}\left[\overbrace{\alpha_{\mathrm{f}}}^{t,e}\overbrace{E_{\mathrm{f}}}^{t,e}\overbrace{V_{\mathrm{f}}}^{t,e} + \overbrace{\alpha_{\mathrm{m}}}^{t,e}\overbrace{E_{\mathrm{m}}}^{t,e}\overbrace{V_{\mathrm{m}}}^{t,e}\right] \tag{5.27}$$

and

$$\overbrace{\alpha_{22}}^{t,e} = \left(1+\overbrace{\nu_{\mathrm{f}}}^{t,e}\right)\overbrace{\alpha_{\mathrm{f}}}^{t,e}\overbrace{V_{\mathrm{f}}}^{t,e} + (1+\overbrace{\nu_{\mathrm{m}}}^{t,e})\overbrace{\alpha_{\mathrm{m}}}^{t,e}\overbrace{V_{\mathrm{m}}}^{t,e} - \overbrace{\alpha_{11}}^{t,e}\overbrace{\upsilon_{12}}^{t,e} \tag{5.28}$$

Note that the coefficients of thermal expansions in the 1 and 2 directions are different for carbon fibers (negative in the fiber direction and positive in the transverse direction).

For random fibers based on Barbero [4], one writes:

$$\overbrace{\alpha}^{t,e} = \frac{\overbrace{\alpha_{11}}^{t,e} + \overbrace{\alpha_{22}}^{t,e}}{2} + \frac{\overbrace{\alpha_{11}}^{t,e} - \overbrace{\alpha_{22}}^{t,e}}{2}\,\frac{\overbrace{E_{11}}^{t,e} - \overbrace{E_{22}}^{t,e}}{\overbrace{E_{11}}^{t,e} + \left(1+2\,\overbrace{\upsilon_{21}}^{t,e}\right)\overbrace{E_{22}}^{t,e}} \tag{5.29}$$

Like the random strand mat case, subscripts 11 and 22 refer to the longitudinal and transverse properties of the equivalent unidirectional composite with same volume fractions.

The longitudinal and transverse moisture absorption coefficients can be calculated as well for a unidirectional composite using Equations (5.30) and (5.31) [4,25]:

$$\overbrace{\beta_{11}}^{t,e} = \frac{\overbrace{E_{\mathrm{m}}}^{t,e}\ \overbrace{\rho_{\mathrm{c}}}^{t,e}}{\overbrace{E_{11}}^{t,e}\ \overbrace{\rho_{\mathrm{m}}}^{t,e}} \overbrace{\beta_{\mathrm{m}}}^{t,e} \tag{5.30}$$

and

$$\overbrace{\beta_{22}}^{t,e} = \overbrace{\beta_{33}}^{t,e} = \left[\left(1+\overbrace{\nu_{\mathrm{m}}}^{t,e}\right) - \frac{\overbrace{E_{\mathrm{m}}}^{t,e}}{\overbrace{E_{11}}^{t,e}}\left(\overbrace{\nu_{12}}^{t,e}\right)\right] \frac{\overbrace{\rho_{\mathrm{c}}}^{t,e}}{\overbrace{\rho_{\mathrm{m}}}^{t,e}} \overbrace{\beta_{\mathrm{m}}}^{t,e} \tag{5.31}$$

where the subscripts c and m are for composite and matrix. It is important to remember that the dependency upon environment is tightly related to the nature of all constituents and not only to the matrix properties; for example and as detailed in Chapter 3, organic fibers absorb water (e.g. aramid) and may experience a significant amount of swelling when inorganic fibers retain their original dimensions (glass, carbon).

For a random composite, the moisture absorption can be calculated using an equation analog to Equation (5.29) [4]:

$$\overbrace{\beta}^{t,e} = \frac{\overbrace{\beta_{11}}^{t,e} + \overbrace{\beta_{22}}^{t,e}}{2} + \frac{\overbrace{\beta_{11}}^{t,e} - \overbrace{\beta_{22}}^{t,e}}{2} \frac{\overbrace{E_{11}}^{t,e} - \overbrace{E_{22}}^{t,e}}{\overbrace{E_{11}}^{t,e} + \left(1+2\,\overbrace{\upsilon_{12}}^{t,e}\right)\overbrace{E_{22}}^{t,e}} \tag{5.32}$$

with the subscripts 11 and 22 referring to the longitudinal and transverse properties of the equivalent unidirectional composite with same volume fractions.

The thermal and electrical conductivities also depend upon the environmental influence on the constituents. Rule of mixtures and Halpin-Tsai equations can be used to predict the conductivities of the composites. Hence, we obtain for the longitudinal thermal or electrical conductivities:

$$\overbrace{k_{11}}^{t,e} = \overbrace{k_{\mathrm{f}}}^{t,e}\ \overbrace{V_{\mathrm{f}}}^{t,e} + \overbrace{k_{\mathrm{m}}}^{t,e}\ \overbrace{V_{\mathrm{m}}}^{t,e} \tag{5.33}$$

and the Halpin-Tsai equations result in:

$$\overbrace{k_{2}}^{t,e} = \overbrace{k_{\mathrm{m}}}^{t,e} \left[\frac{1+\overbrace{\xi\eta V_{\mathrm{f}}}^{t,e}}{1-\overbrace{\eta V_{\mathrm{f}}}^{t,e}}\right] \tag{5.34}$$

with

$$\overbrace{\eta}^{t,e} = \frac{\left(\overbrace{k_{\mathrm{f}}}^{t,e} / \overbrace{k_{\mathrm{m}}}^{t,e}\right) - 1}{\left(\overbrace{k_{\mathrm{f}}}^{t,e} / \overbrace{k_{\mathrm{m}}}^{t,e}\right) + \xi} \tag{5.35}$$

5.2.4 Discussion on the Validity of the Approach

The approach previously introduced is only an approximation valid for small deformations in a linear range according to the definition of Green and Zerna [26]. In an industrial context, micromechanics is only used in a screening phase and calculation results should always be validated in the design stage by experimental data. Indeed, the micromechanical formulas were originally developed for elastic materials. Barbero and Luciano [27,28] point out that micromechanical formulas only hold true for linear viscoelastic materials in the Laplace domain [29]. In other words, Equation (5.9) should really be written in the time domain by performing the inverse Laplace transform:

$$E_1(t) = L^{-1}[E_1(s)] \tag{5.36}$$

with

$$E_1(s) = E_{\mathrm{f}}(s)V_{\mathrm{f}} + E_{\mathrm{m}}(s)V_{\mathrm{m}} \tag{5.37}$$

When exact formulas are needed, Barbero [29] proposes to first model the viscoelastic behavior of the matrix with a simple model, such as a four-parameter model (see Chapter 2), then use the Laplace method from above and back-transform to obtain the final micromechanical formulas in the time domain.

5.3 ENVIRONMENTAL IMPACT ON STRESSES AND STRAINS OF COMPOSITE STRUCTURES: MACROMECHANICS

Using micromechanical models (Section 5.2), we have introduced time and environment dependency into the materials properties. Environmentally dependent ply properties can then be used in macromechanical models in order to anticipate the overall laminate response to mechanical loads and environment.

As a first step (Section 5.3.1), we will now introduce methods valid for linear viscoelastic or linear elastic materials (the impact of non-linear materials properties on the mechanical response of the composite are discussed in Section 5.3.2). The well-established CLT, for example, provides convenient analytical tools to calculate stresses and strains in thin laminates. The CLT assumes that a line perpendicular to the middle surface remains straight after deformation and that the normal strain ε_{zz} is null [4]. These two assumptions are met in the case of thin plates.

By contrast, thick plates and complex geometries generally require the use of finite element methods and common commercial composite software and are introduced in Section 5.5.

5.3.1 Thin Plates – CLT

5.3.1.1 Definitions

Most composites are anisotropic and therefore require calculations in all directions of the space. Therefore the stresses in the material are written in a tensor form:

$$[\sigma] = \begin{bmatrix} \sigma_{11} & \sigma_{12} & \sigma_{13} \\ \sigma_{21} & \sigma_{22} & \sigma_{23} \\ \sigma_{31} & \sigma_{32} & \sigma_{33} \end{bmatrix} \tag{5.38}$$

It can easily be demonstrated [3] that only six of the nine original tensor components are independent. To simplify the equations, contracted notations can therefore be used in which:

σ_{11} becomes σ_1
σ_{22} becomes σ_2
σ_{33} becomes σ_3
σ_{23} (or τ_{23}) becomes σ_4
σ_{31} (or τ_{31}) becomes σ_5 and
σ_{12} (or τ_{12}) becomes σ_6.

We can then re-write Equation (5.38):

$$[\sigma] = \begin{Bmatrix} \sigma_1 \\ \sigma_2 \\ \sigma_3 \\ \sigma_4 \\ \sigma_5 \\ \sigma_6 \end{Bmatrix} \tag{5.39}$$

A similar approach can be taken for the nine-coefficient strain tensor:

$$[\varepsilon] = \begin{bmatrix} \varepsilon_{11} & \varepsilon_{12} & \varepsilon_{13} \\ \varepsilon_{21} & \varepsilon_{22} & \varepsilon_{23} \\ \varepsilon_{31} & \varepsilon_{32} & \varepsilon_{33} \end{bmatrix} \tag{5.40}$$

Indeed, the strain tensor can be contracted:

ε_{11} becoming ε_1
ε_{22} becoming ε_2
ε_{33} becoming ε_3
γ_{23} becoming γ_4
γ_{31} becoming γ_5 and
γ_{12} becoming γ_6

leading to:

$$[\varepsilon] = \begin{Bmatrix} \varepsilon_1 \\ \varepsilon_2 \\ \varepsilon_3 \\ \gamma_{23} \\ \gamma_{31} \\ \gamma_{12} \end{Bmatrix} \tag{5.41}$$

To relate stresses and strains, a generalized Hooke's law can be used in which the strains are related to the stresses through a stiffness matrix C. This approach is only valid if the materials behavior is linear. Should this not be the case, the solution will be in error (see Section 5.3.2).

$$\sigma_i = C_{ij}\varepsilon_j \qquad \text{where } i, j = 1 \ldots 6 \tag{5.42}$$

or

$$\begin{Bmatrix} \sigma_1 \\ \sigma_2 \\ \sigma_3 \\ \sigma_4 \\ \sigma_5 \\ \sigma_6 \end{Bmatrix} = \begin{bmatrix} C_{11} & C_{12} & C_{13} & C_{14} & C_{15} & C_{16} \\ C_{21} & C_{22} & C_{23} & C_{24} & C_{25} & C_{26} \\ C_{31} & C_{32} & C_{33} & C_{34} & C_{35} & C_{36} \\ C_{41} & C_{42} & C_{43} & C_{44} & C_{45} & C_{46} \\ C_{51} & C_{52} & C_{53} & C_{54} & C_{55} & C_{56} \\ C_{61} & C_{62} & C_{63} & C_{64} & C_{65} & C_{66} \end{bmatrix} \begin{Bmatrix} \varepsilon_1 \\ \varepsilon_2 \\ \varepsilon_3 \\ \gamma_{23} \\ \gamma_{31} \\ \gamma_{12} \end{Bmatrix}$$

Inversely, the strains can be calculated using the compliance matrix S:

$$\varepsilon_i = S_{ij}\sigma_j \qquad \text{where } i, j = 1 \ldots 6 \tag{5.43}$$

or

$$\begin{Bmatrix} \varepsilon_1 \\ \varepsilon_2 \\ \varepsilon_3 \\ \gamma_{23} \\ \gamma_{31} \\ \gamma_{12} \end{Bmatrix} = \begin{bmatrix} S_{11} & S_{12} & S_{13} & S_{14} & S_{15} & S_{16} \\ S_{21} & S_{22} & S_{23} & S_{24} & S_{25} & S_{26} \\ S_{31} & S_{32} & S_{33} & S_{34} & S_{35} & S_{36} \\ S_{41} & S_{42} & S_{43} & S_{44} & S_{45} & S_{46} \\ S_{51} & S_{52} & S_{53} & S_{54} & S_{55} & S_{56} \\ S_{61} & S_{62} & S_{63} & S_{64} & S_{65} & S_{66} \end{bmatrix} \begin{Bmatrix} \sigma_1 \\ \sigma_2 \\ \sigma_3 \\ \sigma_4 \\ \sigma_5 \\ \sigma_6 \end{Bmatrix}$$

If no symmetry can be identified in the laminate, the material is anisotropic (or triclinic) and the tensors have 21 independent constants. However, composites selected for industrial purposes generally rely on symmetrical lay-ups. Such symmetries in the composite can help reduce the number of independent constants. Indeed, if the laminate has one plane of symmetry (monoclinic), the number of constant is reduced to 13. For two or three symmetry planes, the material is orthotropic with nine independent constants. If the material is isotropic within the plane (transversely isotropic), this number is reduced to five and finally down to two if the material is perfectly isotropic (infinity of symmetry planes). Common compliance and stiffness matrices reductions are summarized in Table 5.2.

Table 5.2. Compliance and stiffness matrices reductions through symmetry

		Anisotropic (21 independent constants)		Monoclinic (13 independent constants)		Orthotropic (9 independent constants)		Transversely isotropic (5 independent constants)		Isotropic (2 independent constants)	
C_{11}	S_{11}										
C_{12}	S_{12}										
C_{13}	S_{13}									$= C_{12}$	$= S_{12}$
C_{14}	S_{14}			$= 0$	$= 0$	$= 0$	$= 0$	$= 0$	$= 0$	$= 0$	$= 0$
C_{15}	S_{15}			$= 0$	$= 0$	$= 0$	$= 0$	$= 0$	$= 0$	$= 0$	$= 0$
C_{16}	S_{16}					$= 0$	$= 0$	$= 0$	$= 0$	$= 0$	$= 0$
C_{21}	S_{21}	$= C_{12}$	$= S_{12}$	$= C_{12}$	$= S_{12}$	$= C_{12}$	$= S_{12}$	$= C_{12}$	$= S_{12}$	$= C_{12}$	$= S_{12}$
C_{22}	S_{22}									$= C_{11}$	$= S_{11}$
C_{23}	S_{23}									$= C_{12}$	$= S_{12}$
C_{24}	S_{24}			$= 0$	$= 0$	$= 0$	$= 0$	$= 0$	$= 0$	$= 0$	$= 0$
C_{25}	S_{25}			$= 0$	$= 0$	$= 0$	$= 0$	$= 0$	$= 0$	$= 0$	$= 0$
C_{26}	S_{26}					$= 0$	$= 0$	$= 0$	$= 0$	$= 0$	$= 0$
C_{31}	S_{31}	$= C_{13}$	$= S_{13}$	$= C_{13}$	$= S_{13}$	$= C_{13}$	$= S_{13}$	$= C_{13}$	$= S_{13}$	$= C_{12}$	$= S_{12}$
C_{32}	S_{32}	$= C_{23}$	$= S_{23}$	$= C_{23}$	$= S_{23}$	$= C_{23}$	$= S_{23}$	$= C_{23}$	$= S_{23}$	$= C_{12}$	$= S_{12}$
C_{33}	S_{33}									$= C_{11}$	$= S_{11}$
C_{34}	S_{34}			$= 0$	$= 0$	$= 0$	$= 0$	$= 0$	$= 0$	$= 0$	$= 0$
C_{35}	S_{35}			$= 0$	$= 0$	$= 0$	$= 0$	$= 0$	$= 0$	$= 0$	$= 0$
C_{36}	S_{36}					$= 0$	$= 0$	$= 0$	$= 0$	$= 0$	$= 0$
C_{41}	S_{41}	$= C_{14}$	$= S_{14}$	$= 0$	$= 0$	$= 0$	$= 0$	$= 0$	$= 0$	$= 0$	$= 0$
C_{42}	S_{42}	$= C_{24}$	$= S_{24}$	$= 0$	$= 0$	$= 0$	$= 0$	$= 0$	$= 0$	$= 0$	$= 0$
C_{43}	S_{43}	$= C_{34}$	$= S_{34}$	$= 0$	$= 0$	$= 0$	$= 0$	$= 0$	$= 0$	$= 0$	$= 0$

C_{44}	S_{44}									$=(C_{11}-C_{12})/2$	$=(S_{11}-S_{12})/2$
C_{45}	S_{45}					$=0$	$=0$	$=0$	$=0$	$=0$	$=0$
C_{46}	S_{46}			$=0$	$=0$	$=0$	$=0$	$=0$	$=0$	$=0$	$=0$
C_{51}	S_{51}	$=C_{15}$	$=S_{15}$	$=0$	$=0$	$=0$	$=0$	$=0$	$=0$	$=0$	$=0$
C_{52}	S_{52}	$=C_{25}$	$=S_{25}$	$=0$	$=0$	$=0$	$=0$	$=0$	$=0$	$=0$	$=0$
C_{53}	S_{53}	$=C_{35}$	$=S_{35}$	$=0$	$=0$	$=0$	$=0$	$=0$	$=0$	$=0$	$=0$
C_{54}	S_{54}	$=C_{45}$	$=S_{45}$	$=C_{45}$	$=S_{45}$	$=0$	$=0$	$=0$	$=0$	$=0$	$=0$
C_{55}	S_{55}							$=C_{44}$	$=S_{44}$	$=(C_{11}-C_{12})/2$	$=(C_{11}-C_{12})/2$
C_{56}	S_{56}			$=0$	$=0$	$=0$	$=0$	$=0$	$=0$	$=0$	$=0$
C_{61}	S_{61}	$=C_{16}$	$=S_{16}$	$=C_{16}$	$=S_{16}$	$=0$	$=0$	$=0$	$=0$	$=0$	$=0$
C_{62}	S_{62}	$=C_{26}$	$=S_{26}$	$=C_{26}$	$=S_{26}$	$=0$	$=0$	$=0$	$=0$	$=0$	$=0$
C_{63}	S_{63}	$=C_{36}$	$=S_{36}$	$=C_{36}$	$=S_{36}$	$=0$	$=0$	$=0$	$=0$	$=0$	$=0$
C_{64}	S_{64}	$=C_{46}$	$=S_{46}$	$=0$	$=0$	$=0$	$=0$	$=0$	$=0$	$=0$	$=0$
C_{65}	S_{65}	$=C_{56}$	$=S_{56}$	$=0$	$=0$	$=0$	$=0$	$=0$	$=0$	$=0$	$=0$
C_{66}	S_{66}							$=(C_{11}-C_{22})/2$	$=2(S_{11}-S_{22})$	$=(C_{11}-C_{12})/2$	$=(C_{11}-C_{12})/2$

Orthotropic laminates are very widely used. For such materials, time and environmental dependency can be introduced via changes in the modulus and Poisson's ratio. The compliance matrix for the orthotropic composite can then be written:

$$
[\overbrace{S}^{t,e}] = \begin{bmatrix}
1/\overbrace{E_{11}}^{t,e} & -\overbrace{\nu_{21}}^{t,e}/\overbrace{E_{22}}^{t,e} & -\overbrace{\nu_{31}}^{t,e}/\overbrace{E_{33}}^{t,e} & 0 & 0 & 0 \\
-\overbrace{\nu_{12}}^{t,e}/\overbrace{E_{11}}^{t,e} & 1/\overbrace{E_{2}}^{t,e} & -\overbrace{\nu_{32}}^{t,e}/\overbrace{E_{33}}^{t,e} & 0 & 0 & 0 \\
-\overbrace{\nu_{13}}^{t,e}/\overbrace{E_{11}}^{t,e} & -\overbrace{\nu_{23}}^{t,e}/\overbrace{E_{2}}^{t,e} & -1/\overbrace{E_{33}}^{t,e} & 0 & 0 & 0 \\
0 & 0 & 0 & 1/\overbrace{G_{23}}^{t,e} & 0 & 0 \\
0 & 0 & 0 & 0 & 1/\overbrace{G_{13}}^{t,e} & 0 \\
0 & 0 & 0 & 0 & 0 & 1/\overbrace{G_{12}}^{t,e}
\end{bmatrix} \tag{5.44}
$$

where E_{ij} is the tensile modulus in the i-direction, $\nu_{ij} = -\varepsilon_j/\varepsilon_i$ and G_{ij} is the shear modulus in the ij planes.

Inversion of the compliance matrix gives the stiffness matrix Q:

$$
[\overbrace{Q}^{t,e}] = \begin{bmatrix}
\dfrac{1-\overbrace{\nu_{23}}^{t,e}\overbrace{\upsilon_{32}}^{t,e}}{\overbrace{E_2}^{t,e}\overbrace{E_3}^{t,e}\overbrace{\Delta}^{t,e}} & \dfrac{\overbrace{\nu_{12}}^{t,e}+\overbrace{\upsilon_{31}}^{t,e}\overbrace{\upsilon_{13}}^{t,e}}{\overbrace{E_1}^{t,e}\overbrace{E_3}^{t,e}\overbrace{\Delta}^{t,e}} & \dfrac{\overbrace{\nu_{13}}^{t,e}+\overbrace{\upsilon_{12}}^{t,e}\overbrace{\upsilon_{213}}^{t,e}}{\overbrace{E_1}^{t,e}\overbrace{E_2}^{t,e}\overbrace{\Delta}^{t,e}} & 0 & 0 & 0 \\
\dfrac{\overbrace{\nu_{12}}^{t,e}+\overbrace{\upsilon_{31}}^{t,e}\overbrace{\upsilon_{13}}^{t,e}}{\overbrace{E_1}^{t,e}\overbrace{E_3}^{t,e}\overbrace{\Delta}^{t,e}} & \dfrac{1-\overbrace{\upsilon_{13}}^{t,e}\overbrace{\upsilon_{31}}^{t,e}}{\overbrace{E_1}^{t,e}\overbrace{E_3}^{t,e}\overbrace{\Delta}^{t,e}} & \dfrac{\overbrace{\upsilon_{23}}^{t,e}+\overbrace{\upsilon_{21}}^{t,e}\overbrace{\upsilon_{13}}^{t,e}}{\overbrace{E_1}^{t,e}\overbrace{E_2}^{t,e}\overbrace{\Delta}^{t,e}} & 0 & 0 & 0 \\
\dfrac{\overbrace{\nu_{13}}^{t,e}+\overbrace{\upsilon_{12}}^{t,e}\overbrace{\upsilon_{213}}^{t,e}}{\overbrace{E_1}^{t,e}\overbrace{E_2}^{t,e}\overbrace{\Delta}^{t,e}} & \dfrac{\overbrace{\upsilon_{23}}^{t,e}+\overbrace{\upsilon_{21}}^{t,e}\overbrace{\upsilon_{13}}^{t,e}}{\overbrace{E_1}^{t,e}\overbrace{E_2}^{t,e}\overbrace{\Delta}^{t,e}} & \dfrac{1-\overbrace{\upsilon_{12}}^{t,e}\overbrace{\upsilon_{21}}^{t,e}}{\overbrace{E_1}^{t,e}\overbrace{E_2}^{t,e}\overbrace{\Delta}^{t,e}} & 0 & 0 & 0 \\
0 & 0 & 0 & \overbrace{G_{23}}^{t,e} & 0 & 0 \\
0 & 0 & 0 & 0 & \overbrace{G_{31}}^{t,e} & 0 \\
0 & 0 & 0 & 0 & 0 & \overbrace{G_{12}}^{t,e}
\end{bmatrix} \tag{5.45}
$$

Let us now consider the orthotropic lamina in the 1-2 plane. In this case, Equations (5.46) can help simplify the relationships:

$$\sigma_3 = 0,\ \sigma_4 = 0,\ \sigma_5 = 0,\ \varepsilon_3 = S_{13}\sigma_1 + S_{23}\sigma_2,\ \gamma_{23} = 0,\ \gamma_{31} = 0 \tag{5.46}$$

The resulting stress–strain relationships can be established using Equations (5.47) and (5.48):

$$\begin{Bmatrix} \overbrace{\varepsilon_1}^{t,e} \\ \overbrace{\varepsilon_2}^{t,e} \\ \overbrace{\gamma_{12}}^{t,e} \end{Bmatrix} = \begin{bmatrix} \overbrace{S_{11}}^{t,e} & \overbrace{S_{12}}^{t,e} & 0 \\ \overbrace{S_{12}}^{t,e} & \overbrace{S_{22}}^{t,e} & 0 \\ 0 & 0 & \overbrace{S_{66}}^{t,e} \end{bmatrix} \begin{Bmatrix} \sigma_1 \\ \sigma_2 \\ \sigma_6 \end{Bmatrix} = \begin{bmatrix} 1/\overbrace{E_{11}}^{t,e} & -\upsilon_{21}/\overbrace{E_{22}}^{t,e} & 0 \\ -\upsilon_{21}/\overbrace{E_{22}}^{t,e} & 1/\overbrace{E_{22}}^{t,e} & 0 \\ 0 & 0 & 1/\overbrace{G_{12}}^{t,e} \end{bmatrix} \begin{Bmatrix} \sigma_1 \\ \sigma_2 \\ \sigma_6 \end{Bmatrix} \tag{5.47}$$

and

$$\begin{Bmatrix} \overbrace{\sigma_1}^{t,e} \\ \overbrace{\sigma_2}^{t,e} \\ \overbrace{\sigma_6}^{t,e} \end{Bmatrix} = \begin{bmatrix} \overbrace{Q_{11}}^{t,e} & \overbrace{Q_{12}}^{t,e} & 0 \\ \overbrace{Q_{12}}^{t,e} & \overbrace{Q_{22}}^{t,e} & 0 \\ 0 & 0 & \overbrace{Q_{66}}^{t,e} \end{bmatrix} \begin{Bmatrix} \varepsilon_1 \\ \varepsilon_2 \\ \gamma_{12} \end{Bmatrix} = \begin{bmatrix} \dfrac{\overbrace{E_{11}}^{t,e}}{1 - \overbrace{\nu_{12}}^{t,e}\overbrace{\upsilon_{21}}^{t,e}} & \dfrac{\overbrace{\upsilon_{21}}^{t,e}\overbrace{E_{11}}^{t,e}}{1 - \overbrace{\nu_{12}}^{t,e}\overbrace{\upsilon_{21}}^{t,e}} & 0 \\ \dfrac{\overbrace{\upsilon_{21}}^{t,e}\overbrace{E_{11}}^{t,e}}{1 - \overbrace{\nu_{12}}^{t,e}\overbrace{\upsilon_{21}}^{t,e}} & \dfrac{\overbrace{E_{22}}^{t,e}}{1 - \overbrace{\nu_{12}}^{t,e}\overbrace{\upsilon_{21}}^{t,e}} & 0 \\ 0 & 0 & \overbrace{G_{12}}^{t,e} \end{bmatrix} \begin{Bmatrix} \varepsilon_1 \\ \varepsilon_2 \\ \gamma_{12} \end{Bmatrix} \tag{5.48}$$

For multilayered structures, the response has to be calculated in the global coordinate system that rarely coincides with the materials coordinate system. Let us call θ the angle from the x-axis to the 1 axis. The compliance and stiffness matrices have to be transformed to account for the fiber orientations. The transformed compliance and stiffness matrices are obtained using the transformation matrix T and the Reuter's matrix R [30]:

$$[T] = \begin{bmatrix} \cos^2\theta & \sin^2\theta & 2\cos\theta\sin\theta \\ \sin^2\theta & \cos^2\theta & -2\cos\theta\sin\theta \\ -\cos\theta\sin\theta & \cos\theta\sin\theta & \cos^2\theta - \sin^2\theta \end{bmatrix} \tag{5.49}$$

and

$$[R] = \begin{bmatrix} 1 & 0 & 0 \\ 0 & 1 & 0 \\ 0 & 0 & 2 \end{bmatrix} \tag{5.50}$$

The relationships between transformed and untransformed quantities are:

$$[\overline{S}] = [T]^{-1}[S][R][T][R]^{-1} \tag{5.51}$$

$$[\overline{Q}] = [T]^{-1}[Q][R][T][R]^{-1} \tag{5.52}$$

And we can now calculate the stress–strain relationships:

$$\begin{Bmatrix} \overbrace{\varepsilon_x}^{t,e} \\ \overbrace{\varepsilon_y}^{t,e} \\ \overbrace{\gamma_{xy}}^{t,e} \end{Bmatrix} = \begin{bmatrix} \overbrace{\overline{S}}^{t,e} \end{bmatrix} \begin{Bmatrix} \sigma_x \\ \sigma_y \\ \tau_{xy} \end{Bmatrix} \tag{5.53}$$

and

$$\begin{Bmatrix} \overbrace{\sigma_x}^{t,e} \\ \overbrace{\sigma_y}^{t,e} \\ \overbrace{\tau_{xy}}^{t,e} \end{Bmatrix} = \begin{bmatrix} \overbrace{\overline{Q}}^{t,e} \end{bmatrix} \begin{Bmatrix} \varepsilon_x \\ \varepsilon_y \\ \gamma_{xy} \end{Bmatrix} \tag{5.54}$$

5.3.1.2 *Calculation of the laminae macroscopic properties*

We are now able to introduce time and environmental conditions into stress–strain relationships for a lamina. It can also be useful to calculate the apparent properties of the composite part as a whole in the global coordinate system when stressed. Those properties are given for an orthotropic lamina by Jones [3]. We simply introduce time and environment dependence as before:

$$\frac{1}{\underbrace{E_x}_{t,e}} = \frac{1}{\underbrace{E_{11}}_{t,e}}\cos^4\theta + \left(\frac{1}{\underbrace{G_{12}}_{t,e}} - \frac{2\overbrace{\nu_{12}}^{t,e}}{\underbrace{E_{11}}_{t,e}}\right)\sin^2\theta\cos^2\theta + \frac{1}{\underbrace{E_{22}}_{t,e}}\sin^4\theta \tag{5.55}$$

$$\frac{1}{\underbrace{E_y}_{t,e}} = \frac{1}{\underbrace{E_{11}}_{t,e}}\sin^4\theta + \left(\frac{1}{\underbrace{G_{12}}_{t,e}} - \frac{2\overbrace{\nu_{12}}^{t,e}}{\underbrace{E_{11}}_{t,e}}\right)\sin^2\theta\cos^2\theta + \frac{1}{\underbrace{E_{22}}_{t,e}}\cos^4\theta \tag{5.56}$$

$$\frac{1}{\underbrace{G_{xy}}_{t,e}} = 2\left(\frac{2}{\underbrace{E_{11}}_{t,e}} + \frac{2}{\underbrace{E_{22}}_{t,e}} + \frac{4\overbrace{\nu_{12}}^{t,e}}{\underbrace{E_{11}}_{t,e}} - \frac{1}{\underbrace{G_{12}}_{t,e}}\right)\sin^2\theta\cos^2\theta + \frac{1}{\underbrace{G_{12}}_{t,e}}(\sin^4\theta + \cos^4\theta) \tag{5.57}$$

$$\overbrace{\upsilon_{xy}}^{t,e} = \overbrace{E_x}^{t,e}\left[\frac{\overbrace{\upsilon_{12}}^{t,e}}{\overbrace{E_{11}}^{t,e}}(\sin^4\theta+\cos^4\theta)-\left(\frac{1}{\overbrace{E_{11}}^{t,e}}+\frac{1}{\overbrace{E_{22}}^{t,e}}-\frac{1}{\overbrace{G_{12}}^{t,e}}\right)\sin^2\theta\cos^2\theta\right] \tag{5.58}$$

$$\overbrace{\eta_{xy,x}}^{t,e} = \overbrace{E_x}^{t,e}\left[\left(\frac{2}{\overbrace{E_{11}}^{t,e}}+\frac{2\,\overbrace{\upsilon_{12}}^{t,e}}{\overbrace{E_{11}}^{t,e}}-\frac{1}{\overbrace{G_{12}}^{t,e}}\right)\sin\theta\cos^3\theta - \left(\frac{2}{\overbrace{E_{22}}^{t,e}}+\frac{2\,\overbrace{\upsilon_{12}}^{t,e}}{\overbrace{E_{11}}^{t,e}}-\frac{1}{\overbrace{G_{12}}^{t,e}}\right)\sin^3\theta\cos\theta\right] \tag{5.59}$$

$$\overbrace{\eta_{xy,y}}^{t,e} = \overbrace{E_y}^{t,e}\left[\left(\frac{2}{\overbrace{E_{11}}^{t,e}}+\frac{2\,\overbrace{\upsilon_{12}}^{t,e}}{\overbrace{E_{11}}^{t,e}}-\frac{1}{\overbrace{G_{12}}^{t,e}}\right)\sin^3\theta\cos\theta - \left(\frac{2}{\overbrace{E_{22}}^{t,e}}+\frac{2\,\overbrace{\upsilon_{12}}^{t,e}}{\overbrace{E_{11}}^{t,e}}-\frac{1}{\overbrace{G_{12}}^{t,e}}\right)\sin\theta\cos^3\theta\right] \tag{5.60}$$

$\eta_{xy,x}$ and $\eta_{xy,y}$ being the coefficients of mutual influence of the second kind, characterizing the shearing in the xy plane caused by a normal stress in the x- and y-direction respectively [3].

5.3.1.3 Laminate stresses and strains (CLT)

In order to calculate the stresses and strains in each ply of the composite, we will now consider a laminate with N layers (Figure 5.7). For each layer k, Equation (5.54) becomes:

$$\overbrace{\{\sigma\}_k}^{t,e} = \overbrace{\left|\overline{Q}\right|_k}^{t,e}\{\varepsilon\} \tag{5.61}$$

or

$$\begin{Bmatrix}\overbrace{\sigma_x}^{t,e}\\ \overbrace{\sigma_y}^{t,e}\\ \overbrace{\tau_{xy}}^{t,e}\end{Bmatrix}_k = \begin{bmatrix}\overbrace{Q_{11}}^{t,e} & \overbrace{Q_{11}}^{t,e} & \overbrace{Q_{11}}^{t,e}\\ \overbrace{Q_{11}}^{t,e} & \overbrace{Q_{11}}^{t,e} & \overbrace{Q_{11}}^{t,e}\\ \overbrace{Q_{11}}^{t,e} & \overbrace{Q_{11}}^{t,e} & \overbrace{Q_{11}}^{t,e}\end{bmatrix}\left\{\begin{Bmatrix}\varepsilon_x^0\\ \varepsilon_{yx}^0\\ \gamma_{xy}^0\end{Bmatrix}+z\begin{Bmatrix}\kappa_x\\ \kappa_{yx}\\ \kappa_{xy}\end{Bmatrix}\right\} \tag{5.62}$$

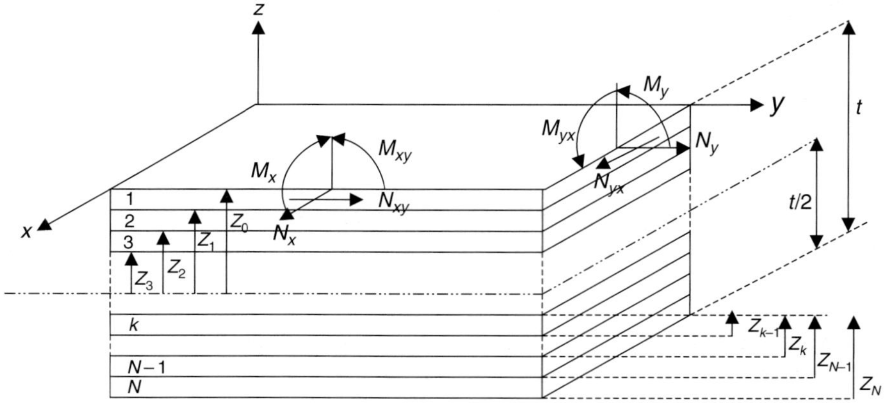

Figure 5.7. Laminate notations.

The resulting forces (N_x, N_y, N_{xy}) and moments (M_x, M_y, M_{xy}) on a thin laminate can be related to the mid-plane strains ε_x^{o}, ε_y^{o}, γ_{xy}^{o} and the curvatures κ_x, κ_y, κ_{xy} by:

$$\overbrace{\begin{Bmatrix} N_x \\ N_y \\ N_{xy} \\ M_x \\ M_y \\ M_{xy} \end{Bmatrix}}^{t,e} = \overbrace{\left[\begin{array}{ccc|ccc} A_{11} & A_{12} & A_{13} & B_{11} & B_{12} & B_{13} \\ A_{21} & A_{22} & A_{23} & B_{21} & B_{22} & B_{23} \\ A_{31} & A_{32} & A_{33} & B_{31} & B_{32} & B_{23} \\ \hline B_{11} & B_{12} & B_{13} & D_{11} & D_{12} & D_{13} \\ B_{21} & B_{22} & B_{23} & D_{12} & D_{22} & D_{23} \\ B_{31} & B_{32} & B_{33} & D_{13} & D_{23} & D_{33} \end{array}\right]}^{t,e} \begin{Bmatrix} \varepsilon_x^0 \\ \varepsilon_y^0 \\ \gamma_{xy}^0 \\ \kappa_x \\ \kappa_y \\ \kappa_{xy} \end{Bmatrix} \tag{5.63}$$

A is the in-plane stiffness matrix:

$$\overbrace{A_{ij}}^{t,e} = \sum_{k=1}^{N} \overbrace{(\overline{Q}_{ij})_k}^{t,e} (z_k - z_{k-1}) = \sum_{k=1}^{N} \overbrace{(\overline{Q}_{ij})_k}^{t,e} t_k \qquad i,\ j = 1,\ 2,\ 6 \tag{5.64}$$

and B is the bending-extension coupling matrix:

$$\overbrace{B_{ij}}^{t,e} = \frac{1}{2} \sum_{k=1}^{N} \overbrace{(\overline{Q}_{ij})_k}^{t,e} (z_k^2 - z_{k-1}^2) = \sum_{k=1}^{N} \overbrace{(\overline{Q}_{ij})_k}^{t,e} t_k \bar{z}_k \qquad i,\ j = 1,\ 2,\ 6 \tag{5.65}$$

where $\bar{z}_k$ is the coordinate along the z-axis of the layer k (Figure 5.7).

Finally D is the bending stiffness matrix:

$$\overbrace{D_{ij}}^{t,e} = \frac{1}{3}\sum_{k=1}^{N}\overbrace{(\overline{Q}_{ij})_k}^{t,e}(z_k^3 - z_{k-1}^3) = \sum_{k=1}^{N}\overbrace{(\overline{Q}_{ij})_k}^{t,e}\left(t_k^{-2}z_k + \frac{t_k^3}{12}\right) \qquad i,\ j = 1,\ 2,\ 6 \tag{5.66}$$

Practically, the coupling matrix B leads to behaviors specific for unbalanced composite structures such as bend-twist coupling. The following case study illustrates the use of such deformation responses for blade technology.

Case study: Bend-twist coupling for blade technology

Non-symmetrical laminates might exhibit combined bending and twisting deformations under load. Such deformations can be permanent, for example, after curing of an unsymmetrical laminate. Figure 5.8 shows an *intelligent* vane prototype for air flow guidance in ventilation systems (EU patent application EP 1 239 151). The vane is a [−45°/45°] long carbon-fiber reinforced PEEK laminate equipped with shape memory alloy wires on both sides. The laminate is characterized by two stable positions. The composite plate can be flipped from one stable position to the other by actuation of the shape memory alloy wires. Indeed, passing current through the proper wire induces shrinkage. The force thus created is sufficient to initiate the necessary twist in the composite. The plate then reaches its second stable position.

Proper design can ensure that such deformations are fully reversible (i.e. elastic). This composite feature can greatly benefit blade technologies, where the optimal twist of the rotating blade varies depending upon the rotation speed. Non-symmetric laminates provide a passive solution for quick response to transient loading, alternative to active pitch control (motor actuated). For example, piezoelectric assisted

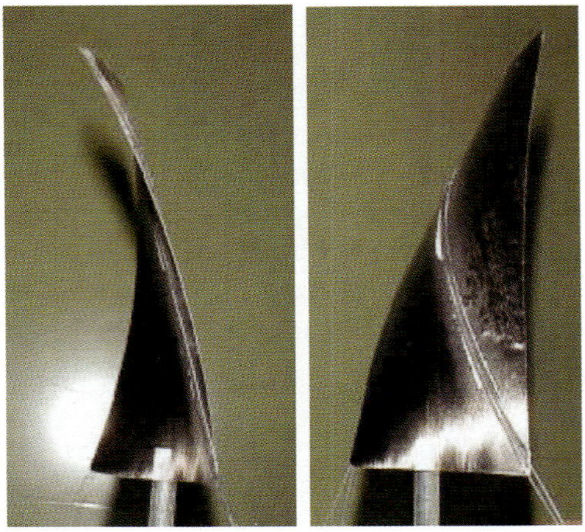

Figure 5.8. Bi-stable [−45°/45°] carbon-fiber PEEK laminate actuated by shape memory alloy wires. (Courtesy of ABB.)

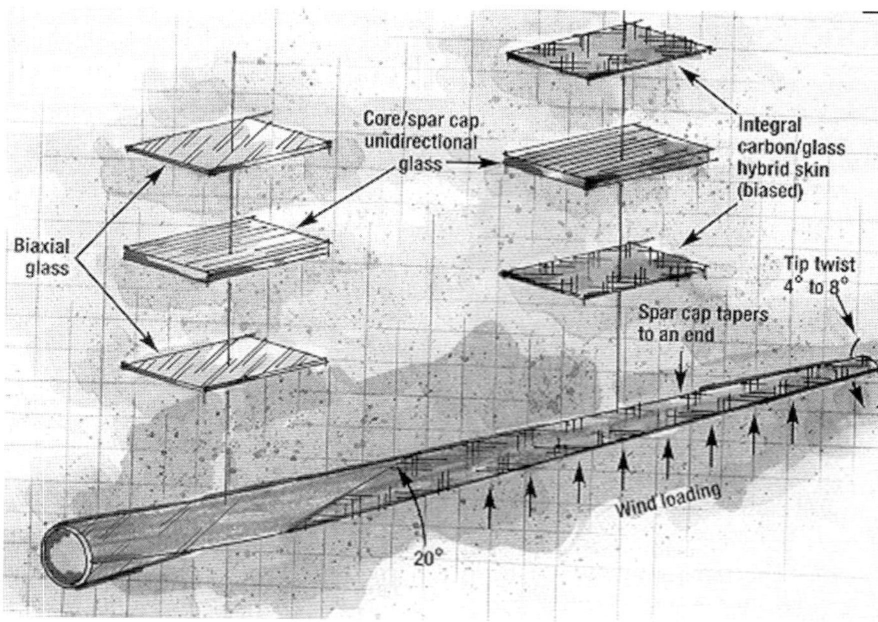

Figure 5.9. Schematic diagram of an adaptive twist-coupled blade [35]. (Copyright 2004, *High Performance Composites*, by K. Fisher Mason, reproduced by permission of Ray Publishing.)

twist coupling was demonstrated to show great potential in improving helicopter blade performance [31–34]. Recent studies also focus on using unbalanced laminates for large windmills in order to optimize the individual response of the blades. Global Energy Concepts LLC, for example, aimed at reducing the cost of energy from a 1.5 MW wind-turbine by 2% using a carbon-glass hybrid adaptive composite blade. The selected lay-up involved a triaxial fabric, a balsa core covered by another layer of triaxial fabric and an external random-mat layer (Figure 5.9). Finite Element calculations using the commercial software ANSYS evidenced that a 20° bias (unbalanced carbon reinforcement) was the best compromise for optimized twist-coupling and sufficient stiffness. The analysis of the 9 m blade predicted a variable twist angle up to 8° translating into a fatigue load reduction between 6 and 10% [35]. However, due to the high level of loads and the large manufacturing dimensions, such technology is still experimental for large windmills.

5.3.1.4 Thermal and moisture stresses

Chapters 2–4 presented examples of the influence of the environment on constituent properties. For example, an increase in temperature was shown to generally reduce the polymers stiffness. So far we have introduced this dependency in the microscopic and macroscopic calculations. The environment will further influence the stresses and strains due to the different anisotropic expansions and contractions of the plies. We therefore need to include this additional aspect into our stress–strain calculations. The equations to perform this operation are well known [3,4]. We just need to introduce time and environment dependency for the properties themselves.

Specifically, the strains can be calculated in the materials and global coordinate systems considering thermal (α) and moisture (β) expansion effects:

$$\begin{Bmatrix} \overbrace{\varepsilon_1}^{t,e} \\ \overbrace{\varepsilon_2}^{t,e} \\ \overbrace{\gamma_{12}}^{t,e} \end{Bmatrix} = \begin{bmatrix} \overbrace{S_{11}}^{t,e} & \overbrace{S_{12}}^{t,e} & 0 \\ \overbrace{S_{12}}^{t,e} & \overbrace{S_{22}}^{t,e} & 0 \\ 0 & 0 & \overbrace{S_{66}}^{t,e} \end{bmatrix} \begin{Bmatrix} \sigma_1 \\ \sigma_2 \\ \sigma_6 \end{Bmatrix} + \begin{Bmatrix} \overbrace{\alpha_{11}}^{t,e} \\ \overbrace{\alpha_{22}}^{t,e} \\ 0 \end{Bmatrix} \Delta T + \begin{Bmatrix} \overbrace{\beta_{11}}^{t,e} \\ \overbrace{\beta_{22}}^{t,e} \\ 0 \end{Bmatrix} \Delta m \tag{5.67}$$

and

$$\begin{Bmatrix} \overbrace{\varepsilon_x}^{t,e} \\ \overbrace{\varepsilon_y}^{t,e} \\ \overbrace{\gamma_{xy}}^{t,e} \end{Bmatrix} = \begin{bmatrix} \overbrace{\overline{S}_{11}}^{t,e} & \overbrace{\overline{S}_{12}}^{t,e} & \overbrace{\overline{S}_{16}}^{t,e} \\ \overbrace{\overline{S}_{12}}^{t,e} & \overbrace{\overline{S}_{22}}^{t,e} & \overbrace{\overline{S}_{26}}^{t,e} \\ \overbrace{\overline{S}_{16}}^{t,e} & \overbrace{\overline{S}_{26}}^{t,e} & \overbrace{\overline{S}_{66}}^{t,e} \end{bmatrix} \begin{Bmatrix} \sigma_x \\ \sigma_y \\ \sigma_{xy} \end{Bmatrix} + \begin{Bmatrix} \overbrace{\alpha_x}^{t,e} \\ \overbrace{\alpha_y}^{t,e} \\ \overbrace{\alpha_{xy}}^{t,e} \end{Bmatrix} \Delta T + \begin{Bmatrix} \overbrace{\beta_x}^{t,e} \\ \overbrace{\beta_y}^{t,e} \\ \overbrace{\beta_{xy}}^{t,e} \end{Bmatrix} \Delta m \tag{5.68}$$

where

$$\begin{Bmatrix} \overbrace{\alpha_x}^{t,e} \\ \overbrace{\alpha_y}^{t,e} \\ \frac{1}{2}\overbrace{\alpha_{xy}}^{t,e} \end{Bmatrix} = [T]^{-1} \begin{Bmatrix} \overbrace{\alpha_{11}}^{t,e} \\ \overbrace{\alpha_{22}}^{t,e} \\ 0 \end{Bmatrix} \tag{5.69}$$

and

$$\begin{Bmatrix} \overbrace{\beta_x}^{t,e} \\ \overbrace{\beta_y}^{t,e} \\ \frac{1}{2}\overbrace{\beta_{xy}}^{t,e} \end{Bmatrix} = [T]^{-1} \begin{Bmatrix} \overbrace{\beta_{11}}^{t,e} \\ \overbrace{\beta_{22}}^{t,e} \\ 0 \end{Bmatrix} \tag{5.70}$$

For example, the environmental dependence of the coefficients of thermal expansion can be introduced in Equation (5.67) with the Equations (5.27) and (5.28):

$$\overbrace{\alpha_{11}}^{t,e} = \frac{1}{\overbrace{E_{11}}^{t,e}} \left[\overbrace{\alpha_f}^{t,e} \overbrace{E_f}^{t,e} \overbrace{V_f}^{t,e} + \overbrace{\alpha_m}^{t,e} \overbrace{E_m}^{t,e} \overbrace{V_m}^{t,e} \right] \tag{5.27}$$

$$\overbrace{\alpha_{22}}^{t,e} = \left(1 + \overbrace{\nu_f}^{t,e}\right) \overbrace{\alpha_f}^{t,e} \overbrace{V_f}^{t,e} + (1 + \overbrace{V_m}^{t,e}) \overbrace{\alpha_m}^{t,e} \overbrace{V_m}^{t,e} - \overbrace{\alpha_{11}}^{t,e} \overbrace{\upsilon_{12}}^{t,e} \tag{5.28}$$

Reciprocally, the stresses can be calculated:

$$\begin{Bmatrix} \overbrace{\sigma_x}^{t,e} \\ \overbrace{\sigma_y}^{t,e} \\ \overbrace{\sigma_{xy}}^{t,e} \end{Bmatrix} = \begin{bmatrix} \overbrace{\overline{Q}_{11}}^{t,e} & \overbrace{\overline{Q}_{12}}^{t,e} & \overbrace{\overline{Q}_{16}}^{t,e} \\ \overbrace{\overline{Q}_{12}}^{t,e} & \overbrace{\overline{Q}_{22}}^{t,e} & \overbrace{\overline{Q}_{26}}^{t,e} \\ \overbrace{\overline{Q}_{16}}^{t,e} & \overbrace{\overline{Q}_{26}}^{t,e} & \overbrace{\overline{Q}_{66}}^{t,e} \end{bmatrix} \left(\begin{Bmatrix} \varepsilon_x \\ \varepsilon_y \\ \gamma_{xy} \end{Bmatrix} - \begin{Bmatrix} \overbrace{\alpha_x}^{t,e} \\ \overbrace{\alpha_y}^{t,e} \\ \overbrace{\alpha_{xy}}^{t,e} \end{Bmatrix} \Delta T - \begin{Bmatrix} \overbrace{\beta_x}^{t,e} \\ \overbrace{\beta_y}^{t,e} \\ \overbrace{\beta_{xy}}^{t,e} \end{Bmatrix} \Delta m \right) \tag{5.71}$$

The resulting forces and moments can finally be evaluated:

$$\begin{Bmatrix} \overbrace{N_x}^{t,e} \\ \overbrace{N_y}^{t,e} \\ \overbrace{N_{xy}}^{t,e} \\ \overbrace{M_x}^{t,e} \\ \overbrace{M_y}^{t,e} \\ \overbrace{M_{xy}}^{t,e} \end{Bmatrix} = \begin{bmatrix} \overbrace{[A]}^{t,e} & \overbrace{[B]}^{t,e} \\ \overbrace{[B]}^{t,e} & \overbrace{[D]}^{t,e} \end{bmatrix} \begin{Bmatrix} \varepsilon_x^0 \\ \varepsilon_y^0 \\ \gamma_{xy}^0 \\ \kappa_x \\ \kappa_y \\ \kappa_{xy} \end{Bmatrix} - \begin{Bmatrix} \overbrace{N_x^T}^{t,e} \\ \overbrace{N_y^T}^{t,e} \\ \overbrace{N_{xy}^T}^{t,e} \\ \overbrace{M_x^T}^{t,e} \\ \overbrace{M_y^T}^{t,e} \\ \overbrace{M_{xy}^T}^{t,e} \end{Bmatrix} \tag{5.72}$$

with

$$\begin{Bmatrix} \overbrace{N_x^T}^{t,e} \\ \overbrace{N_y^T}^{t,e} \\ \overbrace{N_{xy}^T}^{t,e} \end{Bmatrix} = \Delta T \sum_{k=1}^{N} \overbrace{[\overline{Q}]^k}^{t,e} \begin{Bmatrix} \overbrace{\alpha_x}^{t,e} \\ \overbrace{\alpha_y}^{t,e} \\ \overbrace{\alpha_{xy}}^{t,e} \end{Bmatrix}^k t_k + \Delta m \sum_{k=1}^{N} \overbrace{[\overline{Q}]^k}^{t,e} \begin{Bmatrix} \overbrace{\beta_x}^{t,e} \\ \overbrace{\beta_y}^{t,e} \\ \overbrace{\beta_{xy}}^{t,e} \end{Bmatrix}^k t_k \tag{5.73}$$

and

$$\left\{\begin{matrix} \overbrace{M_x^T}^{t,e} \\ \overbrace{M_y^T}^{t,e} \\ \overbrace{M_{xy}^T}^{t,e} \end{matrix}\right\} = \Delta T \sum_{k=1}^{N} \overbrace{[\overline{Q}]^k}^{t,e} \left\{\begin{matrix} \overbrace{\alpha_x}^{t,e} \\ \overbrace{\alpha_y}^{t,e} \\ \overbrace{\alpha_{xy}}^{t,e} \end{matrix}\right\}^k t_k \bar{z}_k + \Delta m \sum_{k=1}^{N} \overbrace{[\overline{Q}]^k}^{t,e} \left\{\begin{matrix} \overbrace{\beta_x}^{t,e} \\ \overbrace{\beta_y}^{t,e} \\ \overbrace{\beta_{xy}}^{t,e} \end{matrix}\right\}^k t_k \bar{z}_k \tag{5.74}$$

5.3.1.5 Shells

We have so far restricted our discussion to thin composite plates. Shells constitute another interesting category of structures extensively covered in the literature [36–38], many composite materials being used for pressure vessels. Shells can be defined as curved thin structures in which through membrane forces (N_x, N_y and N_{xy}) are predominant in the response to transverse loads, bending moments being often negligible.

The membrane forces are a function of the geometry and loading and most cases are tabulated in the literature [39]. For a given load case, let us call N_x^0, N_y^0 and N_{xy}^0 the response of a specific shell obtained from the tabulated values. For example, for a spherical shell of radius r filled with gas with an internal pressure p, we can write:

$$N_x^0 = N_y^0 = \frac{pr}{2} \tag{5.75}$$

Or in the case of a cylindrical gas tank of length L and radius r with an internal pressure p, the meridional force N_x and the hoop force N_y become:

$$N_y^0 = 2N_x^0 = pr \tag{5.76}$$

The environmentally dependent material properties can be introduced by considering a small element of the shell as a flat laminate (approach proposed by Babero [4] for constant properties). This approximation holds for most shells as the thickness to curvature ratio is very small. The values of N_x^0, N_y^0 and N_{xy}^0 can directly be plugged into the CLT equations of Section 5.3.1.3, neglecting the moment effects. Stresses and strains can then be evaluated in the laminated shell.

Analytical calculations can be appropriate in a screening process. Specialized software, however, enables a more precise and quicker calculation of the stresses and strains in a part. The following case study examines composite piping specificities and tool examples embracing stress calculations, environmental considerations as well as composite materials specificities.

Case study: Composite piping

Because of good corrosion resistance, glass-fiber reinforced polymers are often used for piping applications such as sewers, air ventilation and oil transportation. Piping nets can have complex geometries with intersections, diameter changes, joints etc. Furthermore, thermal and pressure loads can lead to important mismatches between pipes having different winding angles. For example, a system involving a small pipe wound at 55°, a larger pipe wound at 75° and reducers all obtained with woven roving reinforcement exhibits a spectrum of axial coefficients of thermal expansion ranging from 28.8 mm/mm/°C for the small pipe to 20.0 mm/mm/°C for the reducers [40]. For comparison, the equivalent thermal expansion of steel is in the range of 15 mm/mm/°C.

To solve those complex challenges, engineers generally revert to finite element calculations. The calculations should always consider the specificities of the composites (plies, anisotropy). Calculations assuming homogeneous isotropic materials are inappropriate for layered structures. Such calculations do not even provide reasonable approximations. Indeed, linear safety factor coefficients are often not sufficient to account for the effects of power laws generally governing polymer composite behavior.

Piping engineers may use the composite shell elements available in commercial general finite element software. Unfortunately, this approach is often time-consuming and not fitted to small projects or rapid tender processes. Specialized commercial software is specifically available for the calculations of glass-fiber reinforced piping systems. Those codes offer libraries of geometrical elements such as supports (anchors, guides, hangars) and fitting options (smooth bends, closely or widely spaced mitered bends, reducers etc.) as well as a pre-selected load range (pressure, temperature, wind, seismic loads etc.) [40]. The calculations consider the non-linear specificities of the material and provide stress and strain information. The displacement information enables an iterative optimization of the support systems, a major and costly challenge in the piping industry.

Bentley's Autopipe software, for example, enables rapid calculations and visualization of pipe stresses and deformations under typical environmental loads such as gravity (Figure 5.10), thermal loads (Figure 5.11), seismic motions (Figures 5.12 and 5.13), wind (Figures 5.14 and 5.15) and water hammering (Figure 5.16). The software enables the input of orthotropic materials properties. Table 5.3 presents the results of typical calculations performed on a glass fiber epoxy 21.91 cm outer diameter piping system. The selected operating environment for the calculations included an internal pressure of 32.07 MPa, an inner temperature of 232.2°C, an outer temperature cycling from −16.0 to 232.2°C, occasional seismic and wind loads as well as a sustained gravity. The results (Table 5.3) show the differences between calculations considering isotropic and orthotropic materials properties. Strain and stresses generally differ by at least 10%. More importantly, under the same load, the isotropic material would fail due to large deformations (exceeding allowable stresses) when the orthotropic pipe was found to remain intact.

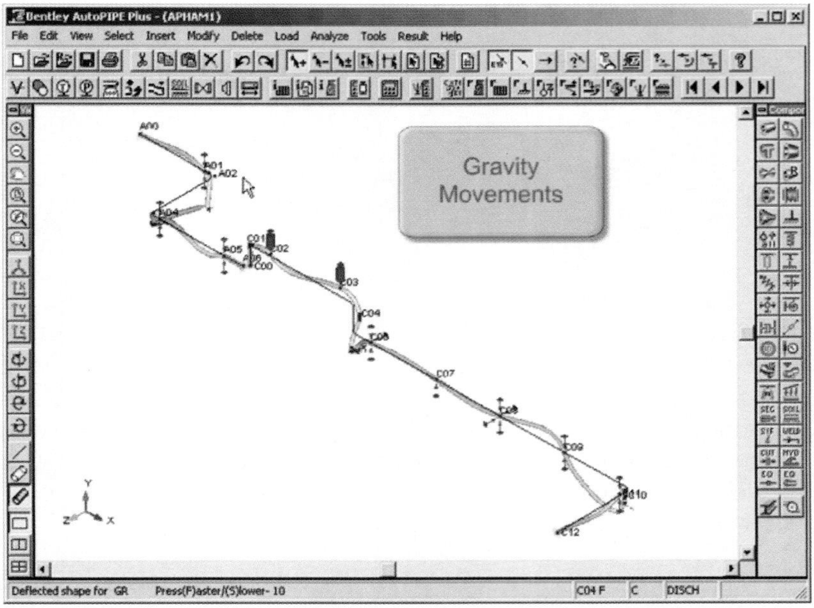

Figure 5.10. Simulation of piping gravity movements obtained with AutoPIPE Plus software. (Courtesy of Bentley Systems, Inc.)

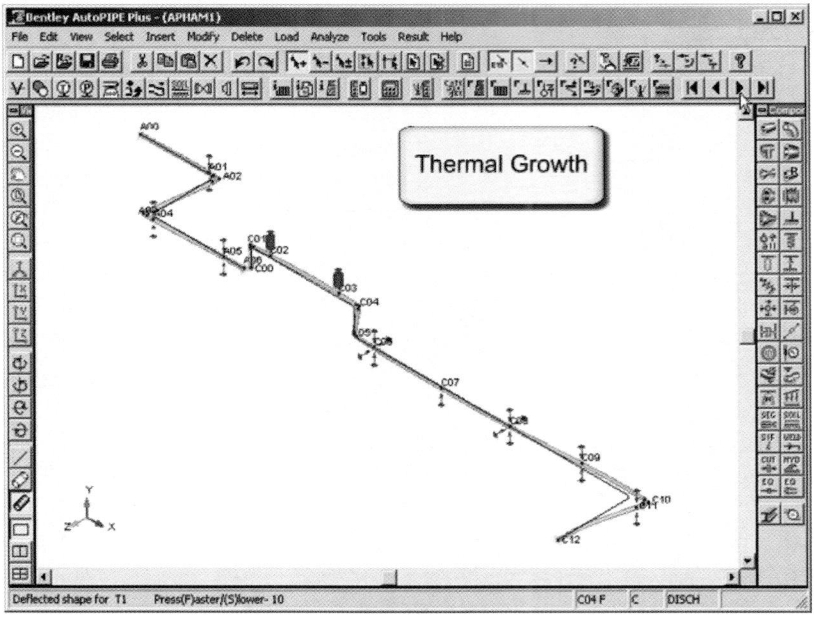

Figure 5.11. Thermal growth simulation of piping system obtained with AutoPIPE software. (Courtesy of Bentley Systems Inc.)

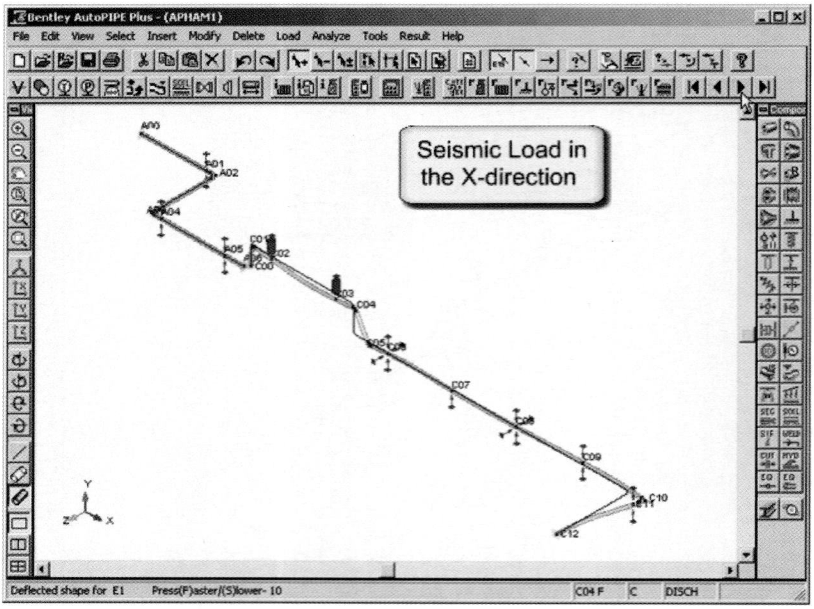

Figure 5.12. Simulation of piping deformation under seismic load in the axial direction obtained with AutoPIPE Plus software. (Courtesy of Bentley Systems, Inc.)

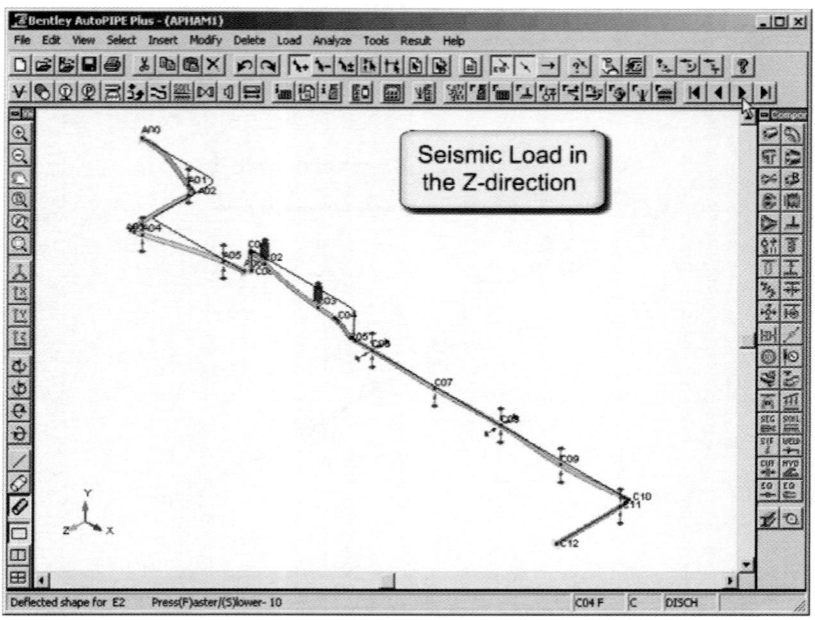

Figure 5.13. Seismic-induced deformation (load in z-direction) simulated with the AutoPIPE Plus software. (Courtesy of Bentley Systems, Inc.)

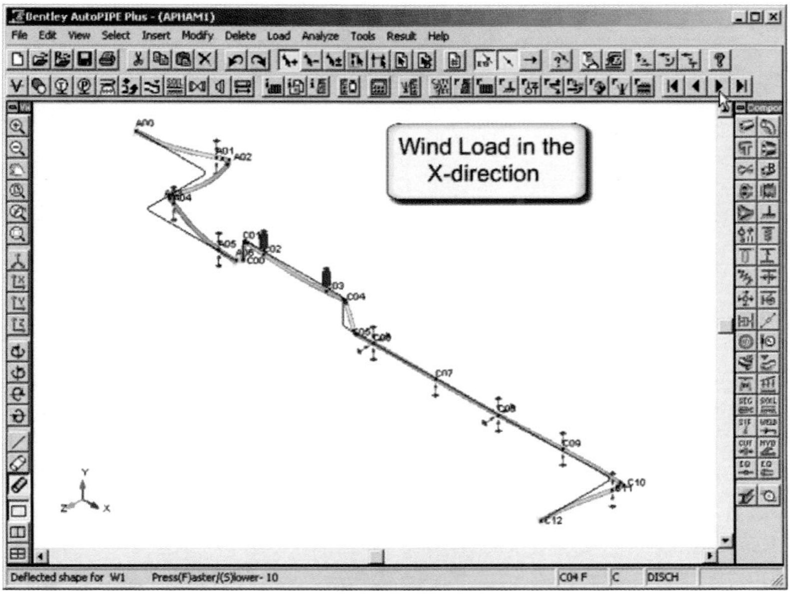

Figure 5.14. Wind-induced piping system deformation (*x*-direction). Simulation with AutoPIPE system. (Courtesy of Bentley Systems, Inc.)

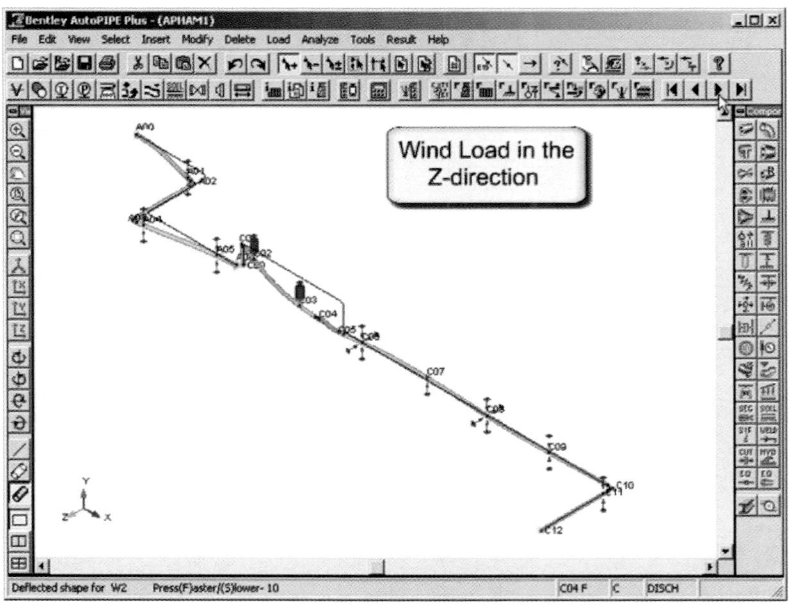

Figure 5.15. Wind-induced piping system deformation (z-direction). Simulation with the AutoPIPE Plus system. (Courtesy of Bentley Systems, Inc.)

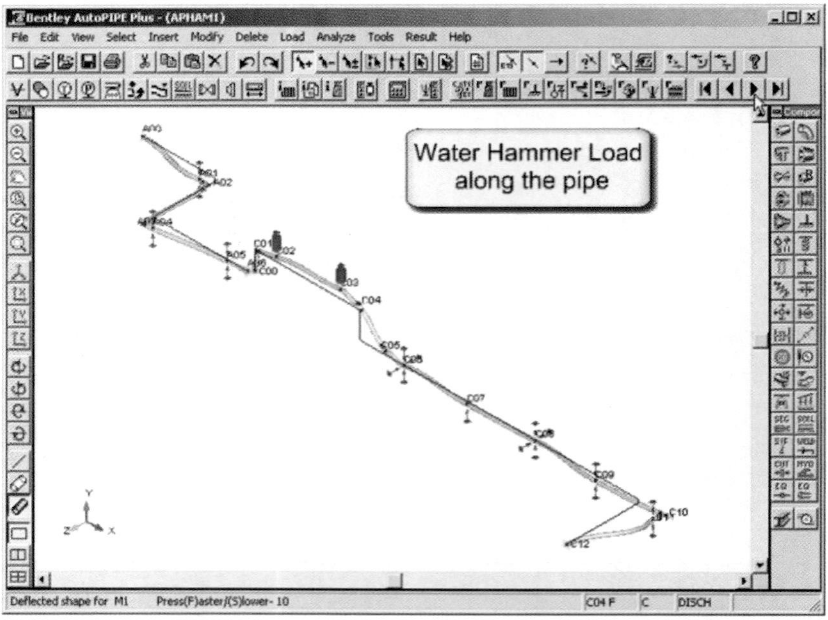

Figure 5.16. Water hammer load induced deformations. Simulation with the AutoPIPE Plus software. (Courtesy of Bentley Systems, Inc.)

Table 5.3. Piping result example: Isotropic versus orthotropic materials properties. Results from AutoPIPE, Bentley Systems, Inc.

	Isotropic pipe material (MPa)	Orthotropic pipe material (MPa)
Maximum sustained stress	33.68	48.68
Allowable sustained stress	99.97	99.97
Maximum displacement stress	234.13	198.32
Allowable displacement stress	231.45	229.62
Maximum occasional stress	81.15	77.10
Allowable occasional stress	132.97	132.97
Maximum hoop stress	44.45	44.45
Maximum allowable stress	99.97	99.97

5.3.2 Impact of Non-linear Viscoelasticity on the Mechanical Properties of Composites

Much like Chapter 2, the present chapter mainly deals with linear viscoelastic behavior. The theory of linear viscoelasticity is well known and rather complete. It proposes established models usable in an industrial context. The field of non-linear viscoelasticity is more complex, though, and far from being fully understood. Several factors contribute to the complexity of the field [41]. For example, non-linear viscoelasticity requires different treatments in the case of fluids and solids. However, Chapter 2 has illustrated the difficulties to distinguish between these two states, especially when temperature and strain rate parameters intervene. To complicate matters further, polymers in the glassy state undergo localized damage even at moderate strain levels (see Chapter 2, Section 2.2.2). Localized processes such as crazing dominate the materials behavior and are difficult to generalize. The presence of crystallinity and reinforcement further reinforce the specificity of the behaviors. Furthermore, unlike the case of linear viscoelasticity, no direct correspondence between creep and relaxation can be established. Finally, Christensen [41] rightfully mentions that the degree of non-linearity not only depends on the current loads and displacements but also on past history.

Theories to determine mechanical properties of non-linear viscoelastic solids were reviewed by Stafford, Lockett and Christiansen [42,43,41] and are mainly based on the Green-Rivlin theory [44]. There is no need to mention that the introduction of non-linear viscoelasticity in finite element calculations is rather complex. Specific studies can serve as starting points in this task [45–48].

These difficulties should, however, not bring the engineer to ignore the non-linearities. The first step in all polymer composite studies is to establish the degree of non-linearity. Simple safety factors can account for minor deviations from linearity. However, if the non-linearities are significant, specific equations can be established based on simple experimental data curve fitting procedures and used in the durability approaches of Chapter 6.

5.4 ENVIRONMENTAL IMPACT ON THE DAMAGE MECHANISMS AND FAILURE OF COMPOSITE STRUCTURES

5.4.1 Composite Failure

Failure is an ambiguous term for most materials including composites. Failure can be defined by the point at which the sample becomes separated (broken) into two or more parts. But failure can also be a loss of stiffness greater than 10%, for example, or the multiplication of partial discharges in high voltage applications. In this part, we will mainly focus on mechanical failure related to strength or strain-to-failure. The failure of multi-layer composites is progressive. For example, in a [0°/90°] carbon-fiber reinforced epoxy under tension, the [90°] ply will break first. The stresses and strains will be re-distributed in the laminate. This progressive ply failure creates a non-linearity in the stress–strain curve (in addition to the potentially nonlinear behavior of the constituents).

The influence of environment on the strength of composites and constituents was detailed in the previous chapters. It was also indicated that there was not much point using micromechanics to predict the effect of environment on the strength degradation of composites. However, it is interesting to look at integrating the effects of property changes of the constituents at the ply level into laminate models.

For that purpose, we need to introduce failure functions for the composite and study their evolution with time and environment. For isotropic materials, several criteria are typically used. The most widely used, at the time of writing, was probably the energy-based Von-Mises criterion that considers the triaxial state of stress:

$$(\sigma_1 - \sigma_2)^2 + (\sigma_2 - \sigma_3)^2 + (\sigma_3 - \sigma_1)^2 = 2\sigma_y^2 \tag{5.77}$$

where the yield stress σ_y can be obtained via tensile tests. The use of Von-Mises criterion can only be allowed under the assumption of isotropy and if the failure modes remain the same under the different axial stresses. The Von-Mises criterion is generally *not applicable* to inhomogeneous, anisotropic (e.g. laminates) and brittle structures.

Most composite materials therefore require the use of different failure criteria. The number of criteria is very large [49] and only the most common ones will be detailed below. It is up to the reader to define the appropriate failure criterion for his own application.

5.4.2 Maximum Stress and Maximum Strain Criteria

5.4.2.1 Maximum stress criterion

The first failure criterion and probably the most intuitive is the maximum stress criterion (Table 5.4). For each ply, failure occurs if one of the stresses exceeds the strength of the material in this direction (F_i). This translates into the following equations (Table 5.4) applicable for each ply in the materials coordinates. Time and environment enter the equations through the stresses and strains calculations derived in Section 5.3, as well as through the strength or strain-to-failure related to the environment (Chapters 2–4). For example $\overbrace{\sigma_{11}}^{t,e}$ can reflect the changes in the stresses resulting from a lower matrix modulus due to moisture absorption, when $\overbrace{X_{1t}}^{t,e}$ can reflect the changes in ply strength due to (for example) the same moisture absorption.

5.4.2.2 Maximum strain criterion

Failure in a ply can also occur if the strain-to-failure is reached in one of the directions. Failure conditions are summarized in Table 5.5.

Table 5.4. Maximum stress criterion

Assumption	Criterion		Consequence
$\sigma_1 > 0$	$\overbrace{\sigma_1}^{t,e} > \overbrace{X_{1t}}^{t,e}$	(5.78)	Failure in the fiber direction $E_{11} = 0$ for the ply
$\sigma_1 < 0$	$\overbrace{-\sigma_1}^{t,e} < \overbrace{X_{1c}}^{t,e}$	(5.79)	Failure in the fiber direction $E_{11} = 0$ for the ply
$\sigma_2 > 0$	$\overbrace{\sigma_2}^{t,e} > \overbrace{X_{2t}}^{t,e}$	(5.80)	Failure in the transverse direction $E_{22} = 0$ for the ply
$\sigma_2 < 0$	$\overbrace{-\sigma_2}^{t,e} > \overbrace{X_{2c}}^{t,e}$	(5.81)	Failure in the transverse direction $E_{22} = 0$ for the ply
	$\left\lvert \overbrace{\sigma_4}^{t,e} \right\rvert > \overbrace{X_4}^{t,e}$	(5.82)	Interlaminar shear failure $G_{23} = 0$ for the ply
	$\left\lvert \overbrace{\sigma_5}^{t,e} \right\rvert > \overbrace{X_5}^{t,e}$	(5.83)	Interlaminar shear failure $G_{31} = 0$ for the ply
	$\left\lvert \overbrace{\sigma_6}^{t,e} \right\rvert > \overbrace{X_6}^{t,e}$	(5.84)	In-plane shear failure $G_{12} = 0$ for the ply

In Chapters 2–4, we have already noted that polymers and polymer matrix composites can exhibit a large amount of non-linearity in their stress–strain responses. Therefore maximum stress and maximum strain criteria are not directly proportional and can lead to different results (Figure 5.17).

For example, the point indicated by x in Figure 5.17 will fail according to the maximum stress criterion. However, this failure would not be predicted by the maximum strain criterion. Reciprocally, the point indicated by o will fail according to the maximum strain criterion, but would be allowed by the maximum stress criterion. It is therefore necessary to test both criteria simultaneously in order to assess ply failure in the laminated structure.

5.4.2.3 Limit of the criteria

Maximum stress and maximum strain criteria apply rather well to situations in which one failure mode dominates clearly. When stresses interact and more than one failure mode is observed, polynomial criteria are more suitable.

Table 5.5. Maximum strain criterion

Assumption	Criterion		Consequence
$\varepsilon_1 > 0$	$\overbrace{\varepsilon_1}^{t,e} > \overbrace{\varepsilon_{1t}}^{t,e}$	(5.85)	Failure in the fiber direction $E_{11} = 0$ for the ply
$\varepsilon_1 < 0$	$\overbrace{-\varepsilon_1}^{t,e} > \overbrace{\varepsilon_{1c}}^{t,e}$	(5.86)	Failure in the fiber direction $E_{11} = 0$ for the ply
$\varepsilon_2 > 0$	$\overbrace{\varepsilon_2}^{t,e} > \overbrace{\varepsilon_{2t}}^{t,e}$	(5.87)	Failure in the transverse direction $E_{22} = 0$ for the ply
$\varepsilon_2 < 0$	$\overbrace{-\varepsilon_2}^{t,e} > \overbrace{\varepsilon_{21c}}^{t,e}$	(5.88)	Failure in the transverse direction $E_{22} = 0$ for the ply
	$\left\lvert\overbrace{\gamma_{23}}^{t,e}\right\rvert > \overbrace{\gamma_{23s}}^{t,e}$	(5.89)	Interlaminar shear failure $G_{23} = 0$ for the ply
	$\left\lvert\overbrace{\gamma_{31}}^{t,e}\right\rvert > \overbrace{\gamma_{31s}}^{t,e}$	(5.90)	Interlaminar shear failure $G_{31} = 0$ for the ply
	$\left\lvert\overbrace{\gamma_{12}}^{t,e}\right\rvert > \overbrace{\gamma_{12s}}^{t,e}$	(5.91)	In-plane shear failure $G_{12} = 0$ for the ply

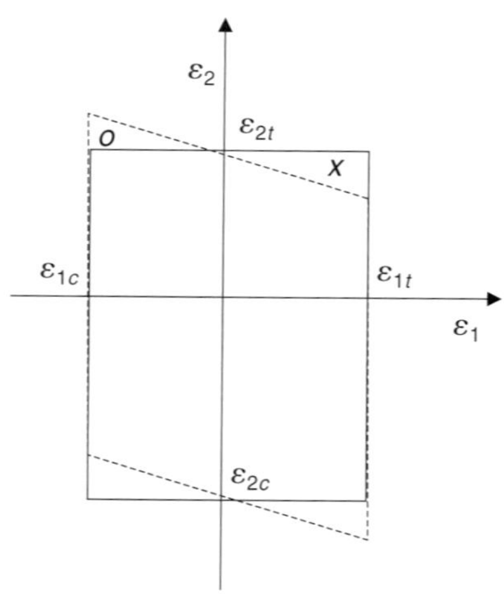

Figure 5.17. Maximum stress (dotted line) versus maximum strain (continuous line) failure envelops.

5.4.3 Polynomial Criteria

Polynomial criteria, such as Tsai-Hill and Tsai-Wu, define failure envelopes, outside of which the material fails. For in-plane stresses in the 1-2 plane, the Tsai-Hill criterion can be written as:

$$\left(\frac{\left(\overbrace{\sigma_1}^{t,e}\right)^2}{\overbrace{X_1}^{t,e}{}^2}-\frac{\left(\overbrace{\sigma_1}^{t,e}\overbrace{\sigma_2}^{t,e}\right)^2}{\overbrace{X_1}^{t,e}{}^2}+\frac{\left(\overbrace{\sigma_2}^{t,e}\right)^2}{\overbrace{X_2}^{t,e}{}^2}+\frac{\left(\overbrace{\sigma_6}^{t,e}\right)^2}{\overbrace{X_6}^{t,e}{}^2}\right)-1\geq 0 \tag{5.92}$$

The Tsai-Wu criterion is a more advanced polynomial criterion, considering the specificities of the materials behavior in tension and compression. The Tsai-Wu criterion can be expressed according to:

$$f_i\sigma_i+f_{ij}\sigma_i\sigma_j=1 \qquad i,\ j=1\ldots 6 \tag{5.93}$$

For an orthotropic lamina under plane stress conditions, Equation (5.93) becomes:

$$\left(\overbrace{f_{11}}^{t,e}\overbrace{\sigma_1^2}^{t,e}+\overbrace{f_{22}}^{t,e}\overbrace{\sigma_2^2}^{t,e}+2\overbrace{f_{12}}^{t,e}\overbrace{\sigma_1}^{t,e}\overbrace{\sigma_2}^{t,e}+\overbrace{f_1}^{t,e}\overbrace{\sigma_1}^{t,e}\right.$$

$$\left.+\overbrace{f_2}^{t,e}\overbrace{\sigma_2}^{t,e}+\overbrace{f_6}^{t,e}\overbrace{\sigma_6}^{t,e}+\overbrace{f_{66}}^{t,e}\overbrace{\sigma_6^2}^{t,e}\right)-1\geq 0 \tag{5.94}$$

where

$$\overbrace{f_1}^{t,e}=\frac{1}{\overbrace{X_{1t}}^{t,e}}+\frac{1}{\overbrace{X_{1c}}^{t,e}} \tag{5.95}$$

and

$$\overbrace{f_{11}}^{t,e}=\frac{-1}{\overbrace{X_{1t}}^{t,e}\overbrace{X_{1c}}^{t,e}} \tag{5.96}$$

in the longitudinal direction and

$$\overbrace{f_2}^{t,e}=\frac{1}{\overbrace{X_{2t}}^{t,e}}+\frac{1}{\overbrace{X_{2c}}^{t,e}} \tag{5.97}$$

$$\overbrace{f_{22}}^{t,e} = \frac{-1}{\underbrace{X_{2t}}^{t,e}\underbrace{X_{2c}}^{t,e}} \tag{5.98}$$

in the transverse direction.

For shear, we write:

$$f_6 = 0 \tag{5.99}$$

and

$$f_{66} = \frac{1}{\underbrace{X_6}^{t,e}{}^2} \tag{5.100}$$

The parameter f_{12} is still to be defined. It represents the interaction between the two normal stresses and can therefore not be obtained from uniaxial experiments. If we fix a biaxial stress state in which the two normal stresses are equal ($\sigma_1 = \sigma_2 = \sigma$) then Equation (5.94) can be solved for f_{12}:

$$f_{12} = \frac{1}{2\sigma^2}\left[1 - \left(\frac{1}{\underbrace{X_{1t}}^{t,e}} + \frac{1}{\underbrace{X_{1c}}^{t,e}} + \frac{1}{\underbrace{X_{2t}}^{t,e}} + \frac{1}{\underbrace{X_{2c}}^{t,e}}\right)\sigma + \left(\frac{1}{\underbrace{X_{1t}}^{t,e}\underbrace{X_{1c}}^{t,e}} + \frac{1}{\underbrace{X_{2t}}^{t,e}\underbrace{X_{2c}}^{t,e}}\right)\sigma^2\right] \tag{5.101}$$

The Tsai-Wu criterion provides reasonable results. Unfortunately, Tsai-Wu and Tsai-Hill criteria alike do not identify the type of failure. It is therefore necessary to compute the maximum stress and maximum strain criteria simultaneously to allow for ply property discount.

5.4.4 Discussion on Recent Failure Criteria

Failure criteria, more recent and advanced than the ones presented in Sections 5.4.2 and 5.4.3, have been developed over the past years. Among those, 19 internationally recognized failure criteria were discussed in the worldwide-failure exercise [50–52]. Table 5.6 summarizes the different approaches represented and the corresponding reference for a detailed description of the method.

This exercise compared the *blind* predictions of the various criteria with experimental results. The results are complex and summarized in three special issues of the journal *Composites Science and Technology* [50–52]. In particular, special recommendations for designers are made in [71]. Comparisons with experimental data led to favor the *Puck and Schürmann*, *Sinoviev et al.*, *Tsai and Liu*, *Cuntze and Freund* and *Bogetti et al.* criteria. However, no failure criterion was found to accurately predict all failure features in all loading cases and most failure criteria are

Table 5.6. Further approaches for failure prediction of composite materials

Contributors [53]	Approach represented [53]	Reference
Chamis C.C., P.K. Gotsis, L. Minnetyan	ICAN (micromechanic based)	[54]
Chamis C.C., P.K. Gotsis, L. Minnetyan	CODSTRAN	[54]
Hart-Smith L.J.	Generalized Tresca theory	[55]
Hart-Smith L.J.	Maximum strain theory	[56]
Eckold G.C.	British standard pressure vessel design codes	[57]
Edge E.C.	British aerospace, in-house design method	[58]
McCartney L.N.	Physically based "damage mechanics"	[59]
Puck A., J. Schürmann	Physically based three-dimensional phenomenological models	[60]
Wolfe W.E., T.S. Butalia	Maximum strain energy method, due to Sandhu	[61]
Sun C.T., J.X. Tao	Linear analysis	[62]
Sun C.T., J.X. Tao	Non-linear FE-based analysis	[62]
Zinoviev P., S.V. Grigoriev, O.V. Labedeva, L.R. Tairova	Development of maximum stress theory	[63]
Tsai S.W., K-S Liu	Interactive matrix and fiber failure theory	[64]
Rotem A.	Interactive matrix and fiber failure theory	[65]
Cuntze R., A. Freund	Failure mode concept (FMC)	[66]
Bogetti T., C. Hoppel, V. Harik, J. Newill, B. Burns	Three-dimensional maximum strain	[67]
Mayes S.J., A.C. Hansen	Multi-continuum micromechanics theory	[68]
Huang Z-M	Bridging model, micromechanics	[69]
Hart-Smith L.J.	Ten-Per-Cent rule	[70]

still in the developmental stage. Facing such uncertainty and for simplicity reasons, we chose to focus in this book on basic criteria and only reference more advanced approaches (Table 5.6), keeping in mind possible prediction discrepancies. We will never stress enough the absolute necessity to validate all predictions or calculations by an extensive experimental plan (see Section 6.4).

5.5 SPECIAL FOCUS: FINITE ELEMENT COMMERCIAL SOFTWARES

We have so far limited our discussion to thin plates and shells. The case of complex geometries will not be described here. Indeed, analytical modeling of thick plates, stiffened panels and beams is dealt with in the literature [72,73,4]. Furthermore, most industrial applications require the use of finite element analysis (FEA).

Several finite element analysis software codes are available on the market. Broadly used ABAQUS and ANSYS solvers can be employed with standard pre- and post-processors or completed by composite specific softwares such as CATIA Composite Design 3 (CPD). This software is currently used by major aircraft manufacturers [74] and covers the full design process including basic and detailed design while considering the product's requirements for finite element analysis and manufacturabililty [75].

The major challenge in the use of commercial FE software codes resides in taking into account the specificities of the composite. For example, modeling thick plates generally requires the use of solid elements. However, the presence of thin layers, such as adhesives can have dramatic effects on the resulting stresses. Figure 5.18 shows a sandwich structure calculated with a p-element FEM program called *StressCheck* developed by ESRD. The honeycomb (assumed anisotropic in the calculation) structure is covered by two orthotropic layers of carbon-fiber reinforced polymer [76]. A metallic insert is attached to the composite structure by an epoxy adhesive (assumed non-linear elastoplastic). Figure 5.18 illustrates the differences between linear and non-linear solutions in response to an in-plane bearing load simulating the presence of a bolt. The results are strikingly different. The linear solution leads to maximum adhesive shear stresses 10% higher than when considering non-linear materials properties. More importantly, the adhesive layer can be clearly identified as the weak point in the assembly. Therefore, the presence

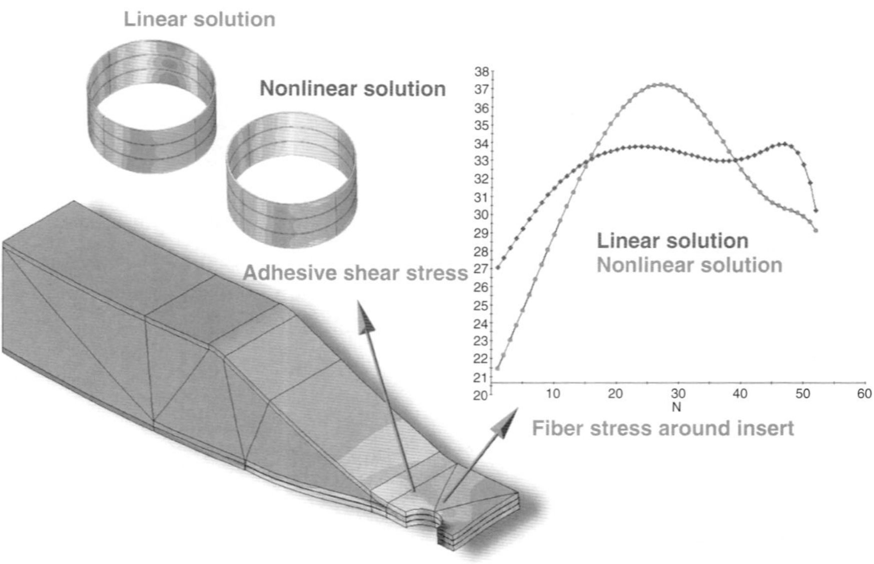

Figure 5.18. Sandwich panel with bonded insert. Maximum shear stress on graph expressed in MPa. (Courtesy of SI-Schweitzer Ingenieurgesellschaft GmbH.)

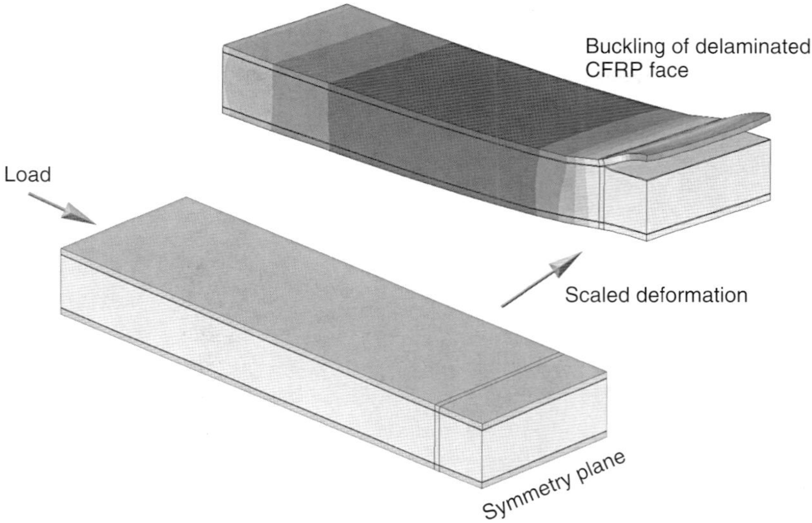

Figure 5.19. Buckling of delaminated carbon-fiber composite face. (Courtesy of SI-Schweitzer Ingenieurgesellschaft GmbH.)

of thin layers cannot be neglected and it is recommended to use mesh and tool calculations enabling geometries and materials properties closest to reality.

Finite element calculations of the composites response under compressive loads are generally more complex. Figure 5.19 illustrates a sandwich structure under compression. The sandwich is made of a honeycomb core between two layers of carbon-fiber reinforced composites bonded by an epoxy layer. To perform the calculations it is necessary to introduce a perturbation in the system by creating an artificial delamination area between the top surface and the core. The compressive loads in this example result in localized buckling of the upper layer (Figure 5.19).

Another composite specificity is related to the manufacturing process. During molding, for example, the fiber orientation might vary depending on the processing parameters. This has led resin supplier BASF to develop the FIBER software bridging the gap between mold-fill simulation software such as MOLDFLOW and commercial finite element analysis software such as ABAQUS or ANSYS. The system integrates non-linearity aspects as well as modified properties based on the calculated fiber orientation considering the molding compound's melt viscosity, fiber content and process parameters (injection speed and holding pressure) [77]. The motivation for the use of such software is illustrated by the example of a beam (LU carrier) under torsional load shown in Figure 5.20. Micrographs reveal strong local differences in fiber orientation (Figure 5.21). Such fiber orientation can be calculated using the BASF specialized software FIBER (Figure 5.22). Subsequent FE analysis run with ABAQUS using the local properties shows excellent agreement

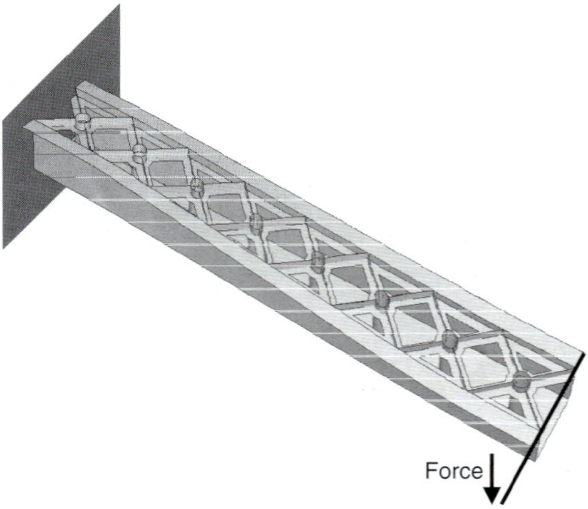

Figure 5.20. Torsional load on short fiber reinforced molded composite beam. (Courtesy of BASF.)

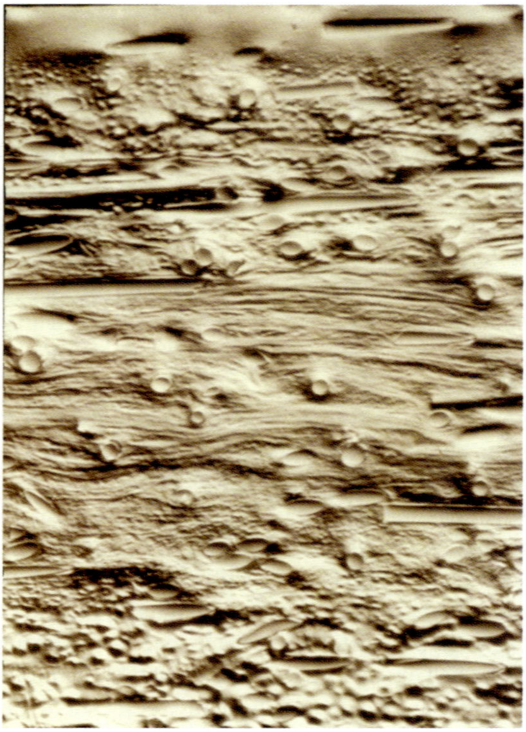

Figure 5.21. Local anisotropy illustrated by different fiber orientation in a short fiber reinforced molded composite sample. (Courtesy of BASF.)

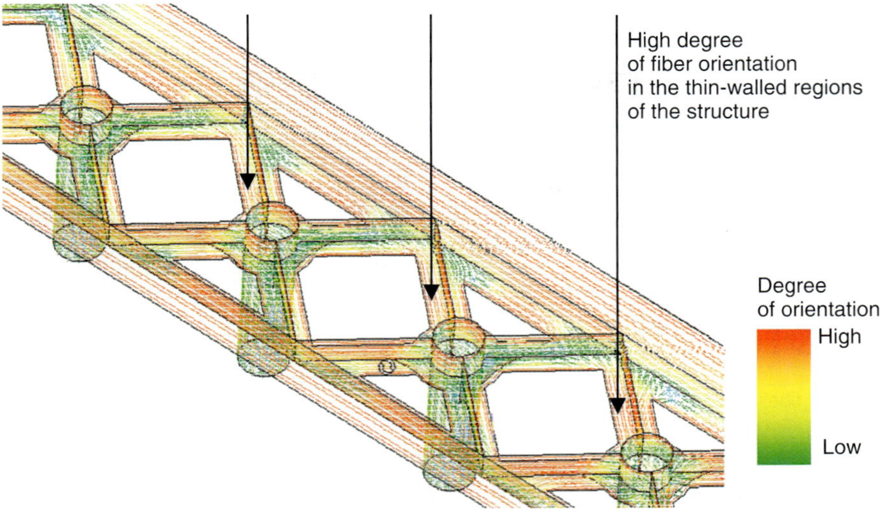

Figure 5.22. Prediction of fiber orientation after molding using BASF FIBER software. (Courtesy of BASF.)

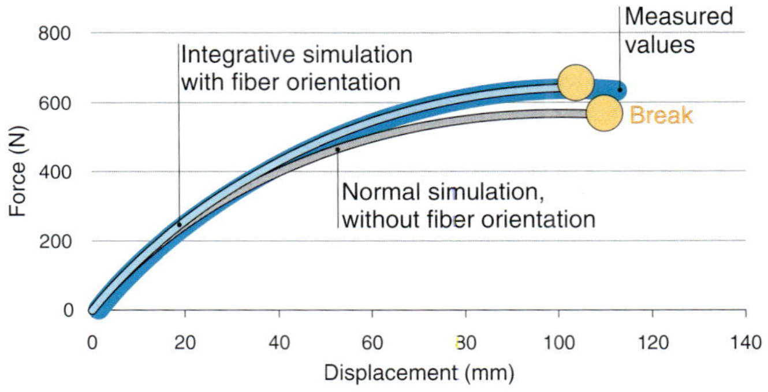

Figure 5.23. Predicted stress–strain curves for beam under torsional load. (Courtesy of BASF.)

with experiments. An ABAQUS calculation without considering the local fiber alignments, as a consequence of the manufacturing process, results in a 6% strength discrepancy (Figure 5.23).

At the other end of the spectrum, software codes such as the Alpha Star Corp GENOA software based on NASA's Composite Durability Structural Analysis program (CODSTRAN) focus on long-term and durability aspects, including local

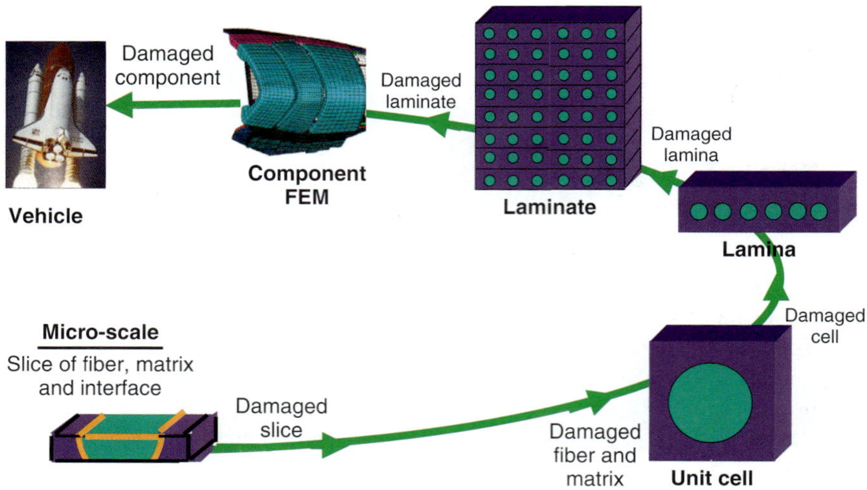

Figure 5.24. GENOA takes a full-scale finite element model and breaks the material properties down to the microscopic level. Materials properties are then updated for the next iteration, reflecting any changes resulting from damage or crack propagation [78]. (Courtesy of Alpha Star.)

damage and progressive failure. Laminate non-linearity is accounted for by incrementally increasing the load and running calculations at each step. The programs generally allow for damage tracking at the microscopic level (such as microbuckling or microcracking) and translate it in terms of macroscopic responses (Figure 5.24).

If scientists and engineers around the world generally agree on stress and strain calculation methods for composite materials, they have not yet reached a consensus on the best failure criteria. The GENOA software, for example, uses 14 different failure criteria. The program can also consider the geometrical specificity of composite material reinforcement such as fiber woven, braided, knitted or stitched composites [78] as well as fiber waviness or void content.

5.6 TESTING

In the previous chapters, we have presented selected test methods related to the environmental parameters of interest. Many test procedures are further available to determine the mechanical properties of the plies or laminates at the macroscopic level. Among those, tensile, compressive, shear and flexural tests on standard and notched specimens provide basic information necessary for the composite design. Care should be taken in performing those tests that experimental conditions are carefully monitored and if necessary controlled. Indeed, parameters such as temperature, moisture or strain rate (see Chapters 2 and 3) can strongly influence the results of quasi-static and dynamic tests.

Major tests are shortly mentioned in following sections. This list is far from exhaustive; entire books are being dedicated to this topic [79–81].

5.6.1 Tensile Testing

Axial tensile testing of long fiber composites is often a challenge. Indeed, the axial load applied by the apparatus is transferred to the specimen as shear. Shear strength in unidirectional composites is typically much lower than axial tensile strength and the specimen tends to fail in the gripping region. The use of dog-boned specimens might lead to shear damage at the fillets at each end of the specimen. These problems can be minimized by increasing the tabbing areas and reducing the specimen thickness (down to 0.4 mm) [81]. Transverse tensile testing of unidirectional polymer composites is generally not a problem and untabbed thicker specimens (up to a 3 mm thickness) can be used.

5.6.2 Compression Testing

Buckling in unidirectional polymer matrix composites, whether at the microscopic or macroscopic scale, is almost inevitable under axial compressive loads. Therefore, axial compression testing requires thick composite specimens as well as an apparatus limiting the buckling of the sample. The compressive strength of the sample in the transverse direction is generally in order of magnitudes lower than in the axial direction and buckling is of lesser concern.

5.6.3 Shear Testing

Shear properties (modulus and strength) can be obtained using thin-walled tube or circular rod torsion experiments. Such experiments, however, require the use of special equipment (torsion machine). The $\pm 45°$ tensile shear tests, two-rail shear and Iosipescu shear tests [81–83] are in-plane experiments alternative to the torsion tests. This last experimental procedure enables the determination of interlaminar shear properties in addition to the in-plane shear information.

5.6.4 Flexural Testing

Bending tests are probably the most common experimental characterization procedure in industry. Indeed, flexural experiments are relatively easy to perform. Unfortunately, such tests do not provide precise information about materials basic properties and failure modes. The specimen simultaneously sees compressive stresses (on the surface where the load is applied), tensile stresses (on the opposite surface of the sample) and shear stresses at the mid-plane (neutral axis). Loading conditions determine the predominant stresses and are therefore the drivers of the failure. Such tests are appropriate only if the materials damage and failure mode under operation correspond to the testing conditions.

Two procedures dominate flexural testing, specifically three- and four-point bending. It is generally recommended to perform four-point bending tests, as a larger portion of the sample is subjected to the maximum bending moment. Furthermore, for a given maximum shear force, specimens undergoing three-point bending experiments are submitted to a concentrated force twice as high [81]. When three-point bending has to be used due to budget constraints, shear effects can usually be minimized by increasing the specimen aspect ratio (length/thickness).

5.6.5 Interface Testing

We have discussed many times the importance of the interface properties in the global response of the composite. Interfaces are keys in defining the state of stresses in the material and ultimately failure. Furthermore, matrix/fiber interfaces are often more sensitive to environmental exposure than the components themselves (e.g. see Section 3.2). However, the experimental determination of the fiber bonding properties and evolution with time still remains a challenge today. Properties can be obtained via interfacial bond test methods [81] that include single fiber tests (embedded single fiber tension/compression, microdebond, single fiber pullout). Unfortunately, such experiments generally result in high levels of scatter in the data, which can often be difficult to interpret. Macroscopic data can also be obtained using the shear methods detailed above.

5.6.6 Fatigue Testing

Fatigue testing can be performed on all the tests presented above by the simple introduction of a cyclic load. Fatigue loads and cycling rates are selected as a function of operational conditions. Fatigue can involve tension–tension, tension–compression, compression–compression and shear–shear cycles. Fatigue is extensively addressed in Chapter 6.

5.6.7 Standardized Tests

The high degrees of inhomogeneity and anisotropy in composite materials often require specific testing procedures. Testing procedures for polymer matrix composites are summarized in the ASTM Standard Guide for Testing Polymer Matrix Composite Materials (ASTM D4762-04). This guide is of course not exhaustive but is a good starting point for the identification of common procedures (Table 5.7).

Table 5.7. Major ASTM norms related to polymer matrix composite testing from ASTM D4762-04

Standard	Designation	Title
General		
ASTM	D5687/D5687M	Guide for Preparation of Flat Composite Panels With Processing Guidelines for Specimen Preparation
ASTM	D618	Practice for Conditioning Plastics for Testing
ASTM	D6856	Guide for Testing Fabric Reinforced Textile Composite Materials
Tension		
ASTM	D3039/3039M	Test Method for Tensile Properties of Polymer Matrix Composite Materials
ASTM	D638	Test Method for Tensile Properties of Plastics
ASTM	D5450/D5450M	Test Method for Transverse Tensile Properties of Hoop Wound Polymer Matrix Composite Cylinders
ASTM	D5766/D5766M	Test Method for Open Hole Tensile Strength of Polymer Matrix Composite Laminates
ASTM	D5083	Test Method for Tensile Properties of Reinforced Thermosetting Plastics Using Straight-Sided Specimens
Compression		
ASTM	D5467/5417M	Test Method for Compressive Properties of Unidirectional Polymer Matrix Composites Using a Sandwich Beam
ASTM	D5449/5449M	Test Method for Transverse Compressive Properties of Hoop Wound Polymer Matrix Composite Cylinders
ASTM	D695	Test Method for Compressive Properties of Rigid Plastics
ASTM	D3410/D3410M	Test Method for Compressive Properties of Polymer Matrix Composite Materials with Unsupported Gage Section by Shear Loading
ASTM	D6484/D6484M	Test Method for Open-Hole Compressive Strength of Polymer Matrix Composite Laminates
ASTM	D6742/D6742M	Practice for Filled-Hole Tension and Compression Testing of Polymer Matrix Composite Laminates
ASTM	D6641/D6641M	Test Method for Determining the Compressive Properties of Polymer Matrix Composite Materials Using the Combined Loading Compression (CLC) Test Fixture
Shear		
ASTM	D3518/D3518M	Test Method for In-Plane Shear Response of Polymer Matrix Composite Materials by Tensile Test of a 45 Laminate
ASTM	D4255/D4255M	Test Method for In-Plane Shear Properties of Polymer Matrix Composite Materials by the Rail Shear Method

(*Continued*)

Table 5.7. *(Continued)*

Standard	Designation	Title
ASTM	D5379/D5379M	Test Method for Shear Properties of Composite Materials by the V-Notched Beam Method
ASTM	D5448/D5448M	Test Method for In-Plane Shear Properties of Hoop Wound Polymer Matrix Composite
ASTM	D3846	Test Method for In-Plane Shear Strength of Reinforced Plastics
Bending		
ASTM	C393	Test Method for Flexural Properties of Sandwich Constructions
ASTM	D6772	Test Method for Flexural Properties of Unreinforced and Reinforced Plastics and Electrical Insulating Materials by Four-Point Bending
ASTM	D6416/D6416M	Test Method for Two-Dimensional Flexural Properties of Simply Supported Sandwich Composite Plates Subjected to a Distributed Load
ASTM	D2344/D2344M	Test Method for Short Beam Strength of Composite Materials and Their Laminates
ASTM	D790	Test Methods for Flexural Properties of Unreinforced and Reinforced Plastics and Electrical Insulating Materials
Fatigue		
ASTM	D3479/D3479M	Test Method for Tension-Tension Fatigue of Polymer Matrix Composite Materials
ASTM	D671	D671 Test Method for Flexural Fatigue of Plastics by Constant-Amplitude-of-Force

5.7 TOOL KIT

Topic	Equation	Assumptions	Importance
Volume fraction	$\overbrace{V_\mathrm{f}}^{t,e} + \overbrace{V_\mathrm{m}}^{t,e} = 1$	Only two phases	
Relationship for isotropic materials	$\overbrace{G}^{t,e} = \dfrac{\overbrace{E}^{t,e}}{2(1+\overbrace{\nu}^{t,e})}$	Isotropic material	
Relationship for anisotropic materials	$\dfrac{\overbrace{\upsilon_{12}}^{t,e}}{\overbrace{E_{11}}^{t,e}} = \dfrac{\overbrace{\nu_{21}}^{t,e}}{\overbrace{E_{22}}^{t,e}}$	Anisotropic in 1-2 plane	

Topic	Equation	Assumptions	Importance
ROM for axial modulus	$\overbrace{E_{11}}^{t,e} = \overbrace{E_{\mathrm{f}}}^{t,e}\overbrace{V_{\mathrm{f}}}^{t,e} + \overbrace{E_{\mathrm{m}}}^{t,e}\overbrace{V_{\mathrm{m}}}^{t,e} = \overbrace{E_{\mathrm{f}}}^{t,e}\overbrace{V_{\mathrm{f}}}^{t,e} + \overbrace{E_{\mathrm{m}}}^{t,e}(1-\overbrace{V_{\mathrm{f}}}^{t,e})$	Perfect bonding/ two-phase-composite	Estimation of the tensile modulus from individual component properties
ROM for transverse modulus	$\frac{1}{\overbrace{E_{22}}^{t,e}} = \frac{\overbrace{V_{\mathrm{m}}}^{t,e}}{\overbrace{E_{\mathrm{m}}}^{t,e}} + \frac{\overbrace{V_{\mathrm{f}}}^{t,e}}{\overbrace{E_{\mathrm{f}}}^{t,e}} = \frac{1-\overbrace{V_{\mathrm{f}}}^{t,e}}{\overbrace{E_{\mathrm{m}}}^{t,e}} + \frac{\overbrace{V_{\mathrm{f}}}^{t,e}}{\overbrace{E_{\mathrm{f}}}^{t,e}}$	Perfect bonding/ two-phase-composite	Estimation of the transverse modulus from individual component properties
Halpin-Tsai	$\overbrace{X}^{t,e} = X_{\mathrm{m}}\left[\frac{1+\xi\overbrace{\eta V_{\mathrm{f}}}^{t,e}}{1-\overbrace{\eta V_{\mathrm{f}}}^{t,e}}\right]$, $\eta = \frac{\left(\frac{\overbrace{X_{\mathrm{f}}}^{t,e}}{\overbrace{X_{\mathrm{m}}}^{t,e}}\right)-1}{\left(\frac{\overbrace{X_{\mathrm{f}}}^{t,e}}{\overbrace{X_{\mathrm{m}}}^{t,e}}\right)+\xi}$		More accurate estimation of the tensile modulus from individual component properties, considers fiber geometry
ROM for Poisson's ratio	$\overbrace{\nu_{12}}^{t,e} = \overbrace{\upsilon_{\mathrm{m}}}^{t,e}\overbrace{V_{\mathrm{m}}}^{t,e} + \sum_i \overbrace{\nu_{\mathrm{f}_i}}^{t,e}\overbrace{V_{\mathrm{f}_i}}^{t,e}$	Perfect bonding/ two-phase-composite	Estimation of the Poisson's ratio from individual component properties
ROM for in-plane shear modulus	$\frac{1}{\overbrace{G_{12}}^{t,e}} = \frac{\overbrace{V_{\mathrm{m}}}^{t,e}}{\overbrace{G_{\mathrm{m}}}^{t,e}} + \frac{\overbrace{V_{\mathrm{f}}}^{t,e}}{\overbrace{G_{\mathrm{f}}}^{t,e}} = \frac{1-\overbrace{V_{\mathrm{f}}}^{t,e}}{\overbrace{G_{\mathrm{m}}}^{t,e}} + \frac{\overbrace{V_{\mathrm{f}}}^{t,e}}{\overbrace{G_{\mathrm{f}}}^{t,e}} = \frac{2\left(1+\overbrace{\nu_{\mathrm{m}}}^{t,e}\right)\left(1-\overbrace{V_{\mathrm{f}}}^{t,e}\right)}{\overbrace{E_{\mathrm{m}}}^{t,e}} + \frac{2\left(1+\overbrace{\nu_{\mathrm{f}}}^{t,e}\right)\overbrace{V_{\mathrm{f}}}^{t,e}}{\overbrace{E_{\mathrm{f}}}^{t,e}}$	Perfect bonding/ two-phase-composite	Estimation of the shear modulus from individual component properties

(*Continued*)

Topic	Equation	Assumptions	Importance
Stress–strain relationship	$\sigma_i = C_{ij}\varepsilon_j \quad \text{where } i,\ j = 1\ldots 6$	Material linear	Stress–strain relationship for anisotropic laminate
Strain–stress relationship	$\varepsilon_i = S_{ij}\sigma_j \quad \text{where } i,\ j = 1\ldots 6$	Material linear	Strain–stress relationship for anisotropic laminate
Transformation matrix	$[T] = \begin{bmatrix} \cos^2\theta & \sin^2\theta & 2\cos\theta\sin\theta \\ \sin^2\theta & \cos^2\theta & -2\cos\theta\sin\theta \\ -\cos\theta\sin\theta & \cos\theta\sin\theta & \cos^2\theta - \sin^2\theta \end{bmatrix}$ $[R] = \begin{bmatrix} 1 & 0 & 0 \\ 0 & 1 & 0 \\ 0 & 0 & 2 \end{bmatrix}$	None	To express quantities in material and global coordinate systems
Transformed compliance and stiffness matrices	$[\overline{S}] = [T]^{-1}[S][R][T][R]^{-1}$ $[\overline{Q}] = [T]^{-1}[Q][R][T][R]^{-1}$	None	To express transformed and compliance matrices in global coordinates
Strain–stress relationship in global coordinates	$\begin{Bmatrix} \overbrace{\varepsilon_x}^{t,e} \\ \overbrace{\varepsilon_y}^{t,e} \\ \overbrace{\gamma_{xy}}^{t,e} \end{Bmatrix} = \begin{bmatrix} \overbrace{\overline{S}}^{t,e} \end{bmatrix} \begin{Bmatrix} \sigma_x \\ \sigma_y \\ \tau_{xy} \end{Bmatrix}$	Material linear	
Strain–stress relationship in global coordinates	$\begin{Bmatrix} \overbrace{\sigma_x}^{t,e} \\ \overbrace{\sigma_y}^{t,e} \\ \overbrace{\tau_{xy}}^{t,e} \end{Bmatrix} = \begin{bmatrix} \overbrace{\overline{Q}}^{t,e} \end{bmatrix} \begin{Bmatrix} \varepsilon_x \\ \varepsilon_y \\ \gamma_{xy} \end{Bmatrix}$	Material linear	
Composite elastic properties in global coordinate system	$\dfrac{1}{\overbrace{E_x}^{t,e}} = \dfrac{1}{\overbrace{E_{11}}^{t,e}}\cos^4\theta + \left(\dfrac{1}{\overbrace{G_{12}}^{t,e}} - \dfrac{2\overbrace{\upsilon_{12}}^{t,e}}{\overbrace{E_{11}}^{t,e}}\right)\sin^2\theta\cos^2\theta + \dfrac{1}{\overbrace{E_{22}}^{t,e}}\sin^4\theta$	Material linear and orthotropic	Enable the calculation of global elastic properties for orthotropic laminates

Topic	Equation	Assumptions	Importance
	$\frac{1}{\overbrace{E_y}^{t,e}} = \frac{1}{\overbrace{E_{11}}^{t,e}}\sin^4\theta + \left(\frac{1}{\overbrace{G_{12}}^{t,e}} - \frac{2\overbrace{\nu_{12}}^{t,e}}{\overbrace{E_{11}}^{t,e}}\right)\sin^2\theta\cos^2\theta + \frac{1}{\overbrace{E_{22}}^{t,e}}\cos^4\theta$		
	$\frac{1}{\overbrace{G_{xy}}^{t,e}} = 2\left(\frac{2}{\overbrace{E_{11}}^{t,e}} + \frac{2}{\overbrace{E_{22}}^{t,e}} + \frac{4\overbrace{\nu_{12}}^{t,e}}{\overbrace{E_{11}}^{t,e}} - \frac{1}{\overbrace{G_{12}}^{t,e}}\right)\sin^2\theta\cos^2\theta + \frac{1}{\overbrace{G_{12}}^{t,e}}\left(\sin^4\theta + \cos^4\theta\right)$		
	$\overbrace{\upsilon_{xy}}^{t,e} = \overbrace{E_x}^{t,e}\left[\frac{\overbrace{\upsilon_{12}}^{t,e}}{\overbrace{E_{11}}^{t,e}}\left(\sin^4\theta + \cos^4\theta\right) - \left(\frac{1}{\overbrace{E_{11}}^{t,e}} + \frac{1}{\overbrace{E_{22}}^{t,e}} - \frac{1}{\overbrace{G_{12}}^{t,e}}\right)\sin^2\theta\cos^2\theta\right]$		
	$\overbrace{\eta_{xy,x}}^{t,e} = \overbrace{E_x}^{t,e}\left[\left(\frac{2}{\overbrace{E_{11}}^{t,e}} + \frac{2\overbrace{\upsilon_{12}}^{t,e}}{\overbrace{E_{11}}^{t,e}} - \frac{1}{\overbrace{G_{12}}^{t,e}}\right)\sin\theta\cos^3\theta - \left(\frac{2}{\overbrace{E_{22}}^{t,e}} + \frac{2\overbrace{\upsilon_{12}}^{t,e}}{\overbrace{E_{11}}^{t,e}} - \frac{1}{\overbrace{G_{12}}^{t,e}}\right)\sin^3\theta\cos\theta\right]$		
	$\overbrace{\eta_{xy,y}}^{t,e} = \overbrace{E_y}^{t,e}\left[\left(\frac{2}{\overbrace{E_{11}}^{t,e}} + \frac{2\overbrace{\upsilon_{12}}^{t,e}}{\overbrace{E_{11}}^{t,e}} - \frac{1}{\overbrace{G_{12}}^{t,e}}\right)\sin^3\theta\cos\theta - \left(\frac{2}{\overbrace{E_{22}}^{t,e}} + \frac{2\overbrace{\upsilon_{12}}^{t,e}}{\overbrace{E_{11}}^{t,e}} - \frac{1}{\overbrace{G_{12}}^{t,e}}\right)\sin\theta\cos^3\theta\right]$		

(*Continued*)

Topic	Equation	Assumptions	Importance
Stress–strain relationship at ply level	$\overbrace{\{\sigma\}_k}^{t,e} = \overbrace{[\overline{Q}]_k}^{t,e} \{\varepsilon\}$	Material linear	
Strain–stress relationships including thermal and moisture stresses	$\begin{Bmatrix} \overbrace{\varepsilon_1}^{t,e} \\ \overbrace{\varepsilon_2}^{t,e} \\ \overbrace{\gamma_{12}}^{t,e} \end{Bmatrix} = \begin{bmatrix} \overbrace{S_{11}}^{t,e} & \overbrace{S_{12}}^{t,e} & 0 \\ \overbrace{S_{12}}^{t,e} & \overbrace{S_{22}}^{t,e} & 0 \\ 0 & 0 & \overbrace{S_{66}}^{t,e} \end{bmatrix} \begin{Bmatrix} \sigma_1 \\ \sigma_2 \\ \sigma_6 \end{Bmatrix} + \begin{Bmatrix} \overbrace{\alpha_{11}}^{t,e} \\ \overbrace{\alpha_{22}}^{t,e} \\ 0 \end{Bmatrix} \Delta T + \begin{Bmatrix} \overbrace{\beta_{11}}^{t,e} \\ \overbrace{\beta_{22}}^{t,e} \\ 0 \end{Bmatrix} \Delta m$	Material linear	
Stress–strain relationships including thermal and moisture stresses	$\begin{Bmatrix} \overbrace{\sigma_x}^{t,e} \\ \overbrace{\sigma_y}^{t,e} \\ \overbrace{\sigma_{xy}}^{t,e} \end{Bmatrix} = \begin{bmatrix} \overbrace{\overline{Q}_{11}}^{t,e} & \overbrace{\overline{Q}_{12}}^{t,e} & \overbrace{\overline{Q}_{16}}^{t,e} \\ \overbrace{\overline{Q}_{12}}^{t,e} & \overbrace{\overline{Q}_{22}}^{t,e} & \overbrace{\overline{Q}_{26}}^{t,e} \\ \overbrace{\overline{Q}_{16}}^{t,e} & \overbrace{\overline{Q}_{26}}^{t,e} & \overbrace{\overline{Q}_{66}}^{t,e} \end{bmatrix} \left(\begin{Bmatrix} \varepsilon_x \\ \varepsilon_y \\ \gamma_{xy} \end{Bmatrix} - \begin{Bmatrix} \overbrace{\alpha_x}^{t,e} \\ \overbrace{\alpha_y}^{t,e} \\ \overbrace{\alpha_{xy}}^{t,e} \end{Bmatrix} \Delta T - \begin{Bmatrix} \overbrace{\beta_x}^{t,e} \\ \overbrace{\beta_y}^{t,e} \\ \overbrace{\beta_{xy}}^{t,e} \end{Bmatrix} \Delta m \right)$	Material linear	
Von-Mises failure criterion	$(\sigma_1 - \sigma_2)^2 + (\sigma_2 - \sigma_3)^2 + (\sigma_3 - \sigma_1)^2 = 2\sigma_y^2$	Isotropic	
Maximum stress failure criterion	$\sigma_i = X_i$ where $i = 1 \ldots 6$	None	Defines failure mode and enables ply property reduction
Maximum strain failure criterion	$\varepsilon_i = \varepsilon_{if}$ where $i = 1 \ldots 6$	None	Defines failure mode and enables ply property reduction

Topic	Equation	Assumptions	Importance
Tsai-Hill failure criterion	$\frac{\left(\overbrace{\sigma_1}^{t,e}\right)^2}{\overbrace{X_1}^{t,e}{}^2} - \frac{\left(\overbrace{\sigma_1}^{t,e}\overbrace{\sigma_2}^{t,e}\right)}{\overbrace{X_1}^{t,e}{}^2} + \frac{\left(\overbrace{\sigma_2}^{t,e}\right)^2}{\overbrace{X_2}^{t,e}{}^2} + \frac{\left(\overbrace{\sigma_6}^{t,e}\right)^2}{\overbrace{X_6}^{t,e}{}^2} - 1 \geq 0$	None	Considers stresses interactions
Tsai-Wu failure criterion	$f_i\sigma_i + f_{ij}\sigma_i\sigma_j = 1 \quad i,\ j = 1\ldots6$	None	Considers stresses interactions and differentiates tension and compression

REFERENCES

1. Brosius, D., Corvette gets leaner with carbon fiber. *High Performance Composites*, March 2004, 33–37.
2. Reifsnider, K.L. and S.W. Case, *Damage Tolerance and Durability of Materials Systems*. John Wiley & Sons, Inc., New York, 2002.
3. Jones, R.M., *Mechanics of Composite Materials*. Taylor & Francis, 1975.
4. Barbero, E.J., *Introduction to Composite Materials Design*. Taylor & Francis, 1999.
5. Reifsnider, K.L., Private communication.
6. Novillo, F.A., M. Fujita, M. Tsuji and S. Kohjiya, Preferential Orientation of Poly(ethylene 2,6-naphthalate) Melt-crystallized on the Friction-transfer Layer of PTFE. *Sen'i Gakkaishi*, 1998, 54, 544–549.
7. Göschel, U., F.H.M. Swartjes, G.W.M. Peters and H.E.H. Meijer, Crystallization in isotactic polypropylene melts during contraction flow: Time-resolved synchrotron WAXD studies. *Polymer*, 2000, 41, 1541–1550.
8. Chamis, C.C. and G.P. Sendeckyij, Critique on theories predicting thermoelastic properties of fibrous composites. *Journal of Composite Materials*, July 1968, 332–358.
9. Halpin, J. and S.W. Tsai, Effects of Environmental Factors on Composite Materials. Air Force Materials Lab – Technical Report AFML-TR 67-423, Department of Defense, USA, 1969.
10. Mahieux, C.A., K.L. Reifsnider and S.W. Case, Property modeling across transition temperatures in PMC's: Part I Tensile properties, *Applied Composite Materials*, July 2001, 8(4), 217–234.
11. Matthews, F.L. and R.D. Rawlings, *Composite Materials: Engineering and Science*. Chapman & Hall, 1994.
12. www.matweb.com.
13. Hashin, Z. and B.W. Rosen, The elastic moduli of fiber-reinforced materials. *Journal of Applied Mechanics*, June 1964, 223–230.

14. Luciano, R. and E.J. Barbero, Formulas for the stiffness of composites with periodic microstructure. *International Journal of Solids and Structures*, 1994, 31, 2933–2944.
15. Tsai, S.W. and H.T. Hahn, *Introduction to Composite Materials*. Technomic, Lancaster, PA, 1980.
16. Bartdorf, S.B., Tensile strength of unidirectionally reinforced composites – I. *Journal of Reinforced Plastic Composites*, 1982, 1, 153–164.
17. Batdorf, S.B., Tensile strength of unidirectionally reinforced composites – II. *Journal of Reinforced Plastic Composites*, 1982, 1, 165–176.
18. Nielsen, L.E., Mechanical properties of particulate-filled systems. *Journal of Composite Materials*, 1967, 1, 100–119.
19. Agarwal, B.D. and L.J. Broutman, *Analysis and Performance of Fiber Composites*, 2nd ed. John Wiley & Sons, New York, 1990.
20. Chamis, C.C., Simplified composite micromechanics equations for hydral, thermal, and mechanical properties. *SAMPE Quarterly*, April 1984, 14–23.
21. Rosen, B.W., Tensile failure of fibrous composites. *American Institute of Aeronautics and Astronautics Student Journal*, 1964, 2, 1985–1991.
22. Budiansky, B. and N.A. Fleck, Compressive kinking of fiber composites: A topical review. *Applied Mechanics Reviews*, 1994, 47(6), 246–250.
23. Xu, Y.L. and K.L. Reifsnider, Micromechanical modeling of composites compressive strength. *Journal of Composite Materials*, 1993, 27(6), 572–587.
24. Schapery, R.A., Thermal expansion coefficients of composite materials based on energy principles. *Journal of Composite Materials*, 1968, 2, 380–404.
25. Tsai, S.W. and H.T. Hahn, *Introduction to Composite Materials*. Technomic Lancaster, PA 1980.
26. Green, A.E. and W. Zerna, *Theoretical Elasticity*. Oxford: Clarendon, 1960, 4.1.
27. Luciano, R. and E.J. Barbero, Analytical expressions for the relaxation moduli of linear viscoelastic composites with periodic microstructure. *ASME Journal of Applied Mechanics*, 1995, 62(3), 786–793.
28. Barbero, E.J. and R. Luciano, Micromechanical formulas for the relaxation tensor of linear viscoelastic composites with transversely isotropic fibers. *International Journal of Solids and Structures*, 1995, 32(13), 1859–1872.
29. Barbero, E.J., Private communication.
30. Reuter, Robert M., Jr., Concise property transformation relations for an anisotropic lamina. *Journal of Composite Materials*, April 1971, 2, 270–272.
31. Büter, A. and E. Bretibach, Adaptive Blade Twist – calculations and experimental results. *Aerospace Science and Technology*, July 2000, 4(5), 309–319.
32. Apache composite blades take off. *Reinforced Plastics*, February 2004, 48(2), 6.
33. Li, J., R. Shen, H. Hua and X. Jin, Bending-torsional coupled dynamic response of axially loaded composite Tomishenko thick-walled beam with closed cross-section. *Composite Structures*, April 2004, 64(1), 23–35.
34. Piatak, D.J., M.W. Nixon and J.B. Kosmatka, Stiffness characteristics of composite rotor blades with elastic couplings. NASA Technical Report 1279, April 1997.
35. Fisher Mason, K., Composite anisotropy lowers wind-energy costs. *High Performance Composites*, November 2004, 44–46.
36. Reddy, J.N., *Mechanics of Laminated Composite Plates and Shells: Theory and Analysis*. 2nd ed. CRC Press, 2003.
37. J. Ye, *Laminated Composite Plates and Shells: 3D Modelling*, Springer-Verlag, 2003.

38. Vinson, J.R., *The Behavior of Shells Composed of Isotropic and Composite Materials* (Solid Mechanics and Its Applications), Kluwer Academic Publishers, 1993.
39. Young, W.C., *Roark's Formulas for Stress and Strains*. McGraw-Hill, New York, 1989.
40. Newberry, A.L., Advanced software for stress analysis of composite pipe systems. *Reinforced Plastics*, October 2002, 46–48.
41. Christensen, R.M., *Theory of Viscoelasticity*, 2nd ed. Dover, 1982.
42. Stafford, R.O., On mathematical forms for the material functions in nonlinear viscoelasticity. *Journal of Mechanics and Physics of Solids*, 1969, 17, 339.
43. Lockett, F.J., *Nonlinear Viscoelastic Solids*. Academic Press, New York, 1972.
44. Green, A.E. and R.S. Rivlin, The mechanics of non-linear materials with memory. *Archive for Rational Mechanics and Analysis*, 1975, 1, 1.
45. Zhang, Y., Z. Xia and F. Ellyin, Nonlinear viscoelastic micromechanical analysis of fibre-reinforced polymer laminates with damage evolution. *International Journal of Solids and Structures*, January 2005, 42(2), 591–604.
46. Leigh Pheonix, S., Modeling the statistical lifetime of glass fiber/polymer matrix composites in tension. *Composite Structures*, January–March 2000, 48(1–3), 19–29.
47. Belytschko, T., W. Kam Liu and B. Moran, *Nonlinear Finite Elements for Continua and Structures*. John Wiley & Sons, 2000.
48. Reddy, J.N., *An Introduction to Nonlinear Finite Element Analysis*. Oxford University Press, 2004.
49. Soden, P.D., A.S. Kaddour and M.J. Hinton, Recommendations for designers and researchers resulting from the world-wide failure exercise. *Composites Science and Technology*, March 2004.
50. Special issue of *Composites Science and Technology*, 1998, 58(7).
51. Special issue of *Composites Science and Technology*, 2002, 62(12/13).
52. Special issue of *Composites Science and Technology*, 2004, 64(3).
53. Kaddour, A.S., M.J. Hinton and P.D. Sodden, A comparison of the predictive capabilities of current failure theories for composite laminates: Additional contributions, *Composites Science and Technology*, 2004, 64, 449–476.
54. Gotsis, P.K., C.C. Chamis and L. Minnetyan, Prediction of composite laminate fracture: Micromechanics and progressive fracture. *Composites Science and Technology*, 1998, 58, 1137–1150.
55. Hart-Smith, L.J., Predictions of a generalized maximum-shear-stress failure criterion for certain fibrous composite laminates. *Composites Science and Technology*, 1998, 58, 1179–1208.
56. Hart-Smith, L.J., Predictions of the original and truncated maximum strain failure models for certain fibrous composite laminates. *Composites Science and Technology*, 1998, 58, 1151–1178.
57. Eckold, G.C., Failure criteria for use in the design environment. *Composites Science and Technology*, 1998, 58, 1095–1106.
58. Edge, E.C., Stress based Grant-Sanders method for predicting failure of composite laminates. *Composites Science and Technology*, 1998, 1043–1044.
59. McCartney, L.N., Predicting transverse crack formation in cross-ply laminate. *Composites Science and Technology*, 1998, 58, 1069–1082.
60. Puck, A. and H. Schürmann, Failure analysis of FRP laminates by means of physically based phenomenological models. *Composites Science and Technology*, 1998, 58, 1045–1068.

61. Wolfe, W.E. and T.S. Butalia, A strain energy based failure criterion for nonlinear analysis of composite laminates subjected to biaxial loacing. *Composites Science and Technology*, 1998, 58, 1107–1124.
62. Sun, C.T. and J.X. Tao, Prediction of failure envelopes and stress strain behaviours of composite laminates. *Composites Science and Technology*, 1998, 58, 1125–1136.
63. Zinoviev, P., S.V. Grigoriev, O.V. Labedeva and L.R. Tairova, Strength of multilayered composites under plane stress state. *Composites Science and Technology*, 1998, 58, 1209–1223.
64. Liu, K.-S. and S.W. Tsai, A progressive quadratic failure criterion of a laminate. *Composites Science and Technology*, 1998, 58, 1023–1032.
65. Rotem, A., Prediction of laminate failure with Rotem failure criterion. *Composites Science and Technology*, 1998, 58, 1083–1094.
66. Cuntze, R.G. and A. Freund, The predictive capability of failure mode concept based strength criteria for multidirectional laminates. *Composites Science and Technology*, 2004, 64, 343–377.
67. Bogetti, T.A., C.P.R. Hoppel, V.M. Harik, J.F. Newill and B.P. Burns, Predicting the nonlinear response and progressive failure of composite laminates. *Composites Science and Technology*, 2004, 64, 329–342.
68. Mayes, S. and A.C. Hansen, Composite laminate failure analysis using multicontinuum theory. *Composites Science and Technology*, 2004, 64, 379–394.
69. Huang, Z.M., A bridging model prediction of the tensile strength of composite laminates subjected to biaxial load. *Composites Science and Technology*, 2004, 64, 395–448.
70. Hart-Smith, L.J., Expanding the capabilities of the ten-percent rule for predicting the strength of fibre-polymer composites. *Composites Science and Technology*, 2002, 62, 1115–1144.
71. Soden, P.D., A.S. Kaddour and M.J. Hinton, Recommendations for designers and researchers resulting from the world-wide failure exercise. *Composites Science and Technology*, 2004, 64, 589–604.
72. Reddy, J.N., *Mechanics of Laminated Composite Plates-Theory and Analysis*. CRC Press, Boca Raton, FL, 1997.
73. Troitsky, M.S., *Stiffened Plates-Bending, Stability, and Vibrations*. Elsevier, New York, 1976.
74. Delsart, L. and Y. Levenez, A new generation of design software for composites. *JEC-Composites*, n8, April 2004.
75. http://www-306.ibm.com/software/applications/plm/catiav5/prods/cpd/.
76. Schweitzer, B., *JEC Composites*, n6, January 2004.
77. Analysis software can predict mechanical behavior. *Reinforced Plastics*, September 2004, 20.
78. Berenberg, B., Virtual testing points way to improved designs. *High Performance Composites*, July 2004, 32.
79. Adams, D.F., L.A. Carlsson and R. Byron Pipes, *Experimental Characterization of Advanced Composite Materials*, 3rd ed. CRC Press, Boca Raton, 2003.
80. Hogg, P.J., K. Schulte and H. Withich, Composites Testing and Standardization. ECCM-Cts2, Technomic Publishing Company, November 1, 1994.
81. Adams, D.F., Test methods for composite materials: Seminar notes, Lancaster, PA, Technomic Publishing, 1996.
82. ASTM D4255/D4255M.
83. Iosipescu, N., New accurate procedure for single shear testing of metals. *Journal of Materials*, 1967, 2(3), 537–566.

6

CYCLING MECHANICAL AND ENVIRONMENTAL LOADS

6.1 INTRODUCTION

Out of the laboratory, composites are rarely submitted to static constant loads. Indeed, the environment generally imposes cycling conditions on the parts. Damage and failure under cycling and static loads can differ drastically. Therefore, the important concept of composite fatigue is defined and discussed in Section 6.2. Fatigue results on composite materials are extremely difficult to generalize and many books already focus on this topic [1–5]. We will therefore restrict our discussion to general trends and try to underline common pitfalls.

Composite products are also submitted to a combination of varying loads. To complicate matters further, the different loads generally interact. The case study composite bridges illustrates the complexity of cycling mechanical and environmental conditions for composites used in outdoor environments and underlines the need for comprehensive lifetime approaches.

Global methodologies for load combination and loading blocks exist and are detailed in Section 6.3. Such methodologies also require experimental validation. Unfortunately, finite budgets rarely allow for testing of all possible load combinations and it is often necessary to reduce the number of experiments. Design of Experiments (DOE) methods, last section of the present text, can assist the scientist in performing this task.

Case study: Composite bridges

The use of polymer composites for civil infrastructure has been experiencing an accelerated growth over the past 10 years. The introduction of polymer composites has been driven by the need for improved structural and environmental stability (such as corrosion resistance) and the potential weight savings leading to faster installation. In the early 1990s, the number of bridges containing polymer-composite parts was around a dozen. This number rose to around 175 vehicular bridges in 2003 and 160 pedestrian bridges [6]. Most bridges are mixing the use of composite, concrete and steel – only few bridges are all composite. Glass fiber vinyl ester composites are widely used in such applications, thanks to corrosion resistance, good stiffness to strength ratio and fire resistance. The large number of design alternatives offered by composites and the use of different manufacturing processes complicate the performance comparison between the different composite candidates, and tends to hinder a more rapid development of composites for civil infrastructure applications.

Composites can be used for deck panels, structural beams or smaller components such as bridge enclosure systems. This later application is not as common as deck or beams, but is nevertheless interesting. For example, the use of a glass reinforced plastic enclosure system manufactured by Fibreforce Composites Ltd for the refurbishment of the bridge connecting Dublin and Belfast proved a reduced corrosion rate of untreated steel in the enclosure to under 0.02 mm per year (or 2 mm over the 100-year bridge's expected lifetime, Figure 6.1) [7].

Decks are usually made of trapezoidal or sinusoidal profiles surrounded by an outer skin panel. Honeycomb structures and chopped strand mats can also be found as sandwich core component of the deck. But the details of the deck can strongly vary from one supplier to another. Variations possibilities are illustrated by Ohio's Salem Ave bridge where four decks from four different manufacturers (Hardcore Composites, Creative Pultrusions Inc., ICI, Composite Deck Solutions LLC) were installed on the same bridge coupled by elastomer joints. The Creative Pultrusion

Figure 6.1. Enclosure curved panels installation. (Courtesy of Fibreforce.)

Figure 6.2. Creative Pultrusion composite deck panels. (Courtesy Creative Pultrusion.)

Figure 6.3. Creative Pultrusion deck on Salem Ave bridge (Ohio). (Courtesy Creative Pultrusion.)

deck during transportation and installation is shown in Figures 6.2 and 6.3. The four-deck solutions varied in design choice (profiles), materials (from all polymer composite to hybrid polymer/concrete) and manufacturing methods (from pultrusion to vacuum infusion process) [6].

The project revealed major challenges in matching steel support girders and composite deck coefficients of thermal expansions as well as avoiding overlay cracks and debonding. Asphalt overlay problems are recurrent for composite bridge technology and finding the proper thickness is still greatly empirical. In the case of the Salem Ave bridge, those challenges led to the removal of two of the composite decks.

The use of polymer composites for bridges is a rather recent development still needing validation, especially in terms of lifetime prediction accuracy. Indeed, mistakes in this area could lead to extremely large number of fatalities. The bridges designs are usually *stiffness controlled* and life predictions focus essentially on changes in performance. Experience has shown that quasi-static deformations were usually well predicted from finite element analysis based on material properties obtained from laboratory experiments. On the other hand, time-dependent degradation was more difficult to anticipate showing the need for a deeper material understanding.

The problem is complex. Bridges are exposed to a combination of environmental loads including UV exposure, rain, heat and cold temperatures, vibrations, impact and wear. For example, Portland's Broadway, all composite tilting bridge (Oregon), is likely to be one of the largest and most frequently traveled composite bridge deck in the world. The fiber reinforced plastic bridge deck developed by Martin Marietta, nominally 12.7 cm deep, weighing approximately 73 kg/m^2 and primarily consisting of continuous glass fibers in a polyester resin containing a UV inhibitor (Figures 6.4 and 6.5) experiences an average of 30 000 vehicles per day [8] inducing significant vibrations and wear to the deck.

In order to evaluate short- and long-term load and environmental effects, most bridges undergo two series of investigation. Indeed, short-term monitoring is usually performed during commissioning, when a calculated number of large trucks are circulated and parked on the bridge. Strain gages are commonly used on decks and beams to measure deformations. For example, commissioning of the New York State Route 248 Bennetts Creek crossing involved the loading of the bridge with four fully loaded ten-wheel dump trucks. The maximum recorded strain was 5.2 μm and the maximum deflection at mid-span was less than 3.5 mm (against 8.8 mm allowable) [6].

Long-term monitoring is additionally performed to ensure the control of the integrity of the structure. Cracks in the pavement coating are natural indicators of the deck panel motion. Such damage can result from mismatches in thermal expansions as well as excessive strains. In the case of hybrid materials combination (e.g. steel-composite), excessive strains can also result from loosening of mechanical fastening systems.

Long-term monitoring methods vary for the different projects but generally rely upon strain gages, thermistors and optical sensors (Bragg grating fiber optic or sapphire wire chemical fiber sensors) to register changes and potential damage in the structure. Continuous monitoring of the New York State bridge, for example, showed stable maximum strain data over time. Unfortunately, this bridge alike many showed extensive wear of the polymer concrete coating.

Figure 6.4. Most traveled composite deck bridge (Broadway Bridge, Portland, Oregon). (Courtesy of Martin Marietta.)

6.2 ENVIRONMENTAL AND MECHANICAL CYCLING VERSUS STATIC LOADING

6.2.1 Definitions

Repeated application of a load or strain on a composite can lead to failure even for an applied load much below the static failure limit of the material. Repeated loading is generally referred to as *Fatigue* loading. The loads can be mechanical or environmental (such as applying and removing a thermal load onto the material).

The term *Fatigue* is rather generic and can be ambiguous. Indeed, it is sometimes difficult to distinguish between mechanical fatigue and vibrations. The term *fatigue* can even be extended to a part undergoing constant load conditions over long

Figure 6.5. Installed composite decks. (Broadway Bridge, Portland, Oregon). (Courtesy of Martin Marietta.)

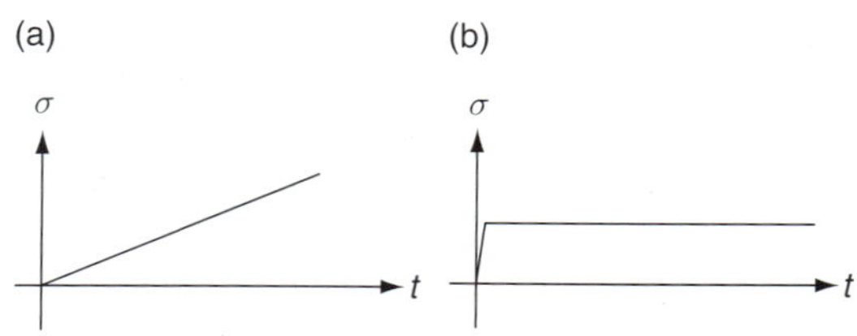

Figure 6.6. (a) *Quasi-static* loading. (b) *Static* loading.

periods of time (static fatigue). Considering such ambiguities, we propose to set some definitions and notations that will be used in the present chapter.

The failure of a material under an increasing load with a *fast* loading rate (failure within a few hours) is referred to as *quasi-static* loading. This is, for example, a tensile test or a temperature ramping until failure (Figure 6.6(a)).

On the other hand, long-term degradation of a composite under mechanical or environmental loads will be called *static* loading (Figure 6.6(b)). The resulting failure will be referred to as *stress-rupture*.

Mechanical and environmental *fatigue* will refer to the materials behavior under cyclic application of mechanical and environmental loads respectively. Vibrations

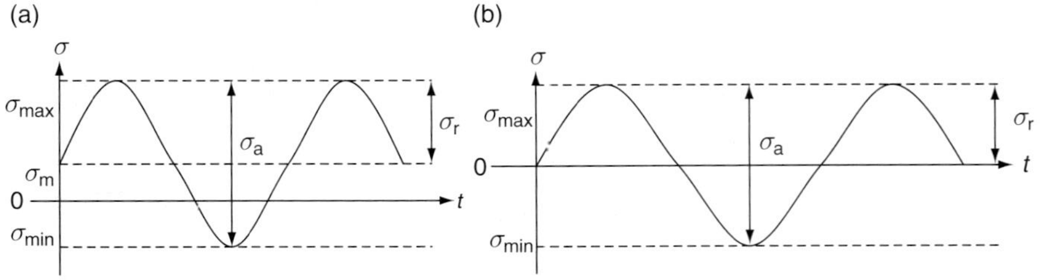

Figure 6.7. (a) Repeated stress cycles. (b) Reversed stress cycles.

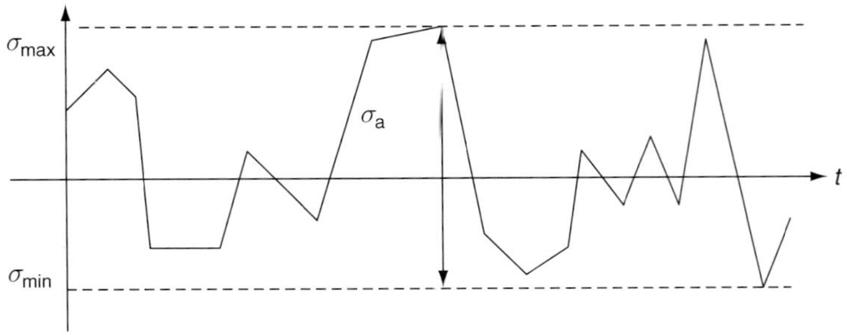

Figure 6.8. Random cycling.

are usually a rapid cycling, with low excitation amplitude. According to our definition, vibrations are therefore a special case of *fatigue*. In-depth analysis of vibrations in composites is beyond the scope of this book. The interested reader can revert to the literature [9–11].

Three examples of loading are given in Figures 6.7 and 6.8. Random loading is often representing the load cases of real composite parts. For accelerated testing purposes, reversed stress cycles (Figure 6.7(b)) and repeated stress cycles (Figure 6.7(a)) are usually used. In such tests, it is important to ensure that the experimental loading rates correspond to the real loading case and do not induce (or hide) degradation seen in operation.

Following the model of metallic materials, we define four characteristic quantities:

(1) The load (here stress) amplitude (σ_a):

$$\sigma_a = \frac{\sigma_{\max} - \sigma_{\min}}{2} \tag{6.1}$$

(2) The load range (σ_r):

$$\sigma_r = 2\sigma_a \tag{6.2}$$

(3) The mean stress (σ_m, mainly useful for periodic cycling):

$$\sigma_m = \frac{\sigma_{max} + \sigma_{min}}{2} \tag{6.3}$$

(4) Most importantly the load ratio, R:

$$R = \frac{\sigma_{min}}{\sigma_{max}} \tag{6.4}$$

By conventions, tensile mechanical loads will be positive.

Most fatigue notations and equations were initially developed for mechanical loads. However, they can be applied to any (environmental) load. For example, the relative humidity varies significantly over the year. Figure 6.9 shows the RH variations in Switzerland. Such variations can seriously impact composites used in an outdoor environment (Chapter 3). The quantities defined in Equations (6.1)–(6.4) can be calculated for the RH. Indeed, based on Chapter 3, we can anticipate that the presence of humidity will trigger changes in the material state and that moisture level cycling (coupled with temperature) will most likely result in fatigue degradation of the composite. For linear behavior and by analogy with mechanical fatigue, the stress amplitude, the load range and mean stress become the RH amplitude, the RH range and the mean RH. An analog R ratio is the ratio of the minimum RH to the maximum RH. A more general approach, applicable to non-linear materials responses is to calculate the stresses induced by the humidity variations (Chapters 3 and 5) and use Equations (6.1–6.4) *as is*.

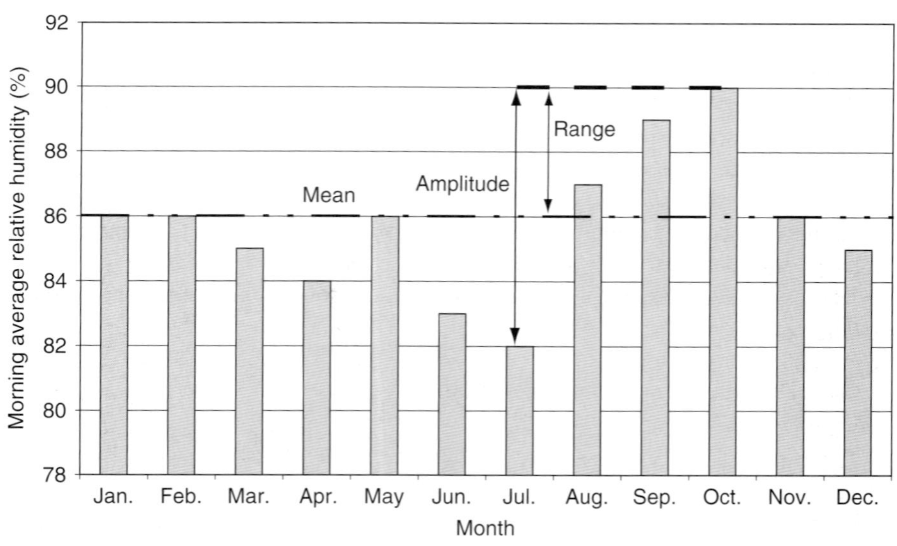

Figure 6.9. Relative humidity in Switzerland (morning data) over the year. (Data from the Washington Post [12].)

6.2.2 Mechanical Fatigue in Composite Materials

6.2.2.1 *Statistical nature of polymer matrix composite failure under cycling loads*

The results of mechanical fatigue are usually presented as an S–N curve (Wohler curve): S representing the stress amplitude and N the number of cycles to failure. Note that this convention is odd, as the parameter being varied (the stress level) appears on the x-axis.

The cycles are usually plotted on a logarithmic scale. Many materials show a fatigue limit defined by a given stress amplitude under which the sample does not fail anymore.

One common means to describe the fatigue failure of homogeneous materials containing flaws is fracture mechanics. In this approach, brittle failure is described using the crack growth rate. The crack growth rate in the region of stable crack growth often follows the form:

$$\frac{\mathrm{d}a}{\mathrm{d}N} = A(\Delta K)^m \tag{6.5}$$

where A and m are constants for a given material. Integration of this equation leads to a useful expression of the number of cycles to failure (N_f) [13]:

$$N_\mathrm{f} = \frac{1}{A\pi^{m/2}(\Delta\sigma)^m} \int_{a_0}^{a_\mathrm{c}} \frac{\mathrm{d}a}{Y^m a^{m/2}} \tag{6.6}$$

where Y depends on crack and specimen geometry and may be determined using stress analysis tool, a_0 being the original crack length and a_c the critical crack length.

Equations (6.5) and (6.6) are only valid for elastic and brittle materials. These equations are in most cases, therefore, not exactly applicable to polymer composites.

Damage accumulation in composites under fatigue is a complex process. Failure, especially at low loads, is rarely the result of the initiation and the propagation of one single crack. This is due to the diverse nature of composite materials and degradation processes in composite materials (fiber failure, microbuckling, matrix cracking, interface degradation and debonding, delamination between plies). These defects will not only accumulate but also often interact during cycling.

The number of cycles to failure under pure mechanical load depends upon many parameters, such as the nature of the composite constituents, the fiber length, the volume fraction and fiber orientations, the lamination sequence, the residual stresses, the environment, the presence of original defects, voids, notches etc.

Therefore, results and models are difficult to generalize. In the previous chapters (Chapters 2–5), many of the effects of the environment on the composite could be extrapolated from the weighted combination of the effects of the environment on the constituents and the interface. For mechanical fatigue, however, we quickly reach the limits of this approach. The role of the interface in fatigue becomes dominant, modifying the stress transfer between fibers and matrix. The role of the interface is further reinforced as it often acts as a damage initiation site.

Statistics play a preponderant role in describing the fatigue failure of polymer matrix composites. Local defects statistically distributed in the material will act as initiation sites. When damage grows locally, stresses are redistributed. Due to the inhomogeneous and viscoelastic characteristics of the composite, this new state of stress is difficult to simulate and anticipate. In 90° plies, matrix cracks might propagate and lead to a stiffness reduction. This damage might not alter the overall strength of the composite in the axial direction; however, this progressive degradation gives an early warning for failure and increases the predictability (in the sense that changes in the material state can be detected by appropriate sensors). In 0° plies, failure might occur without early warning (called *catastrophic failure* or *sudden death*). This process depends upon the statistical strength distribution of the fibers in the composite. A Weibull distribution describes accurately the probability of survival of fibers in a bundle (neglecting the matrix). Indeed, the stress carried by a bundle of fibers can be written as [14]:

$$\sigma = E_{\mathrm{f}}\varepsilon \exp\left[-\frac{l}{l_0}\left(\frac{\varepsilon}{\varepsilon_0}\right)^m\right] = E_{\mathrm{f}}\varepsilon R(\varepsilon) \tag{6.7}$$

where E_{f} is the fiber's modulus, ε is the strain, l is the fiber length and ε_0 is the average strain to failure of a fiber of length l_0. R is conventionally called the *reliability*. Using Reifsnider and Case's words, σ can be viewed as the *stress carried by an individual unbroken fiber multiplied by the fraction of unbroken fibers* [14]. Reifsnider and Case propose to describe the distribution of the probability of survival of different laminates using Weibull distributions (Figure 6.10).

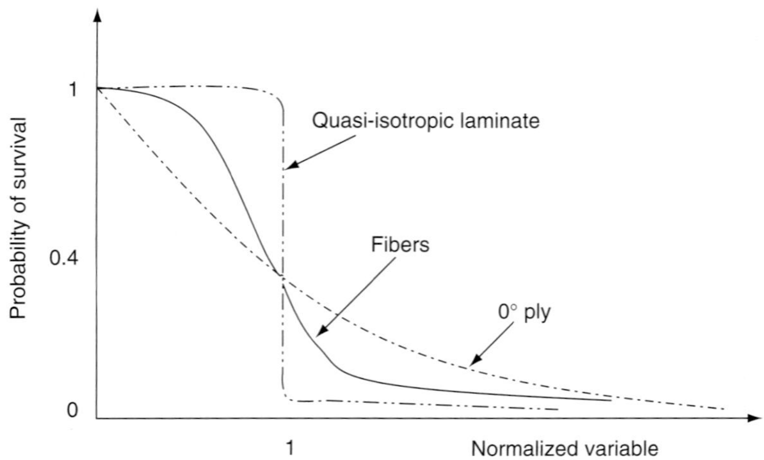

Figure 6.10. Weibull survival distribution.

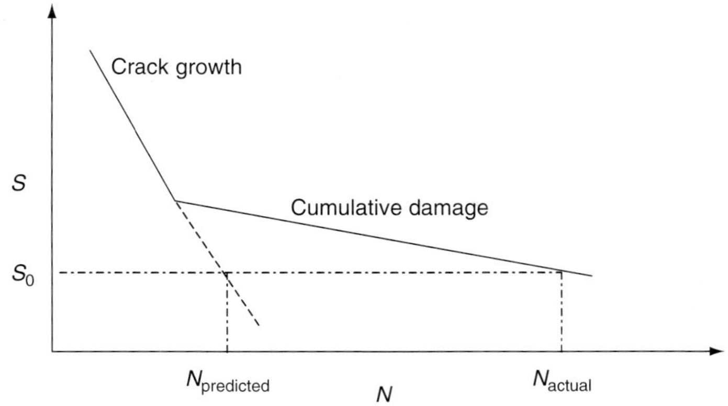

Figure 6.11. Example of dual damage mode in S–N curve.

The quasi-isotropic laminate has the smallest statistical spread when the 0° ply shows the greatest distribution width. It is worth noting that rupture of one fiber can be followed by the almost simultaneous failure of the remaining fibers, especially in the case of small bundles.

Generally speaking, cyclic composite failure results from cumulative damage or crack growth. These mechanisms are translated by different slopes of the S–N curve [14]. Most composites, however, are only dominated by one mechanism (unique slope).

Imagine the consequences of only detecting the cumulative damage portion of the S–N curve. This would lead to the prediction of lives at high stress levels much longer than the actual lives. On the other hand, detecting only crack growth would lead to very costly over designs (Figure 6.11).

The consequences are clear. Fatigue life can only be assessed by a sufficiently large experimental set. A common rule of thumb recommends the testing of 20 samples for a given condition. A more solid recommendation would be to use DOE (see Section 6.4) to ensure the confidence in the results.

6.2.2.2 Factors influencing the fatigue life

Beyond statistical considerations, it is interesting to review the many factors affecting the damage accumulation and the lifetime of polymer matrix composites under cyclic mechanical loads.

6.2.2.2.1 Constituents

(a) Fibers: The damage mechanisms depend of course on the type of fibers used. In most industrial applications, the cyclic load is well below the static strength of the composite. For unidirectional composites containing high volume fractions of brittle fibers (such as glass, boron or carbon), the behavior of the composite is said to be *fiber-dominated*. However, during cycling, the matrix will still undergo changes under the effect of the cyclic load. Molecular rearrangement of parts of the polymer chains (Chapter 2–4) leads to fairly uneven stress distributions within the material. This stress distribution combined with the statistical presence of defects in the fibers can lead to failure even at loads well under the static strength of the composite.

Kerr and Haskins [15] studied the effect of fatigue on different composites. Figure 6.12 shows the S–N curves for two materials with different fiber reinforcement (boron or graphite) tested in exactly similar conditions (room temperature, $R = 0.1$) with identical lay-up sequence. The differences in the S–N curves are striking and can originate only from the nature of the fibers. The differences in the fibers stiffness within the ±45° layers may induce different stress distributions which will in turn affect the damage mechanisms such as the crack development.

(b) Matrix: The role of the matrix in fatigue is of increasing importance in the transverse direction as well as in the case of short fiber composites. The fatigue resistance of a polymer can be improved by many ways [16] such as increasing the molecular weight and narrowing down the molecular weight distribution, avoiding chemical changes during cycling, favoring energy

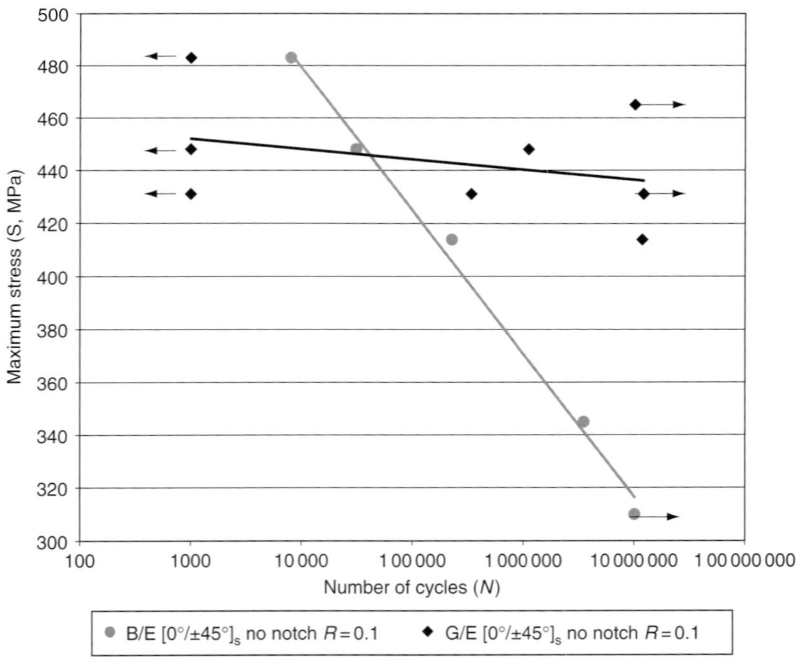

Figure 6.12. Effect of the fiber type on the S–N curves. Arrows to the left indicate lifetime shorter than 1000 cycles. Arrows to the left indicate run-outs (experiments were interrupted prior to sample failure). (Data from Kerr and Haskins [15].)

absorption via elastic/inelastic/viscoelastic deformations and morphological changes or avoiding cycling stresses operations close to a transition temperature.

(c) Constituent combination and interfaces: Once combined, the constituents can have very different behaviors. Studies on the influence of the fiber volume fraction on the fatigue life of composite materials are rather inconclusive [3]. The lifetime is influenced by conflicting mechanisms: some slowing down, others accelerating the degradation in the material. For example, the lifetime of composites is generally increased by enhancing the bonding strength between fibers and matrix. Too strong bonds can however be detrimental, as the induced brittleness can negatively impact the materials lifetime. Environmental exposure such as moisture absorption (Chapter 3) that contributes to bonding degradation can therefore significantly affect the fatigue life of polymer matrix.

6.2.2.2.2 The composite lay-up and reinforcement geometry

Long unidirectional fiber composites under axial cyclic loads generally yield longer lifetimes than off-axis samples. Kerr and Haskin's data (Figure 6.13 [15]) confirms the superior lifetime of unidirectional composites compared with $[0°/\pm45°]_s$ samples. Failure of unidirectional composites is influenced by the reinforcement

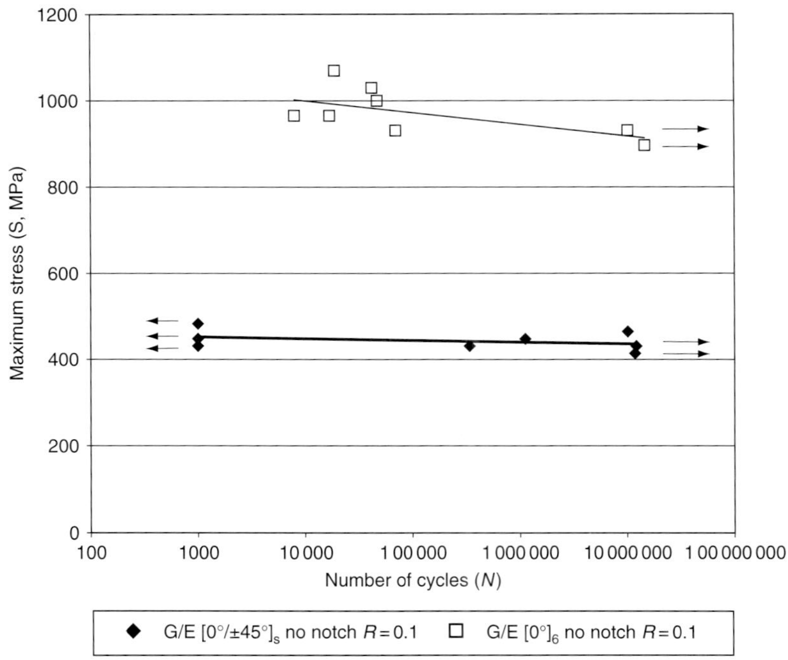

Figure 6.13. Effect of lay-up configuration on S–N curve. (Data from Kerr and Haskins [15].)

content. Indeed, high volume fractions of very stiff fibers often lead to failure by longitudinal splitting (catastrophic failure).

Short fiber composites are generally dominated by the mechanical response of the matrix and typically yield shorter lifetimes. Their behavior in fatigue is closer to homogeneous materials with the preponderance of localized failure. Unfortunately, no general rule permits to anticipate composites response as a function of reinforcement content. Indeed, the lifetime of the composite might increase or decrease with increasing fiber content, some polymers showing shorter lifetime when reinforced [3].

6.2.2.2.3 The loading conditions

Industrial composites are rarely submitted to a single stress direction. Flexural, axial, transverse and compressive forces can act simultaneously or sequentially. Such combinations are complex. For example, compressive stresses can help *close* matrix cracks but can also create other damage in the material such as buckling (see Section 6.2.3). An in-depth discussion of the failure criteria that consider multiaxial stresses can be found in the literature [17].

Strain rate and number of load reversals are prime factors for the failure of polymer matrix composite and can greatly influence damage accumulation and failure modes. This point is best illustrated by an example. The data from Kerr and Haskins [15] drawn in Figure 6.14 show a same material tested at two load

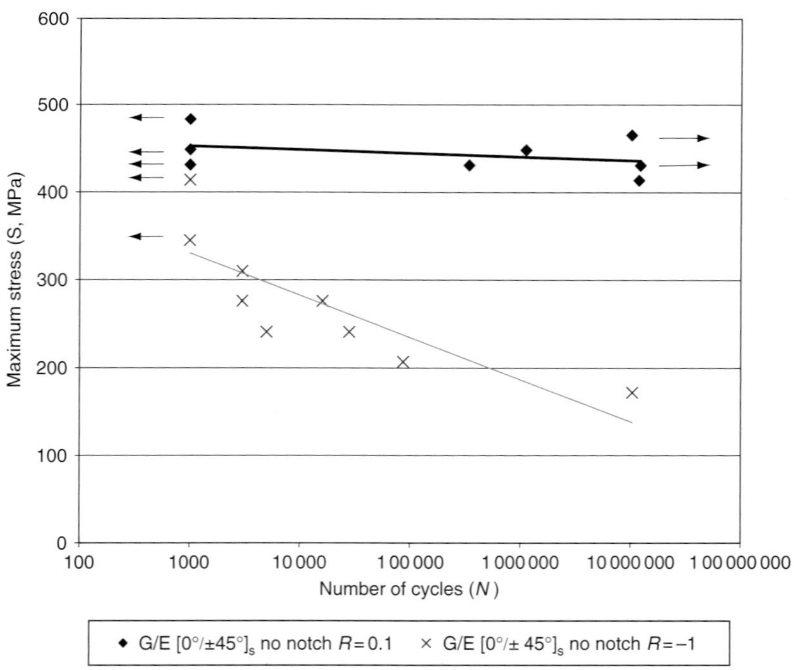

Figure 6.14. Effect of *R* ratio on the S–N curves. (Data from Kerr and Haskins [15].)

ratio levels (0.1 and −1). The full load reversal was found very damaging for the material. Indeed, the composite lifetime for a given load level is much smaller at $R = -1$ than at $R = 0.1$. Furthermore, the slopes of the curves are very different illustrating an accelerated damage process for $R = -1$.

6.2.2.2.4 The environment

The extreme sensitivity of polymer matrix composites to the environment was established in Chapters 2–4. Heat (which can also be generated by the material itself when stressed), moisture, acid, radiation exposure can significantly accelerate the fatigue degradation processes.

6.2.2.2.5 The initial state

The lifetime of composites under cyclic loading is very sensitive to the original state of the material. Internal or external stress concentrations will alter the composites response. At the macroscopic scale, for example, the presence of a notch in the samples can (but not always) completely deteriorate the composite lifetime (Figure 6.15). Residual thermal stresses after manufacturing are not to be forgotten as they can lengthen or shorten the lifetime of the composite. Finally, due to the highly statistical nature of the fatigue damage process, the presence of defects

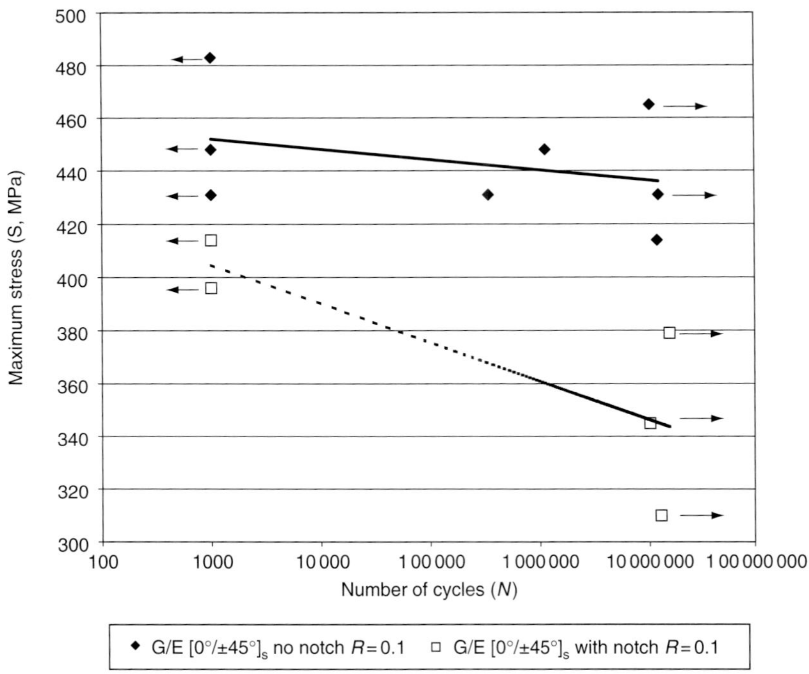

Figure 6.15. Influence of notch on S–N curve. (Data from Kerr and Haskins [15].)

in the constituents or at the interface will also be key in determining the lifetime of the part.

6.2.3 Stress Rupture

The phenomenon of stress rupture or *static fatigue* is not as widely studied and understood as dynamic fatigue [18–20]. However, this phenomenon cannot be disregarded. Indeed, the lifetime of a unidirectional carbon-fiber composite under static bending load, for example, can vary from a few seconds to several months depending on the environmental conditions.

Static end-loaded bending experiments illustrate important degradation processes in stress rupture. This experimental method, relying on a simple fixture (Figure 6.16), was developed in order to investigate the possible use of carbon-fiber reinforced polyphenylene sulfide (PPS) composites for piping application [21,22].

Specimens are bent in the fixture and can easily be placed in an oven or in a liquid bath. Most samples break after a certain exposure time (stress rupture). The lifetime varies significantly with maximum applied strain and oven temperature.

Result examples on unidirectional carbon-fiber polyphenylene sulfide (AS4/PPS) composites are shown in Figures 6.17, 6.18 and 6.19. As intuitively expected, high strain levels and high temperatures tend to reduce the materials lifetime. More unexpected are failures observed at temperatures as low as 40°C hinting toward a potential long-term room temperature failure under end-loaded static conditions.

At higher strain levels, failure occurred on the AS4/PPS system without warning (no visible damage accumulation) by the rapid propagation of a single fatal microbuckle located at the center of the specimen.

At lower strain levels, however, investigations of the failure mechanisms evidenced a fairly complex damage accumulation process. Figure 6.20 shows the sequence of damage on the bottom surface (compression side) of the unidirectional

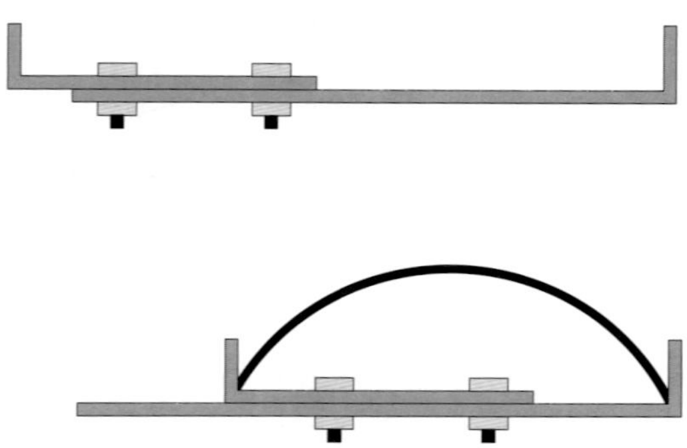

Figure 6.16. End-loaded compression bending fixture [22]. (Copyright 1998, *Journal of Composite Materials*, C.A. Mahieux et al., reproduced by permission of Sage Publications.)

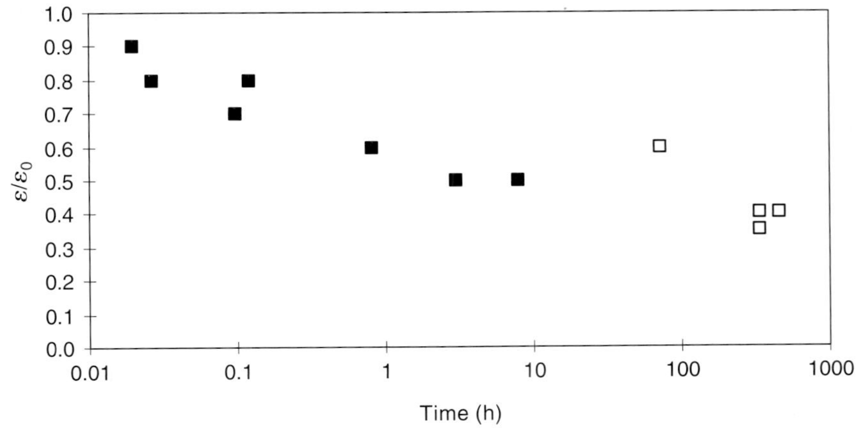

Figure 6.17. Maximum applied strain/tensile strain-to-failure ratio versus time-to-failure ratio at 90°C (the empty squares indicate run-out experiments) [21]. (Copyright 1998, *Journal of Applied Composites*, B.E. Russell et al., reproduced by permission of Sage Publications.)

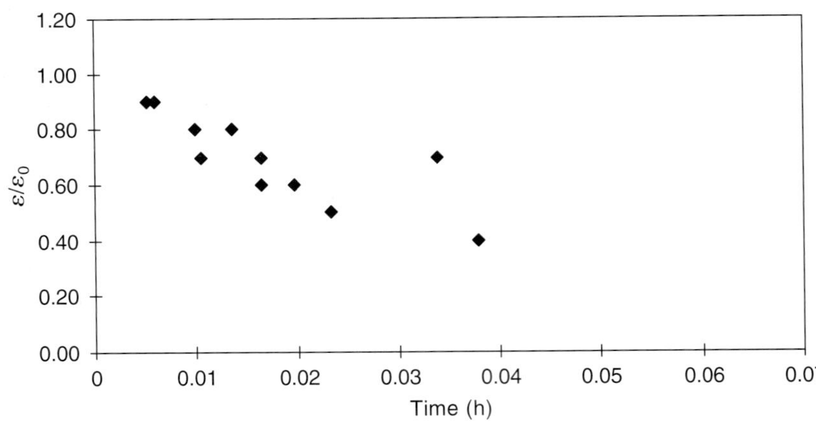

Figure 6.18. Maximum applied strain/tensile strain-to-failure ratio versus time-to-failure ratio at 120°C (the empty squares indicate run-out experiments) [21]. (Copyright 1998, *Journal of Applied Composites*, B.E. Russell et al., reproduced by permission of Sage Publications.)

composite under moderate load. The first damage appeared on the edges of the sample but not at the point of highest strain level (which is located in the middle of the sample length), illustrating the statistical nature of the damage initiation site location (Figure 6.20 (a)). Electron microscopy helped unravel the nature of the damage and designates microbuckling as the predominant degradation process (Figure 6.21). Microbuckles are kink-bands characteristic of compressive loading, in which the fibers deviate from their original axial direction (Figure 6.22); the fibers

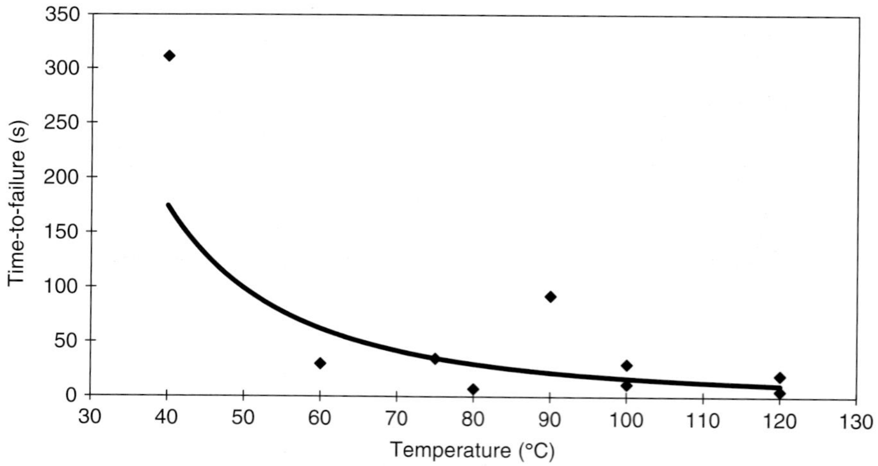

Figure 6.19. Time-to-failure ratio versus temperature for specimens bent at 90% of their strain-to-failure ratio [23]. (Copyright 1997, ASME 97, C.A. Mahieux and K.L. Reifsnider, reproduced by permission of ASME International.)

may deform or rupture. In the end-loaded experiments at moderate loads under discussion, initial microbuckling was followed by the initiation of new kink-bands on both sides (edges) of the sample (Figure 6.20 (b–e)). The rather regular spacing between microbuckles tend to corroborate an assumption of stress re-distribution and interaction between damage sites.

However, despite such damage accumulation, the sample has not yet lost all of its load carrying ability. Final failure was observed to occur according to a different process in which a microbuckle located at the center of the specimen and that was experiencing so far a stable (slow) growth suddenly propagated across the width of the composite and induced failure of the unidirectional composite. This last phase is characteristic of a material *sudden death* or catastrophic failure (Figure 6.20 (f)).

Static fatigue shows striking similarities with cyclic fatigue such as the possibility of different damage accumulation mechanisms and the statistical nature of the damage. Though specific models can be established to fit those experimental data, observations and empirical approaches are necessary for the study of polymer composites undergoing such loads.

6.2.4 Environmental Cycling

Environmental cycling is a complex mechanism as mechanical cycling. However, the literature is rather sparse on the topic and once again, models are difficult to generalize. Environmental cycling was already approached in the present book with freeze thaw, thermal spiking, electrical effects and will therefore not be repeated

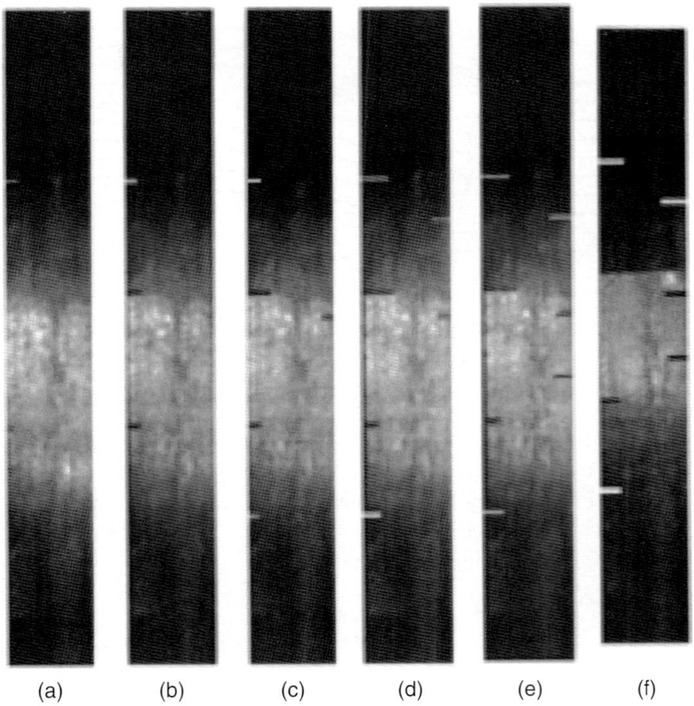

Figure 6.20. Underneath of the bent specimen in oven (sequence of events). (a) at 60 s, (b) at 75 s, (c) at 80 s, (d) at 85 s, (e) at 90 s, (f) at 92 s. Data from Mahieux et al. [22]. (Copyright 1998, *Journal of Composite Materials*, C.A. Mahieux et al., reproduced by permission of Sage Publications.)

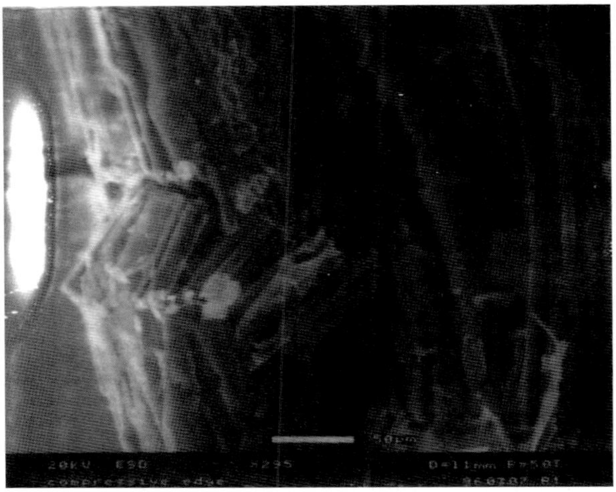

Figure 6.21. Microbuckling in end-loaded experiments [22]. (Copyright 1998, *Journal of Composite Materials*, C.A. Mahieux et al., reproduced by permission of Sage Publications.)

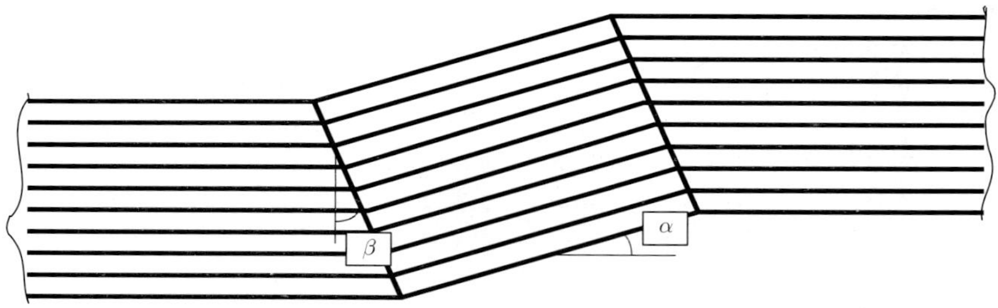

Figure 6.22. Schematic diagram of a microbuckle. α and β are the characteristic angles [22]. (Copyright 1998, *Journal of Composite Materials*, C.A. Mahieux et al., reproduced by permission of Sage Publications.)

here. We will direct the reader to Chapters 2–4. Nevertheless, it is worth mentioning again that the cyclic application of an environmental load is generally more damaging than the constant application of a load of identical amplitude.

6.2.5 Practical Complexity

Dealing with dynamic load is complex. This complexity is illustrated by the testimony of R. Schmidt, blade designer at Aerodyn.

Case study: Stress analysis for wind turbine rotor blades (by R. Schmidt)
The engineering office Aerodyn was established in 1983 and since hen has devoted its engineering exclusively to the development of wind energy converters (WECs) and their components. Complete WECs and single components have been developed and designed for more than 30 international manufacturers, including the required type approval documentation. During this period, more than 70 converter concepts with rated powers between 5 and 5000 kW have been implemented. Amongst the numerous components, more than 40 rotor blades have been designed by Aerodyn.

A rotor blade of a modern horizontal axis wind turbine is an application of fiber reinforced composites comprising a wide range of design aspects to be considered (see Chapter 1). The blades of a typical three-bladed horizontal axis turbine can be regarded as built-in cantilever beams mounted to the rotor hub. In their function to transform the translatory air movement into rotational energy, they are subject to high cycle loading in the form of combined thrust forces, bending moments and torsional moments when the turbine is operating. In addition to the cyclic loading due to gravity forces and aerodynamic forces in a turbulent wind field, the blades experience extreme loads due to emergency stop procedures and in extreme wind conditions, when the turbine is at standstill. In an operating lifetime of 20 years, the rotor blades experience more than 10^8 load cycles. The magnitude of the loads for given wind conditions depends on the aerodynamic shape of the rotor blades, the turbine control strategy and the mass and stiffness properties of the blade structure as well as other factors. Since the wind turbine is a rather flexible system with rotating components, a careful tuning of component and drive train natural frequencies and rotational frequencies is important in order to avoid unwanted vibrations and hence loading. Apart from the natural frequencies, blade deflection, load carrying capacity, fatigue strength and buckling stability as well as the blade root interface to the rotor hub are important aspects in the structural design of rotor blades.

In the design process, computational tools based on analytical methods and the FEM are used in a number of design loops. The fatigue and extreme loads on the rotor blades and the whole turbine are calculated by subjecting an aero-elastic model of the turbine to a turbulent wind field and simulating various operating and extreme load cases in the time domain. In this model, the mass and stiffness distribution of the blades is considered in the form of beam elements. Calculation outputs are extreme loads and fatigue load time series for the desired number of blade sections. Each load case consists of a combination of the following five components: edgewise and flatwise forces and moments, torsion and centrifugal forces.

In a typical structural layout of a rotor blade, the direction-dependent mechanical properties of fiber reinforced composite materials are used in an efficient lightweight structure. A light sandwich shell with biaxial or multiaxial skins providing the required aerodynamic shape and some torsional stiffness is combined with an either partly integrated or separately manufactured spar structure to carry the loads. The spar structure consists of unidirectionally reinforced (UD) spar caps and multiaxially reinforced sandwich spar webs. Most commonly, the blade shell is made up of two halves, which are bonded either to the separate spar structure or to spar webs in case of spar caps integrated into the shell. The shell halves are also bonded together at nose and tail. A typical cross-section of a blade structure is shown in Figure 6.23.

At the root end of the rotor blade, there is usually a cylindrical interface for a bolt connection to the blade bearing and the cast iron rotor hub. This interface is provided either in the form of metallic inserts integrated or bonded into the blade root laminate, metallic flanges bonded and/or bolted to the root laminate or in the form of pre-tensioned T-bolts. In order to transfer the loads from the spar cap into the bolt circle, a build-up of multiaxial reinforcement is used in the blade root. A typical cross-section of a T-bolt connection is shown in Figure 6.24.

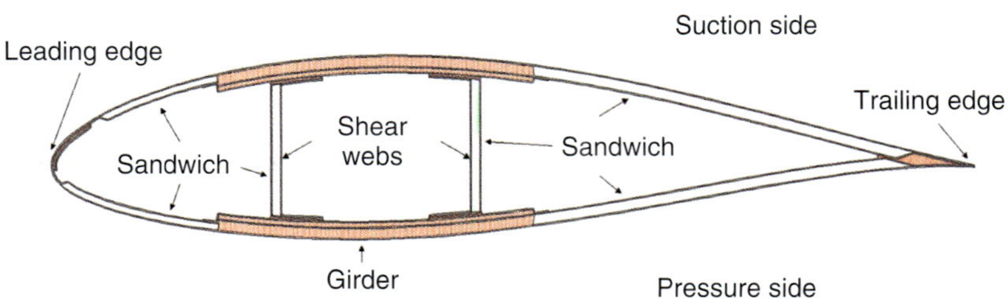

Figure 6.23. Blade cross-section. (Courtesy of Aerodyn.)

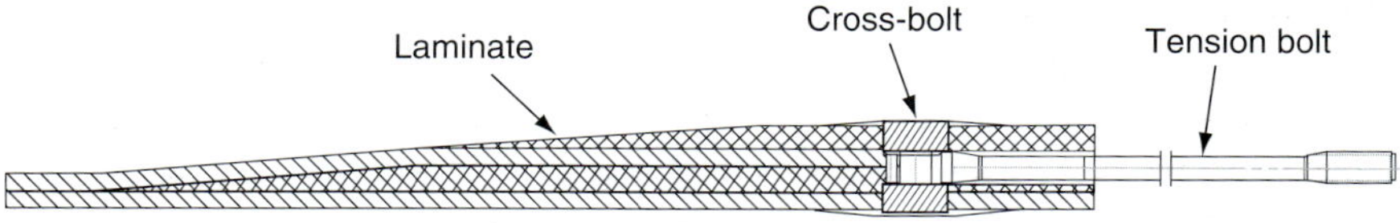

Figure 6.24. T-bolt connection. (Courtesy of Aeordyn.)

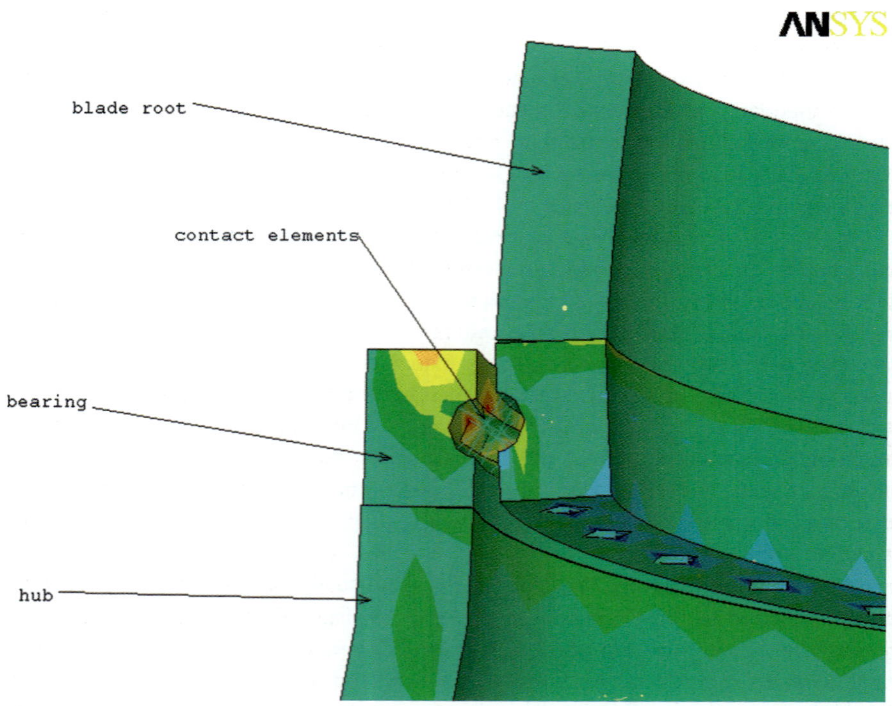

Figure 6.25. Partial FEM model. (Courtesy of Aerodyn.)

In the design of the blade connection, the stiffness of blade root, bearing and hub have to be considered as well as the component geometries, since they have an influence on the loading on the whole connection and the bolts and on the gaping behavior of the connection. For the design of the blade connection, the components are modeled in FEM-volume models featuring contact elements and analysed with non-linear calculations. A partial FEM model including rotor hub, bearing and blade root is shown in Figure 6.25.

Depending on the design details of the blade connection, it may be required to perform static and dynamic component tests. Test results may be used not only to prove the connection but also to improve and tune the FEM models.

The blade structure is pre-dimensioned in a first design loop with a custom beam model based on two-dimensional cross-sections. With this model, tensile, compressive and shear stresses and strains are calculated. Fatigue damage calculations are performed on the basis of load time series and strains are determined with the analytical model. A current limitation in the fatigue design is the lack of understanding of the fatigue behavior of composite laminates under combined loading. This requires conservative design approaches, especially in the root region of the blade with a complex geometry and laminate lay-up. In the absence of proven failure mechanisms for complex dynamic loading, design experience and component tests for verification are vital for a successful design.

Methods for buckling stability calculations are integrated into the analytical model. Natural frequency and deflection calculations are performed with simple FEM beam models. On a typical horizontal axis wind turbine with upwind rotor, the maximum blade deflection is limited by the requirement for a minimum tower clearance. On the less common turbines with downwind rotor, a very large deflection may be desirable in extreme winds in order to reduce the rotor projected area and hence the extreme loads. Another design approach is to tune the reaction of the blade structure to wind gusts by the so-called *bend-twist coupling* in order to influence the aerodynamic behavior of the blade and to reduce turbine loading. This means, the laminate lay-up is designed in a way that the blade twists in addition to bending under gust loading and the aerodynamic inflow angles are changed favourably. Research and development work has been and is being carried out in this area, but systems employing such technology have not been implemented in a large scale.

Using the mass and stiffness distribution from the first design loop, new turbine loads are calculated. In further design loops, global FEM shell models of the blade structure and detail volume models are used for design optimization. For efficient modeling of the blade geometry, structure and blade loads, a special pre-processor is used which also allows efficient and reliable checking of the model. The global FEM model gives an insight into the spatial stress distribution and local stress concentrations in the blade structure. Based on this, the structure is analysed with suitable criteria for fiber and inter-fiber failure. Figure 6.26 shows a typical global FEM model of a rotor blade. Further, buckling stability is investigated with linear or non-linear analysis and blade deformations are determined.

Local aspects, such as the shear stress distribution in bonds, are analysed with local volume models. Figure 6.27 shows a detailed volume model of a blade cross-sections with local bonds. In the design of the bonds, there are also some limitations. The fatigue behavior of thick bonds common in rotor blades is not fully understood, neither the influence of aging of the bonding material on the fatigue behavior. More analysis and mechanical tests are required in this area.

Finally, the load carrying capacity of the blade and the calculation models are verified by means of full-scale blade tests. The blade's natural frequencies are measured and, during a static load test, deflections and strains are recorded. The design life of the blade is sometimes simulated in a full-scale fatigue test. In order to shorten the test time to a few months, the load level has to be increased.

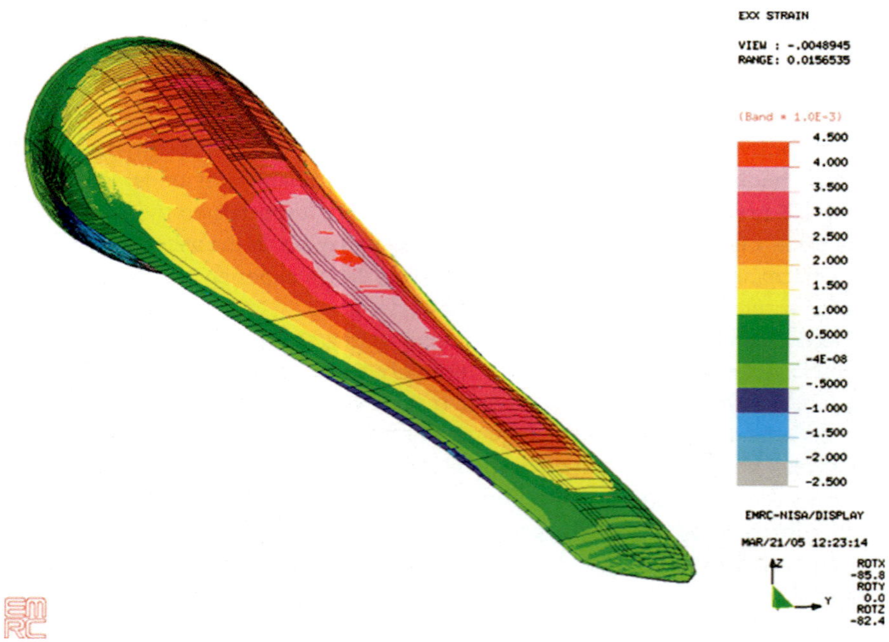

Figure 6.26. Blade FEM global model. (Courtesy of Aerodyn.)

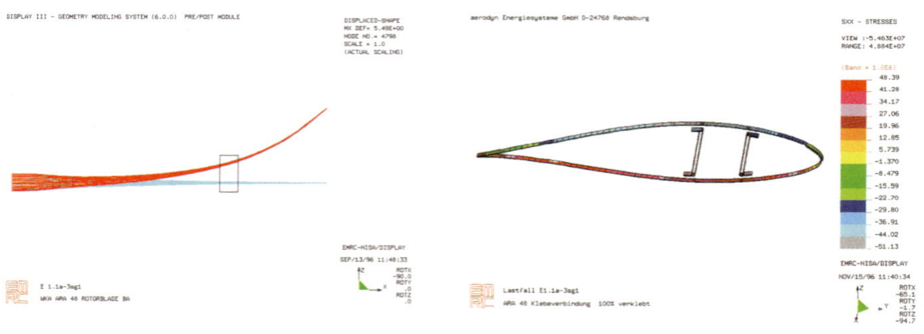

Figure 6.27. Detailed volume model with detailed bonds. (Courtesy of Aerodyn.)

It is required to understand the influence of this and the lack of aging on the damage development for an adequate interpretation of the test results. Such full-scale dynamic test will not give a full understanding of the fatigue behavior of the blade structure; However, it will give a good insight into the behavior of the blade structure under dynamic loading. A static blade test is shown in Figure 6.28.

This complex case study clearly evidenced the urgent need for analytical load combination schemes, topic of the following section.

Figure 6.28. Blade test. (Courtesy of Aerodyn.)

6.3 SEQUENTIAL AND COMBINED LOADING

6.3.1 Approaches

Miner's rule is often used to account for the sequential loading of metals at different stress levels [24]. The fatigue damage accumulation can be calculated using *block* summation:

$$\sum_{i=1}^{k} \frac{n_i}{N_i} = 1 \tag{6.8}$$

where n_i is the number of cycles spent at S_i and N_i the lifetime of the sample under S_i.

Unfortunately, Miner's rule was found very inexact for composite materials [3,25,26]. For example up to a 60% discrepancy was found between Miner-based lifetime predictions and experimental data for carbon-fiber reinforced polymer T800/5242 laminates under fatigue.

Indeed, composite materials are very sensitive to the order of loading sequences, as well as to the frequency of loading (strain rate influence, see Chapter 2). Therefore, a failure function accounting for the non-linearity of the damage accumulation has to be considered in order to calculate the composite remaining and ultimate lifetime.

A common pitfall consists of using Miner's rule and adding a very large safety coefficient. This approach is extremely dangerous considering the typical non-linearity of the composites response. Indeed, multiplying by a single factor can have difficultly in covering the effects of a power law. Therefore, an alternate approach is proposed for composite materials.

Let us for now consider the example of a composite bridge. The bridge is exposed over its lifetime to a large number (N) of parameters such as varying temperature, cyclic and static mechanical loads, moisture and water, ultraviolet radiations etc. Testing the material in all individual and interacting conditions would mean performing N^N experiments. This would induce costs and timeframes unrealistic for most companies. Tools are therefore needed to decrease the number of required experiments and enable partial virtual design [14]. The durability approach presented thereafter makes use of such tools and is perfectly adapted to *block* loading situations.

6.3.2 Durability Concept

Very few global durability schemes are today available for composites. Most models rely only on specific configurations and difficulty can be generalized. We will focus here on a durability assessment method that can be used for the widest range of materials (not only composites) and that encompasses the effects of combined mechanical and environmental loads.

This method that originally relied on a proprietary code (MRLife^{TM}) was introduced by Reifsnider and Stinchomb [27]. This approach is not unique but enables a comprehensive life prediction and a measurable damage progression. It was successfully applied to the life prediction of many industrial projects ranging from aerospace to truck tires applications. For more details on the methodology, it is recommended to further consult reference [14]. Our purpose here will be to demystify the approach by providing a short summary of the basic methodology and tools, illustrated by a basic application example.

6.3.2.1 Critical element

The first step in this method is to determine the critical element. Its definition is based on the critical element theory [27–29]. The critical element is defined as the last element that will fail in the sample. Let us illustrate this concept with a simple example such as a [0°/90°] laminate under tensile axial load: the 90° ply first experiences matrix cracking. However, the 0° ply still provides strength to the laminate. In other words, failure of the 90° ply does not necessarily imply direct failure of the laminate. The 90° ply is not a critical element. However, the degradation of the laminate will lead to a stress redistribution in the 0° ply (see ply-discount method, Chapter 5). The degradation in the 90° ply leads to a change in the state of material of the 0° ply. Therefore, the 90° ply will be called *a sub-critical element*. Then under the new state of stress and material state, the 0° ply might fail. After the failure of this 0° ply, the laminate fails. In other words, failure of the 0° ply necessarily implies failure of the laminate (and vice versa). The 0° ply is therefore the critical element.

Though very practical, it is not always possible to find a *small* critical element. In the case of a random fiber single thick composite plate, for example, the critical element is the entire part.

6.3.2.2 Failure functions

The second step of the durability approach is the determination of a single failure function for the critical element [27–29]. This step is probably the most complex in the life prediction process.

This failure function depends upon the materials failure mode and is generally defined experimentally. Despite the broad range of damage types in composites, experience shows that the critical element mainly experiences one failure type for a given (mechanical and environmental) load interval. As a first approximation, data from the literature can be used to estimate the failure mode that the material will experience. It is, however, strongly recommended to perform thorough validation experiments after screening.

Examples of failure functions were already introduced in Chapter 5 with maximum strain, maximum stress, Tsai-Wu and Tsai-Hill criteria. We will now generalize our discussion and use Fa as the notation for any failure function representing the failure mode. Fa is therefore a function of the state of stress and the strength of the critical element [14]:

$$Fa = Fa\left(\frac{\sigma_{ij}}{X_{ij}}\right) \qquad 0 \leq Fa \leq 1 \tag{6.9}$$

Both stresses and strengths change with time and environmental conditions (Chapters 2–5). We can explicit the relationship between time and environmental exposure in the failure function using the notations introduced in Chapter 5:

$$\overbrace{Fa}^{t,e} = Fa\left(\frac{\overbrace{\sigma_{ij}}^{t,e}}{\overbrace{X_{ij}}^{t,e}}\right) \qquad 0 \leq \overbrace{Fa}^{t,e} \leq 1 \tag{6.10}$$

The failure function Fa therefore reflects the combined degradation processes and considers differing degradation rates.

6.3.2.3 Strength as a damage metric

The present methodology considers strength as a damage metric. It assumes that all material changes and degradation relevant to the composites lifetime are translated in term of stiffness and strength variations. Therefore, the failure function Fa can be related to the materials remaining strength Fr [27] by:

$$Fr = 1 - \int_0^{\tau_1} (1 - Fa)\left(\frac{\sigma_{ij}(\tau)}{X_{ij}(\tau)}\right) j\tau^{j-1}\mathrm{d}\tau \tag{6.11}$$

where j is a material parameter. A j value equal to 1 describes a degradation rate constant over time when a j smaller than unity indicates faster degradation at the beginning of the loading. Finally a j greater than 1 indicates an increase in the degradation rate with time characteristic of sudden death.

In the case of the quasi-static experiments, the characteristic time is the creep rupture life (as a function of stress level) and in the case of fatigue:

$$\tau = \frac{n}{N} \tag{6.12}$$

where n is the number of cycles and N the number of cycles to failure.

This allows Reifsnider and Case to rewrite the remaining strength equation (Equation (6.11)) in a more friendly form [14], namely:

$$Fr = 1 - \int_0^{t} \left(1 - Fa\left(\frac{\sigma_{ij}}{X_{ij}}\right)\right) j\left(\frac{n}{N}\right)^{j-1} \mathrm{d}\left(\frac{n}{N}\right) \tag{6.13}$$

If the element is not damaged, Fa equals 0 and the remaining strength is maximum. When damage accumulates, Fa increases and the remaining strength is reduced. Failure is reached when Fr and Fa become equal:

$$\text{Failure when } Fa = Fr \tag{6.14}$$

6.3.2.4 Practical implications

Using strength as a damage metric has clear practical implications raised by Reifsnider et al. [14] on equivalences possibilities, illustrated by Figure 6.29. Namely, exposure to n_1 cycles at a given level 1 is equivalent to n_2 cycles at a level 2 from a materials perspective. Figure 6.29 illustrates this concept and clearly indicates the deviations from linearity, which reminds us of the non-applicability of Miner's rule.

Major consequences of Equation (6.13) from an industrial perspective are twofold. First, damage and lifetime can be anticipated under different conditions. Second, and more importantly, damage can also be *tracked and measured* at all times. Influence of the environment (mechanical loads, creep and relaxation, plas-

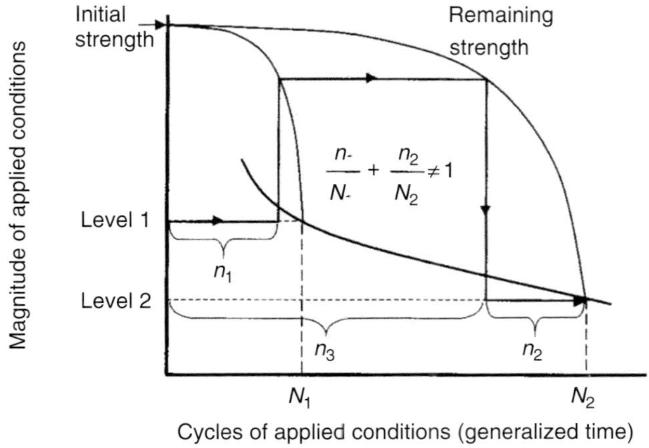

Figure 6.29. Concept of remaining strength as a damage metric [14]. (Copyright 2002, *Damage Tolerance and Durability of Material Systems*, by K.L. Reifsnider and S.W. Case, reprinted by permission of John Wiley & Sons.)

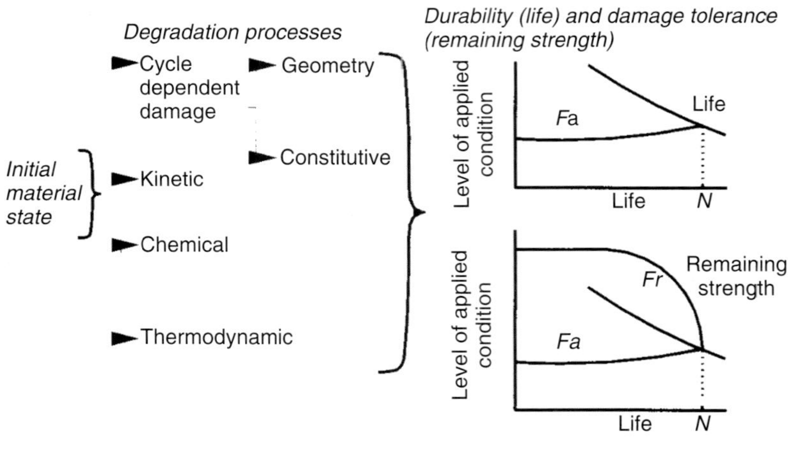

Figure 6.30. Damage tolerance and durability in composite systems that degrade by multiple, interacting progressive degradation processes under mechanical, thermal and chemical applied environments [14]. (Copyright 2002, *Damage Tolerance and Durability of Material Systems*, by K.L. Reifsnider and S.W. Case, reprinted by permission of John Wiley & Sons.)

ticization trough moisture absorption, UV induced degradation etc.) can enter the equation via the failure function (Figure 6.30):

$$Fr = 1 - \int_0^t \left(1 - Fa\left(\frac{\overbrace{\sigma_{ij}}^{t,e}}{\overbrace{X_{ij}}^{t,e}}\right)\right) j\left(\frac{n}{N}\right)^{j-1} \mathrm{d}\left(\frac{n}{N}\right) \tag{6.15}$$

For mechanical fatigue, the applied load will typically be sinusoidal (laboratory experiment). To deal with the variation and to combine the effects of the different loads, an incremental approach is proposed by Case et al. [30]. From Case [30], for a constant Fa applied, one can write:

$$Fr = 1 - (1 - Fa)\tau^{j} \tag{6.16}$$

If we consider a first-loading process at a constant Fa_1 for a time τ_1 resulting in a remaining strength Fr followed by a different loading condition Fa_2, we can write the time that would have been necessary to cause an equivalent amount of damage at this second load level (pseudo time, τ_2^0):

$$\tau_2^0 = \left(\frac{1 - Fr}{1 - Fa_2}\right)^{\frac{1}{j}} \tag{6.17}$$

Over an interval of loading the remaining strength can be written as:

$$\Delta Fr = -(1 - Fa_2)\left\{\left(\frac{t_2^0 + \Delta t}{\hat{\tau}_2}\right)^{j} - \left(\frac{t_2^0}{\hat{\tau}_2}\right)^{j}\right\} \tag{6.18}$$

where t_2^0 is the unnormalized pseudo time:

$$t_2^0 = \hat{\tau}_2 \tau_2^0 \tag{6.19}$$

Finally the remaining strength can be expressed as:

$$Fr = 1 - \sum_{i=1}^{i=N_{\text{blocks}}} \Delta Fr_i \tag{6.20}$$

Using Equation (6.20), it is now possible to use an incremental approach to the problem.

6.3.3 Example

With an almost infinite number of load combinations, durability assessments for real-life cases are complex. However, experience enables the selection of major and most severe contributors to the composite lifetime. Those few parameters and their interaction are generally studied with great detail during the design phase. For example, the stress rupture fixture presented in Section 6.2.3 was designed to accelerate the stress rupture behavior of a composite flexible riser tensile armor for oil rising. Indeed, weight savings and corrosion resistance are great drivers for the development of the use of composites for oil and gas exploration. Before analyzing this example, the following case study relates recent developments in this area.

Case study: Composites for the oil and gas industry
Off-shore applications are traditionally divided into surface water, shallow water (<500 m), deep water (between 500 and 1500 m) and ultradeep water applications (depth in excess of 1500 m). The Petrobras platform Roncador in the Campos Basin in Brazil reaching a 2000 m depth is an example of the most challenging ultradeep oil rising.

Thanks to light weight and corrosion resistance, polymer-based materials are used in all parts of over and under-water structures, ranging from simple parts such as glass fiber reinforced polymer gratings (Figure 6.31) or stairs to more complex applications such as underwater risers.

It is evaluated that around 70% of the undiscovered oil and gas reserves lay in deep and ultradeep waters [31]. However, to allow for a cost-efficient exploitation of such resources, piping equipment resisting internal and external pressure as well as axial load created by the kilometers of pipe length have yet to be developed.

Several oil exploitation methods exist but can be grouped into two main categories: temporary and (semi)-permanent structures. Semi-permanent structures involve the use of platforms. In the case of deep-water rising (500 m and deeper), spars platforms (vertical floating cylinders) or tension leg platforms (TLP, Figure 6.32) have attractive attributes. For this later platform type, the mooring system, aimed at limiting vertical motion, is made of high stiffness tendons attached

Figure 6.31. Platform composite grating. (Courtesy of American Grating.)

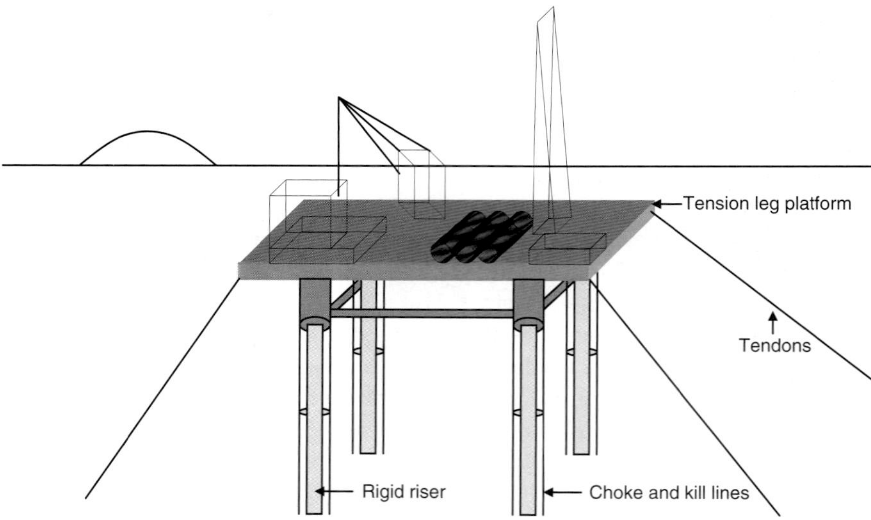

Figure 6.32. Schematic diagram of a tension leg platform.

at each corner of the platform. Rigid vertical risers are used to raise oil from the seafloor to the production wellheads located on the platform deck. Choke and kill lines are additional smaller diameter pipes running along the rigid riser, the purpose of which is to prevent blowout in case of sudden gas leak [32].

Composite materials can advantageously be used in the different parts of the platform. Current and potential composite applications for topside and underwater equipment are reviewed thereafter.

(a) TLP topside: The economical benefits obtained by the use of composite components for a TLP topside, such as fire resistant reinforced polymer gratings, pipes, storage tanks, were studied by Vennett et al. [33] (Figure 6.33). The drastic expected 45% weight reduction (1200 ton reduction compared to the steel structure) on a typical TLP would translate into a 10.5 million dollar saving for the platform.
(b) Underwater equipment (Choke and kill lines, rigid risers): More than a cost saving opportunity, composites can also be an enabling technology for deep-water oil rising. The use of composites for underwater equipment is more recent, though, and was mainly motivated by the need to explore deeper waters to extract oil. Aerospatiale and the Institut Francais du Petrole pioneered in this field by installing Choke and Kill lines in 1979 in three North Sea projects. The French company in cooperation with Aerospatiale also engaged in joint industrial projects (JIP) in the early 1980s aimed at establishing the feasibility of composite risers. It is however much later, in 1995, that the interest in composite materials for rigid risers really exploded.

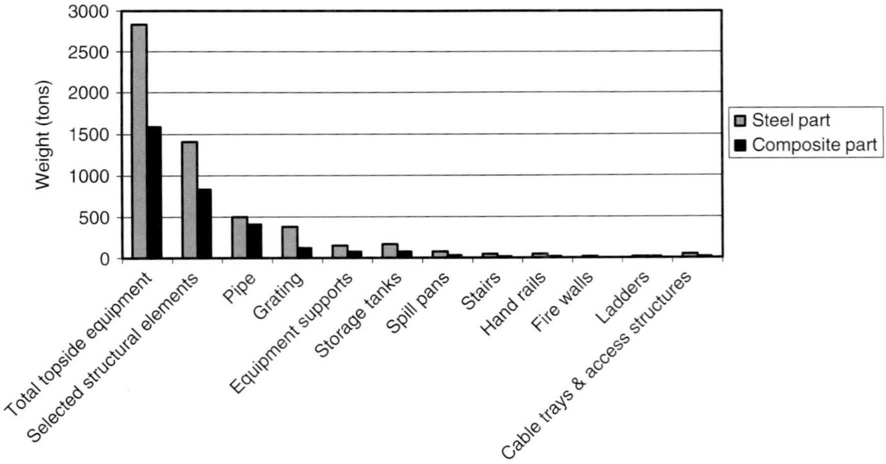

Figure 6.33. Topside weights: Steel versus composites. (Data from Vennett et al. [33].)

In the late 1990s, two main design concepts emerged for rigid composite risers. They resulted from the efforts of two industrial consortiums, one lead by ABB Vetco Grey and the second lead by Conoco/Kvaerner. Both groups had to develop solutions to a long list of challenges such as composite/metal joining, higher performance fluid liners, low weight structures combined with internal/external pressure resistance, systematic fatigue inspection methods and cost-effective semi-mass manufacturing.

Indeed, providing a sealed joint between the composite riser and the metallic end fitting still remains a challenging task. The joint should be mechanically sound and ensure that no fluids are going in or out of the pipe. Gluing the composite and the metal piece together is an impractical solution due to difficulties in the lifetime assessment of the adhesive. Instead, the companies concentrated their efforts toward geometric traps, in which the composite is placed in tight contact with the grooved metal fitting either by winding the composite or by injecting epoxy into the system.

No less challenging was the development of high performance inner liners, destined to serve as fluid barriers. Indeed, the liners must protect the composite structural elements that are more prone to fluid diffusion than traditional steel structures. Inner liners can be exposed to relatively high temperatures (90 – 120°C) combined with aggressive environments such as H_2S, CO_2 and chloride derivatives. The fluid barrier should have extremely high diffusion resistance to water and hydrocarboneous fluids. Differing liner solutions based on either metallic (titanium) or polymer (elastomeric) materials were selected.

In addition to the environmental loads, the risers can be subjected to large constant and cycling mechanical stresses. The rigid risers are traditionally designed to resist the external pressure created by the surrounding water as well as internal pressures in the range of 40–100 MPa for choke and kill lines. Wavy conditions also impose fatigue conditions on the riser and specifications require resistance to occasional but unusually high loads such as the Hundred Year Wave in the shallow region of the pipe.

Typical specifications require a minimum of 150-year lifetime and we understand from the present book that durability assessment on composite structures in such complex environments is not trivial. After manufacturing and in-operation, inspections are therefore crucial for off-shore structures. Directly embedding sensors, such as optical fibers, in the composite pipe offer possibilities for continuously monitoring strains, stresses and fiber failure evolution.

An additional challenge is the selection of a realistic manufacturing process, offering the possibility of an integration in an existing traditional plant. Considering the kilometers of length of risers that should be manufactured, the required manufacturing process is close to a semi-mass fabrication method. Automated methods such as filament winding are therefore prime candidates for the large-scale manufacturing of rigid risers.

Like for most composite applications, the use of composites is actually greater than the use of metals (one-to-one comparison). The cost savings are obtained on the system when considering the weight savings. However, cost acts as a serious limiting factor for the choice of materials and manufacturing processes. Though the composite riser technology is achievable from a technical standpoint, the financial viability still has to be proven.

The lack of design guidelines related to composite in this field further hinders the development of composite rigid risers. Due to their specificities, traditional (metal) norms cannot be directly applied to composite materials. Efforts are ongoing in this field to establish composite norms. To date a recommended practice for composite risers (DNV-RP-F202) is available to the industry.

Despite all those difficulties, a composite drilling riser joint developed by Norske Conoco AS and Kvaerner Oilfield Products was used in 2001 for the drilling of three wells on the Statoil-operated Heidrun field in the Norvegian North Sea (Figure 6.34). The 38.1 cm diameter, 15 m long carbon fiber/epoxy composite riser weighed half of an equivalent steel pipe. The riser joint was manufactured by Spencer Composites Corp. by filament winding. The stringent specifications required a 195 MNm^2 minimum bending stiffness, a nominal operating pressure of 6.3 MPa and impact resistance of 250 kJ for joints above 50 m water depth and 50 kJ below 50 m, a minimum burst pressure of 83 MPa and an axial load capacity of 1.36×10^6 tons [32]. The actual burst pressure achieved on prototypes was 110 MPa. Combination of laboratory experiments and analysis concluded to an anticipated lifetime of 150 years (still to be validated).

Unfortunately, commercialization did not occur rapidly after this first successful prototype. At the time of writing, Spencer Corp. produced composite riser joints for use in the ConocoPhillips Magnolia TLP in the Gulf of Mexico (moored in 1433 m of water, deepest in the world to date). The 3.2 m long, 34.3 m outer-diameter pipes were obtained by filament winding around an inner steel liner. The burst pressure obtained was in excess of 145 MPa for an operating pressure of 69 MPa and an axial load capacity of 1.14×10^6 tons [32].

Figure 6.34. Rigid riser. (Courtesy of Aker Kvaerner.)

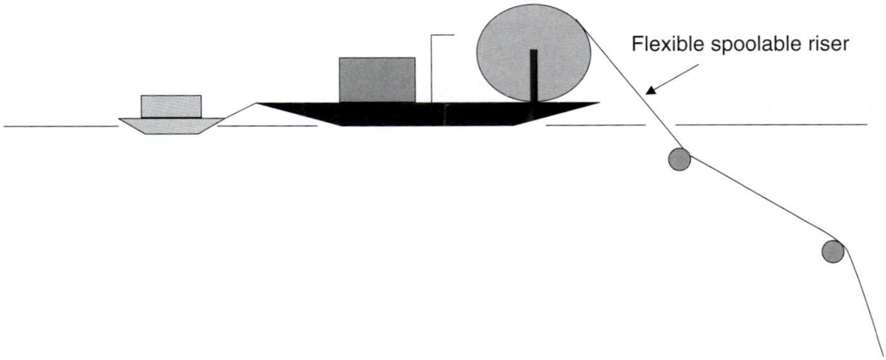

Figure 6.35. Boat with reeled riser.

An alternative and more economic way of performing deep-water oil exploitation is the use of mobile equipment. This method is an attractive candidate for deep and limited oil supplies such as the off-shore oil sources located in the west coast of Africa. A flexible production riser can be reeled on a boat, serving as a mobile platform, and released into the water at the wanted location. The use of a smaller boat is then required to ensure the periodic oil transport to the ground (Figure 6.35).

Flexible risers must fulfill all the requirements listed above for rigid pipes, with an additional stringent requirement on flexibility. Indeed, due to the transportation system, the riser sees very large bending strains. The structure should not only accept such large deformations but also be structurally resistant in order to sustain internal and external pressures. For traditional structures, this is usually achieved by using a multiarmored structure: the riser is made of concentric pipes, free to slide against each other. Unfortunately, such complex structures result in large weights and using traditional flexible risers in deep waters becomes quickly challenging as the pipe tends to collide under its own weight.

Composites would therefore be an enabling technology for deep-water exploitation using spoolable risers. Wellstream Inc. successfully introduced at the beginning of the twenty-first century a hybrid flexible riser (Figure 6.36) where the structural (axial load) steel armor was replaced by unidirectional carbon-fiber polyphenylene sulfide (AS4/PPS). Other companies are trying to achieve all composite structures but a technically viable solution has not yet been reached.

In the frame of the development of the composite tensile armor for flexible risers, a durability study was initiated on carbon-fiber reinforced polyphenylene

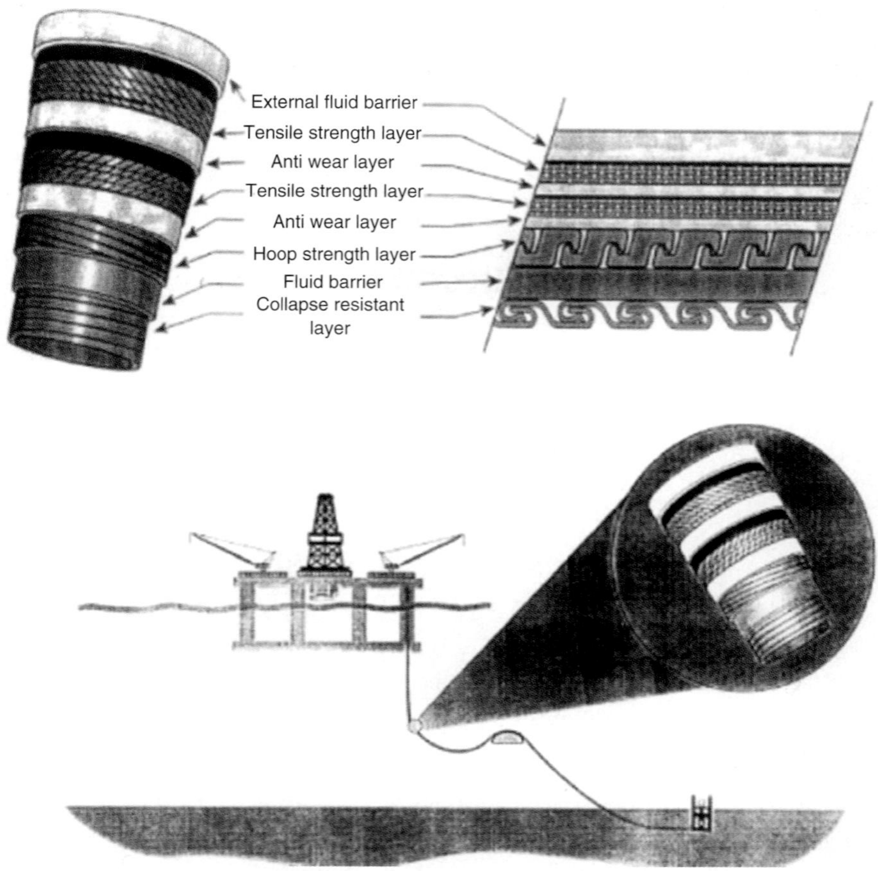

Figure 6.36. Multilayered composite flexible riser [34]. (Copyright 1999, *Composite Materials for Offshore Operations*, 2nd ed., by S.S. Wang et al., reproduced by permission of the American Bureau of Shipping.)

sulfide. Extensive end-loaded stress rupture experiments were performed under various environmental conditions (see Section 6.2.3). A simple example of load combination procedure related to this investigation is detailed thereafter.

The purpose of the load combination scheme thereafter is to predict the lifetime of a composite part under conditions 1 and 2 combined, based on individual data of the composite part life under conditions 1 only and under condition 2 only. For example, let us consider a material undergoing end-loaded bending fatigue (condition 1) at elevated temperature (condition 2). Our goal is to analytically predict the lifetime of a unidirectional composite under cyclic end-loaded bending at elevated temperature (combined conditions 1 and 2) knowing the materials response under bending at room temperature (condition 1) and under stress rupture (condition 2).

Step 1: Critical element

The first step is to determine the critical element. We are considering for simplicity purposes a unidirectional carbon-fiber reinforced polyphenylene sulfide (AS4/PPS) sample. Therefore, the critical element is the sample itself.

Step 2: Failure function

The second step consists in defining the failure function. The loading is sinusoidal (Figure 6.37), therefore independent of damage formation, the failure function *Fa* varies with time:

$$Fa = \frac{Fa_{\max} - Fa_{\min}}{2} \sin\left(\frac{2\pi t}{P} - \frac{\pi}{2}\right) + \frac{Fa_{\max} + Fa_{\min}}{2} \tag{6.21}$$

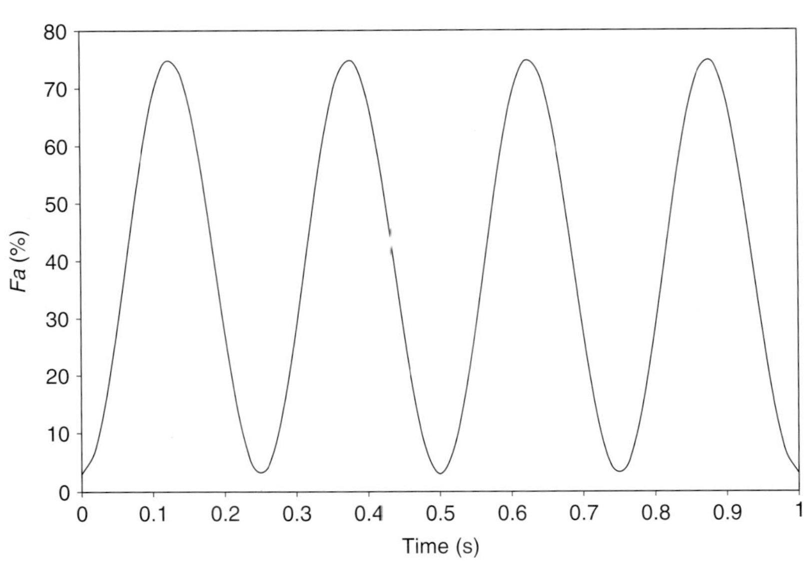

Figure 6.37. Sinusoidal variations of the failure function *Fa* (with $Fa_{max} = 75\%$).

where P is the period. On the loading portion $(Fa_{min} < Fa < Fa_{max})$ the influence of creep on the remaining strength of the specimen can be computed assuming the failure function to be constant and equal to its average value over the half-period of time. The half-period is then divided into an increasing number of steps where the average failure function can be written as:

$$Fa_{average} = \frac{aT}{2\pi\Delta t}\left(\cos\left(\frac{2\pi t}{T} - \frac{\pi}{2}\right) - \cos\left(\frac{2\pi t - \Delta t}{T} - \frac{\pi}{2}\right)\right) + b \qquad (6.22)$$

where

$$a = \frac{Fa_{max} - Fa_{min}}{2} \qquad (6.23)$$

and

$$b = \frac{Fa_{max} + Fa_{min}}{2}$$

until convergence of the remaining strength is obtained.

At the load reversal, the strength reduction due to fatigue for a half period is applied. Then this process is repeated upon the unloading portion of the failure function.

In this analysis, the failure function Fa is arbitrarily chosen to be related to the applied maximum strain by:

$$Fa = \frac{\varepsilon_{max}}{\varepsilon_0} \qquad (6.24)$$

where ε_0 is the ultimate strain-to-failure.

Step 3: Modeling of condition 1 (room temperature fatigue)
Data about fatigue experiments in end-loaded bending fatigue have now to be considered. Such data is available in the literature [35]. The fatigue apparatus used to generate the data is shown in Figure 6.38.

The unidirectional AS4/amorphous PPS composite samples were tested in fatigue. The applied displacement varied sinusoidally from a small compression (3% of strain-to-failure) to a maximum value reached at the center of the specimen (ε_{max}) with a period T $(T = 0.25\,\text{s}$ (4 Hz)). The normalized strain (Figure 6.39) was taken as the maximum strain at the mid-length of the specimen divided by the ultimate strain to failure at room temperature. The experiments included room temperature fatigue for various ε_{max}. The specimens were tested at very high strain levels. At room temperature, no failure was recorded below 90% of the ultimate strain-to-failure (the experiments were stopped after 100 000 cycles that will define a run-out). The results are shown in Figure 6.39.

Observations of the compression side and the failure surface of the broken specimens were made with an environmental scanning electron microscope. The

Figure 6.38. End-loaded fatigue fixture from Jackson et al. [35,36]. (Copyright 2001, *Applied Composite Materials*, by C.A. Mahieux et al., with kind permission of Springer Science and Business Media.)

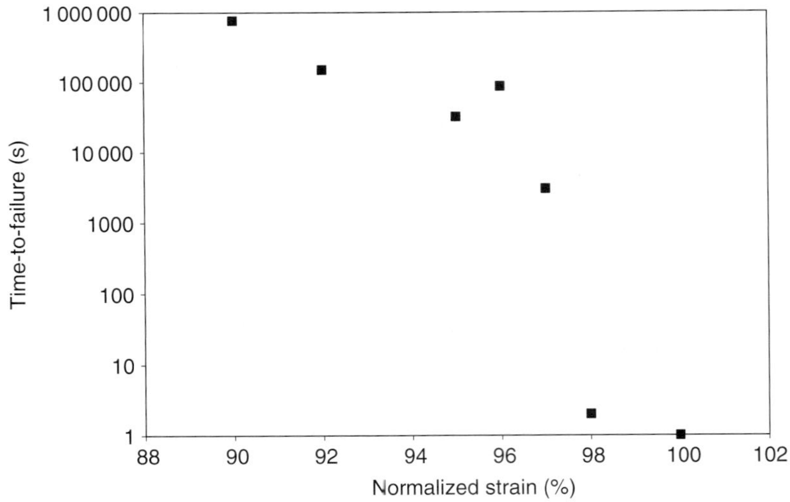

Figure 6.39. Room temperature end-loaded fatigue experiments.

compression side was severely damaged by the cyclic load. Figure 6.40 shows the presence of a microbuckle initiating from the edge. Figure 6.41 shows a bundle of fibers that macroscopically buckled and broke. The fibers are clearly crushed, resulting from a compressive cyclic load. This type of damage was seen only on specimens that underwent fatigue at room temperature.

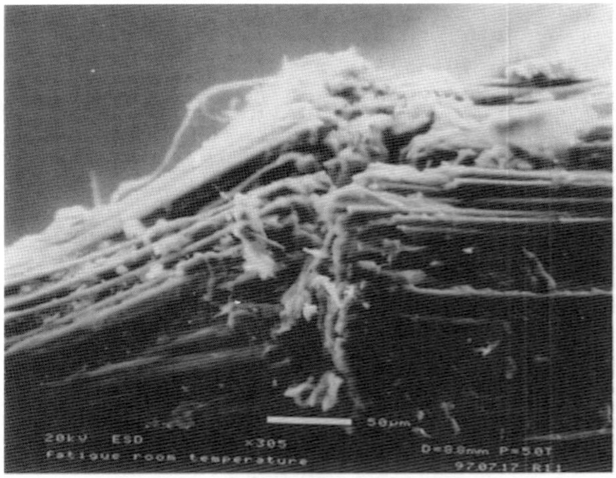

Figure 6.40. SEM picture. Room temperature bending fatigue. Microbuckle on the compression side [35]. (Copyright 2001, *Applied Composite Materials*, by C.A. Mahieux et al., with kind permission of Springer Science and Business Media.)

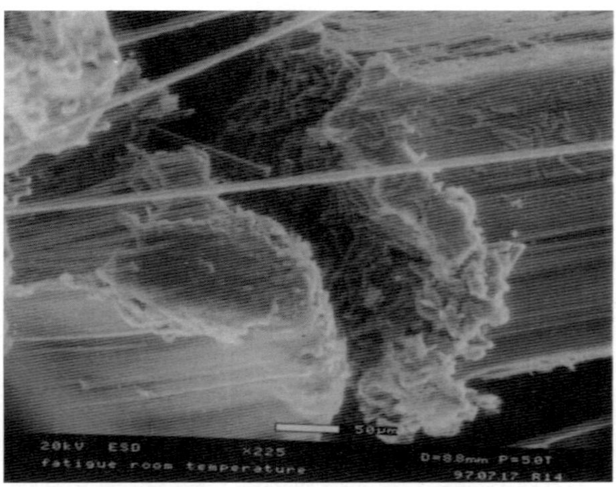

Figure 6.41. SEM picture. Room temperature bending fatigue. Damage on the compression side [35]. (Copyright 2001, *Applied Composite Materials*, by C.A. Mahieux et al., with kind permission of Springer Science and Business Media.)

Step 4: Modeling of condition 2 (Elevated temperature static loading)
We now need to model damage and failure of the sample under static loads at elevated temperature. Stress-rupture experiments shown in the example in Section 6.2.3 exhibited a degradation characterized by microbuckling. In order to model this process, a rapid literature search for this failure mode indicates that Budiansky [37]

proposes the following constitutive equation, which can be used to derive the lifetime:

$$\frac{\dot{\gamma}}{\dot{\gamma}_{\text{ref}}} = \left(\frac{\tau}{\tau_{\text{ref}}}\right)^{M} \tag{6.25}$$

where M is a material constant, $\dot{\gamma}_{\text{ref}}$ is a reference value of the creep rate produced by the reference shear stress τ_{ref}. However, we have seen in Chapter 2 that the shear stress τ in the material depends on temperature. Therefore, we write:

$$\frac{\dot{\gamma}}{\dot{\gamma}_{\text{ref}}} = \left(\frac{\tau(T)}{\tau_{\text{ref}}(T)}\right)^{M} \tag{6.26}$$

We now need to explicit the temperature dependency. For this, we assume that the dependence of the reference shear stress follows the behavior of the modulus in temperature, which can be described by (Equation (2.31)).

$$\tau_{\text{ref}}(T) = \sum_{i=1}^{N} \mu_i \exp\left(-\left(\frac{T}{T_i}\right)^{m_i}\right) \tag{6.27}$$

From Equation (6.26) we obtain

$$\frac{\mathrm{d}\gamma}{\mathrm{d}t} = \dot{\gamma}_{\text{ref}} \left(\frac{\tau(T)}{\tau_{\text{ref}}(T)}\right)^{M} \tag{6.28}$$

or

$$\mathrm{d}t = \frac{1}{\dot{\gamma}_{\text{ref}}} \left(\frac{\tau(T)}{\tau_{\text{ref}}(T)}\right)^{-M} \mathrm{d}\gamma \tag{6.29}$$

Knowing that for small γ:

$$\sigma = \frac{\tau}{\gamma} \tag{6.30}$$

and integrating for a time varying between 0 and t and the microbuckle's angle varying from the initial misalignment of the fiber $\overline{\phi}$ to a greater angle $\alpha = \overline{\phi} + \phi$, where ϕ is the additional angular displacement, we get:

$$\int_0^t \mathrm{d}t = \frac{1}{\dot{\gamma}_{\text{ref}}} \times \left(\frac{\tau_{\text{ref}}(T)}{\sigma(T)}\right)^{M} \int_{\overline{\phi}}^{\overline{\phi}+\phi} \gamma^{-M} \mathrm{d}\gamma \tag{6.31}$$

or

$$t = \frac{1}{M-1} \times \frac{1}{\dot{\gamma}_{\text{ref}}} \times \left(\frac{\tau_{\text{ref}}(T)}{\sigma(T)}\right)^{M} \times \left(\frac{1}{\overline{\phi}^{M-1}} - \frac{1}{(\overline{\phi}+\phi)^{M-1}}\right) \tag{6.32}$$

Assuming that failure occurs when the fibers reach 90°, we can write the time-to-failure according to Equation (6.33):

$$t_f = \frac{1}{M-1} \times \frac{1}{\dot{\gamma}_{ref}} \times \left(\frac{\tau_{ref}(T)}{\sigma(T)}\right)^M \times \left(\frac{1}{\phi^{M-1}} - \frac{1}{\left(\frac{\pi}{2}\right)^{M-1}}\right) \tag{6.33}$$

or

$$t_f = \frac{1}{M-1} \times \frac{1}{\dot{\gamma}_{ref}} \times \left(\frac{\sum_{i=1}^{N} \mu_i \exp\left(-\left(\frac{T}{T_i}\right)^{m_i}\right)}{E_{11}(T) \times \varepsilon_{max}}\right)^M \times \left(\frac{1}{\phi^{M-1}} - \frac{1}{\left(\frac{\pi}{2}\right)^{M-1}}\right) \tag{6.34}$$

The longitudinal modulus ($E_{11}(T)$) can be calculated, for example, according to the model given in Chapter 2, Section 2.4.1.5.

In the example already explained in this chapter (Section 6.2.3) we can find experimental data for such a composite under *static* end-loaded bending at various temperatures (Figures 6.17, 6.18 and 6.19). Fitting parameters such as μ can be adjusted on one dataset (and one only). Other curves are predicted then checked with experimental data (datasets at other strain level or other temperature).

Such model derived from Equation (6.24) was found to fit very well all the data generated in stress-rupture. It includes the effects of applied strain level and temperature (stress-rupture, condition 1) but excludes fatigue. This model is of course not unique and any curve fit representing the entire set of data could be used instead.

An explicit relationship between time-to-failure, strain and temperature has now been established for the static experiments (Equation (6.34)). However, in order to use Equation (6.18) and predict the life of the composite under cyclic loading at elevated temperatures, we also need data to characterize the strength degradation of the composite during the end-loaded static experiments.

Such data can be found in the literature [35]. Remaining-strength experiments were conducted on the amorphous composite: a series of specimens were bent at 38% of their ultimate tensile strain-to-failure. They were taken out of the oven before failure. Quasi-static tensile remaining strength experiments were performed in tension at a loading rate of 22.7 kg (50 lbs) per second and the ultimate load-to-failure was recorded. The same experiments were performed with specimens bent at 57% of their ultimate strain-to-failure. As intuitively expected, the remaining strength decreased as the oven exposure time increased (Figures 6.42 and 6.43).

We now have to determine the parameter j, which can be determined only experimentally. Remaining strength experiments run at two different strain levels (in

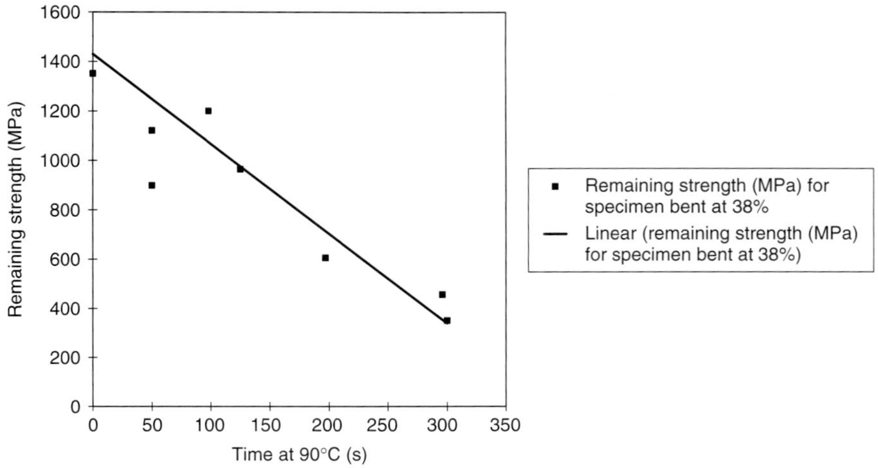

Figure 6.42. Remaining strength. Stress rupture experiments at 90°C and 38% strain-to-failure

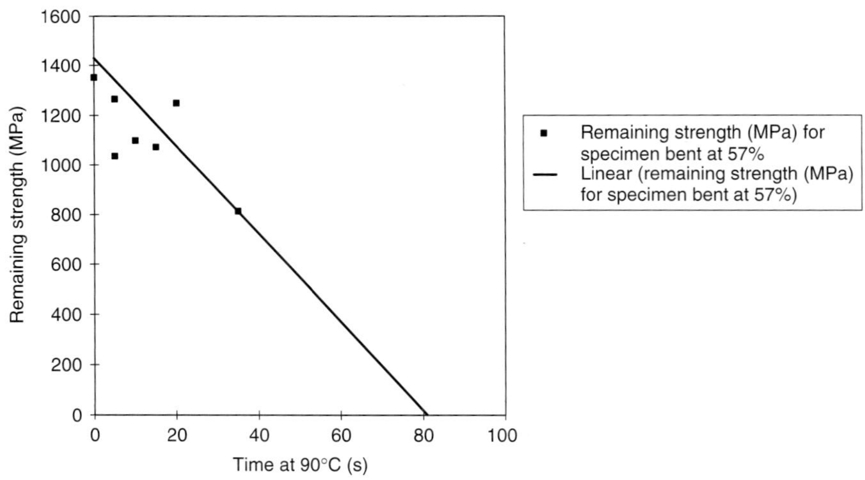

Figure 6.43. Remaining strength. Stress rupture experiments at 90°C and 57% strain-to-failure.

this example 38 and 57% of the maximum strain-to-failure) enable the computation of the j_{rupture} value according to Equation (6.35):

$$\sigma = 1 - (1 - Fa)\left(\frac{t}{t_{\text{r}}}\right)^{j_{\text{rupture}}} \tag{6.35}$$

Both fits led to a value of 0.66 for j_{rupture}.

For fatigue, it is also possible to relate the failure function to the number of cycles to failure. Indeed, during the bending experiments, the applied displacement varied sinusoidally from a small compression ($Fa_{\min} = 0.03$) to a maximum value reached at the center of the specimen ($Fa_{\max}$) with a period $T(T = 0.25\,\mathrm{s})$. Performing experiments for different values of $Fa_{\max}$ lead to:

$$Fa = 0.9941 \cdot N_{\mathrm{f}}^{-0.0045} \tag{6.36}$$

where N_{f} is the number of cycles to failure ($N_{\mathrm{f}} \geq 1$).

Step 5: Validation of analytical combination of conditions 1 and 2

The next step is to analytically combine the results of the room-temperature fatigue experiments in bending with the quasi-static experiments at elevated temperatures in bending to predict the life of unidirectional carbon-fiber polymer matrix composites under bending fatigue at elevated temperatures. We proceed by calculating *Fr* (Equation (6.20)) and *Fa* (Equation (6.24)) for each load increment (reversal) until $Fr = Fa$ (rupture).

Analytical predictions are plotted against the literature data in Figures 6.44–6.46:

- isostrain experiments at 75% for various temperatures (Figure 6.44)
- isostrain experiments at 90% for various temperatures (Figure 6.45)
- isotemperature experiments at 95°C for various strain levels (Figure 6.46).

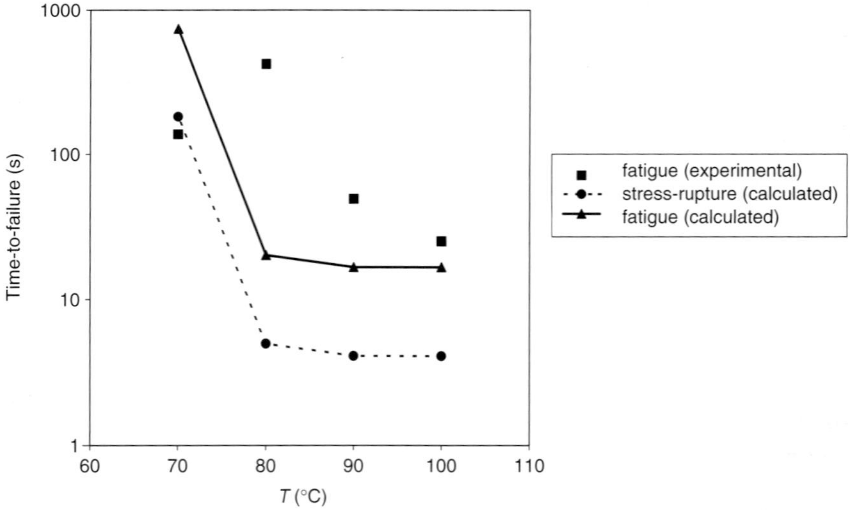

Figure 6.44. Isostrain experiments and theoretical results at 75% for various temperatures [35]. (Copyright 2001, *Applied Composite Materials*, by C.A. Mahieux et al., with kind permission of Springer Science and Business Media.)

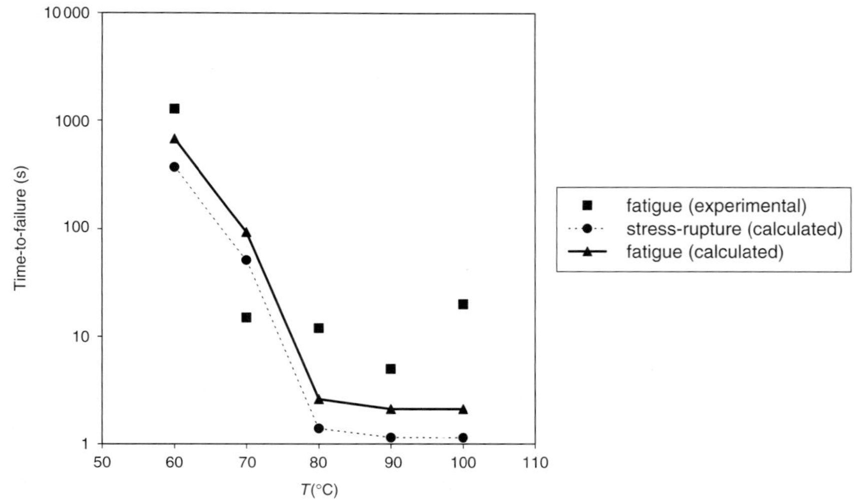

Figure 6.45. Isostrain experiments and theoretical results at 90% for various temperatures [35]. (Copyright 2001, *Applied Composite Materials*, by C.A. Mahieux et al., with kind permission of Springer Science and Business Media.)

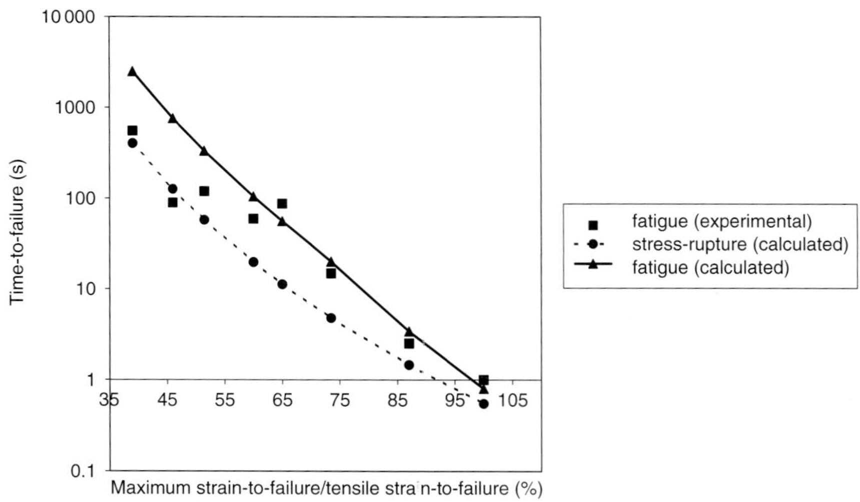

Figure 6.46. Isotemperature experiments and theoretical results at 90°C for various strain levels [35]. (Copyright 2001, *Applied Composite Materials*, by C.A. Mahieux et al., with kind permission of Springer Science and Business Media.)

The life of the composite in bending fatigue is in the present case longer than the life in bending static-load experiments, which is not a typical result. The durability scheme results reflect this concept and predict longer lives for the fatigued specimen.

In order to validate the model, it is necessary to check the damage and failure mode consistency. Cumulative damage is clearly driven by compression and by the formation of microbuckles. Typical *static* bending damage at elevated temperature is shown in Figure 6.22, damage under room temperature cyclic bending is shown in Figure 6.40 and elevated temperature cyclic bending conditions

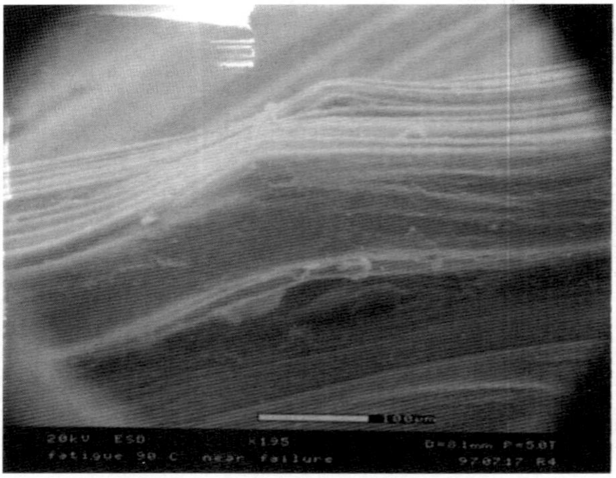

Figure 6.47. SEM picture. 90°C bending fatigue. Microbuckle on the compression side [35]. (Copyright 2001, *Applied Composite Materials*, by C.A. Mahieux et al., with kind permission of Springer Science and Business Media.)

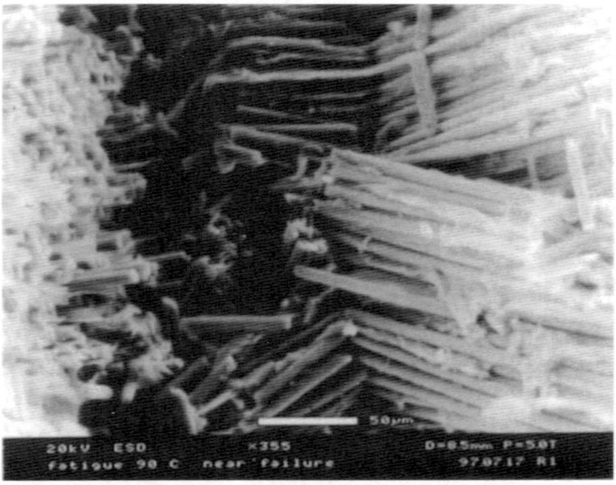

Figure 6.48. SEM picture. 90°C bending fatigue. Microbuckle on the compression side [35]. (Copyright 2001, *Applied Composite Materials*, by C.A. Mahieux et al., with kind permission of Springer Science and Business Media.)

are shown in Figures 6.47 and 6.48. On the other hand, the failure modes are characteristic of more rapid damage (Figure 6.49 for stress rupture, Figure 6.50 for bending fatigue at room temperature and Figure 6.51 for bending fatigue at elevated temperatures) [35].

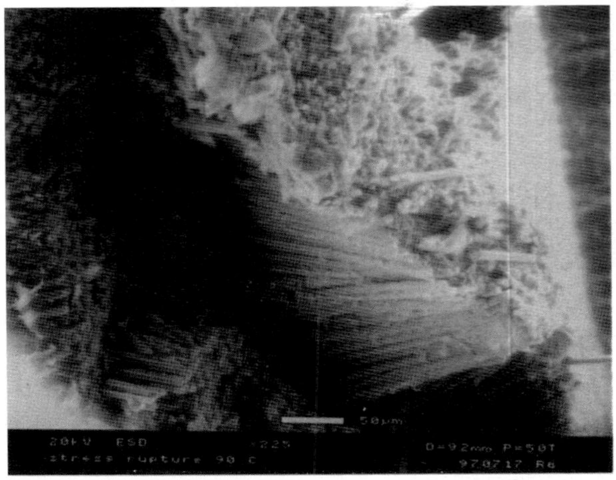

Figure 6.49. SEM picture. Stress rupture at 90°C. Failure surface [35]. (Copyright 2001, *Applied Composite Materials*, by C.A. Mahieux et al., with kind permission of Springer Science and Business Media.)

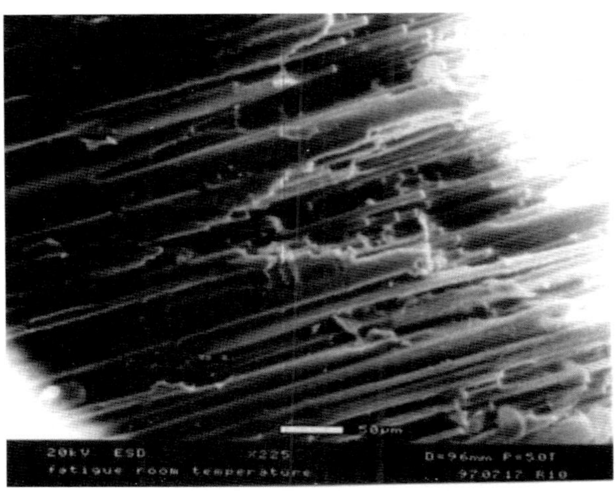

Figure 6.50. SEM picture. Room temperature bending fatigue. Failure surface [35]. (Copyright 2001, *Applied Composite Materials*, by C.A. Mahieux et al., with kind permission of Springer Science and Business Media.)

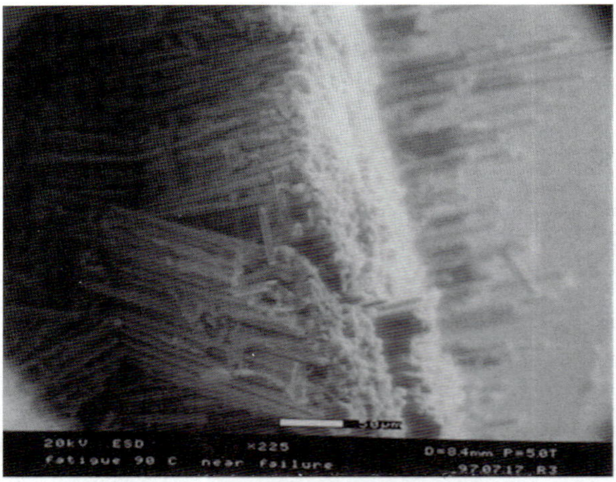

Figure 6.51. SEM picture. Bending fatigue at 90°C. Failure surface [35]. (Copyright 2001, *Applied Composite Materials*, by C.A. Mahieux et al., with kind permission of Springer Science and Business Media.)

6.4 SPECIAL FOCUS – TESTING: DESIGN OF EXPERIMENTS FOR COMPOSITES

6.4.1 Introduction

The behavior and lifetime of polymer composites is influenced by a very large number of parameters. Common test methods were summarized in Chapters 2–5 and concrete industrial examples such as the impressive set-up for full-scale windmill blades fatigue testing (Chapter 1) or the large-scale airplane wing tests (Figures 6.52 and 6.53) were presented along the book.

Unfortunately, long-term fatigue and full-scale testing are often very costly. Budget and time constraints generally do not allow for the testing of all possible environmental and mechanical load combinations. It is therefore necessary to find ways to reduce the number of experiments without overseeing major degradation processes.

The purpose of the present section is to provide the reader with an overview of a method that can help reduce the number of experiments in the determination of environmental effects on composite degradation and lifetime. This simple introduction does not intend to give an in-depth summary of the DOE method and more information on this topic can be found in the literature [38,39].

Design of experiments can be used in any situation involving processes/systems, in which inputs are transformed into outputs (Figure 6.54). Design of experiments can therefore be used for diverse applications ranging from administrative to manufacturing processes. This broad range of applications and a strong reliance upon statistical significance made DOE a key tool in the six-sigma methodology, statistically based quality method for the improvement of company performances.

Figure 6.52. A 340 wing section test. (Courtesy of Airbus.)

Figure 6.53. Static loading of a carbon-fiber demonstrator wing. (Courtesy of Airbus.)

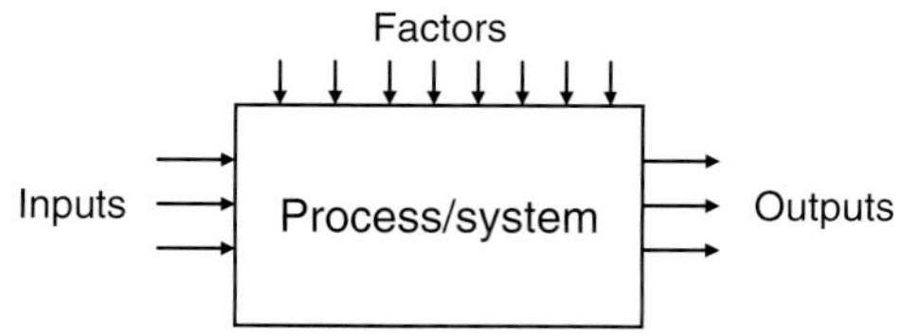

Figure 6.54. Inputs, outputs, factors and processes.

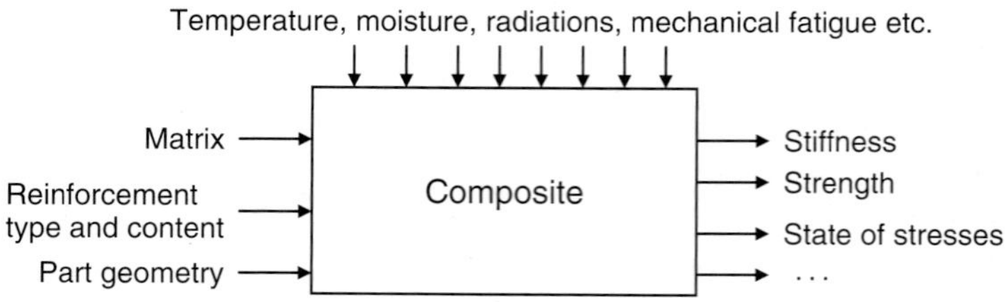

Figure 6.55. Composite example.

In our case, the system is the polymer composite exposed to a given environment (Figure 6.55). Along this book, we have seen that the degradation process is influenced by many variables (called *factors*). Indeed, temperature, gas, electrical field, radiations and mechanical loads are some of the many factors influencing the materials state. In the DOE, we will vary intentionally the variables and identify the key factors to the materials response of interest (mainly degradation in our case). From this knowledge, the number of experiments to be performed will be greatly reduced with a controlled resolution on the results.

6.4.2 Selecting the Proper Design

Once the system has been identified and inputs, outputs and factors listed, the proper experimental design has to be selected. Let us consider a composite exposed to temperature (T), ultra-violet radiation (UV) and a static mechanical load (P). For simplification purposes, we will consider that the different factors are discrete and can take only high and low values. For example, the temperature can be high (90°C) or low (0°C), the UV radiation can be on or off, the mechanical load can be 10 or 0 MPa. By convention, we will note -1 the low level (0°C, no UV, no load) and $+1$ the high level (90°C, UV on, 10 MPa). We would like to assess the effects of the environmental parameters on the axial tensile stiffness of the material. It is very likely that the composite under temperature (at no load and no UV radiation) reacts differently than the composite under temperature and load. We therefore have to test the effects of the three main factors (main effects) as well as the interactions of the second (temperature/load, UV/load, temperature/UV) and third order (temperature/load/UV). Testing of all load combinations results in $2^3 = 8$ experimental situations. Furthermore, to obtain statistically valid information, a *minimum* of two replications (two repetitions) is necessary. This means that for this very simple case, a minimum of 16 samples should be tested. Increasing the number of replication is a recommended practice, as it will increase the level of confidence. A replication is, however, costly as it does involve not only a simple repetition of the measurement but a complete replication of the experimental conditions.

The list of experiments is summarized in Table 6.1. This complete set of experiments (2^3 full factorial design) provides exhaustive informations on main effects and all interactions. The resolution of our results is therefore maximal.

Unfortunately, full factorial experiments can bear high costs. In an industrial context, budget and time usually define the number of experiments that can realistically be performed. Given these two constraints, when it is not possible to use a full factorial design, the engineer must select experiments providing the maximum amount of useful information.

To revert to fractional designs is the solution. Several fractional designs are available leading to different design resolution. Indeed, reducing the number of experiments also means systematically changing several factors simultaneously (confounding). In such situations, it is not always possible to distinguish between the effects of the different factors. Naturally, it is always recommended to minimize confounding and specialized softwares such as Minitab, which generally assists one in choosing the proper design.

Generally, runs one to four in Table 6.1 are always performed. Therefore, main effects are not confounded with other main effects. However, if only those four runs are executed, main effects are then confounded with two- and three-factor interactions, and two factor interactions are confounded with each other. By convention, such experimental plan is a resolution III design. This notion is best illustrated by our example. Table 6.2 clearly evidences that the effect of the load, for example, is confounded with the temperature/UV interaction. In other words,

Table 6.1. Run, factors and interactions

Run (standard order)	Factors (normal notation)			Factors (DOE notation)		
	T (°C)	UV	P (MPa)	A	B	C
1	0	off	0	+1	+1	+1
2	90	off	0	+1	−1	−1
3	0	on	0	−1	+1	−1
4	0	off	10	−1	−1	+1
5	90	on	0	+1	+1	−1
6	0	on	10	−1	+1	+1
7	90	off	10	+1	−1	+1
8	90	on	10	−1	−1	−1

Table 6.2. Resolution III design example, A = B × C, B = A × C and C = A × B

Runs	A	B	C	B × C	A × C	A × B	A × B × C
1	+	+	+	+	+	+	+
2	+	−	−	+	−	−	+
3	−	+	−	−	+	−	+
4	−	−	+	−	−	+	+

a hypothetically observed stiffness decrease could not be attributed with a high confidence level to the 10 MPa load, as it could also be the consequence of the combined UV/90°C temperature exposure.

For a larger number of parameters, we can also chose to run more experiments in which no main effects are confounded with each other or with any two-factor interactions, but where two-factor interactions may be confounded with each other. We will then obtain a resolution IV design. Finally, in a resolution V design, confounding is limited to two-factor with three-factor interactions.

Considering the specificity of composite materials, where we have assessed a large level of interactions between factors and after the screening step, it is recommended to rely on resolution V designs.

6.4.3 Conducting the Experiments

Once the proper design is selected, it is necessary to ensure that the measurement system is appropriate. For this, it is recommended to perform a Gauge R&R [38], in which reproducibility and repeatability are verified. Reproducibility experiments judge the ability of different operators to obtain statistically comparable results for the same sample. On the other hand, repeatability experiments measure the performance of the measurement process in giving results statistically similar when multiple measurements are performed on the same sample by the same operator. In order to further eliminate noise influence and bias, it is additionally recommended to randomize the experiments. For example, the runs 1–4 of Table 6.2 should be performed in a random order (Table 6.3). We have also mentioned that runs necessitate replication in order to extract statistical significance to the results. Therefore, measurements under the conditions of run 1 in Table 6.2 should be repeated at least twice (Table 6.3).

To complicate matters further, it sometimes occurs that not all experiments can be performed on the same day. It also happens that a large number of experiments necessitate the use of two different batches of material. Such situations require the use of the blocking technique [38], which considers this potential additional source of variation as an additional parameter. Here too, commercial softwares facilitate the *block* consideration by offering blocking options while setting up the DOE.

Table 6.3. Hypothetical results of a full factorial design with three factors, two replicates

Standard order	Run order	Temperature	UV	Static load	Stiffness (GPa)
3	1	−1	+1	−1	30
7	2	−1	+1	+1	20
15	3	−1	+1	+1	22
9	4	−1	−1	−1	35
2	5	+1	−1	−1	15
10	6	+1	−1	−1	17
6	7	+1	−1	+1	5
1	8	−1	−1	−1	34
13	9	−1	−1	+1	24
12	10	+1	+1	−1	12
16	11	+1	+1	+1	5
4	12	+1	+1	−1	13
8	13	+1	+1	+1	7
14	14	+1	−1	+1	6
5	15	−1	−1	+1	23
11	16	−1	+1	−1	28

6.4.4 Analyzing the Experiments

Once the experiments were performed under the conditions listed above, the results can be analyzed. The main purpose of the analysis is to identify the main contributors to the materials response and reduce the number of parameters that will later be studied in further detail.

Let us consider our typical unidirectional composite exposed to temperature, UV radiation and static load. The results of a full factorial design of experiment performed on the material for the three factors with two replicates are summarized in Table 6.3.

Traditional DOE methods offer statistical and graphical tools to enable the analysis of such results. Indeed, *main effects* plots are basic graphical representations indicating the general direction of variation of the composite response as a function of the level of the individual factors (slope). They also provide an indication of the magnitude of the variation of the mean value of the response between the two levels. In our example, the composite stiffness is negatively affected by increased temperature, load and UV exposure (Figure 6.56). Main effect plots identify temperature as most influencing the composite response and UV radiation the least.

Interaction plots provide additional information on the second-order interaction between the factors. If the lines are parallel, little interaction occurs between the factors. In our example, only load and UV seem to converge and indicate a possible interaction (Figure 6.57). This observation is confirmed by the statistical analysis of the data that indicates a p value below traditional limits of statistical irrelevance [38].

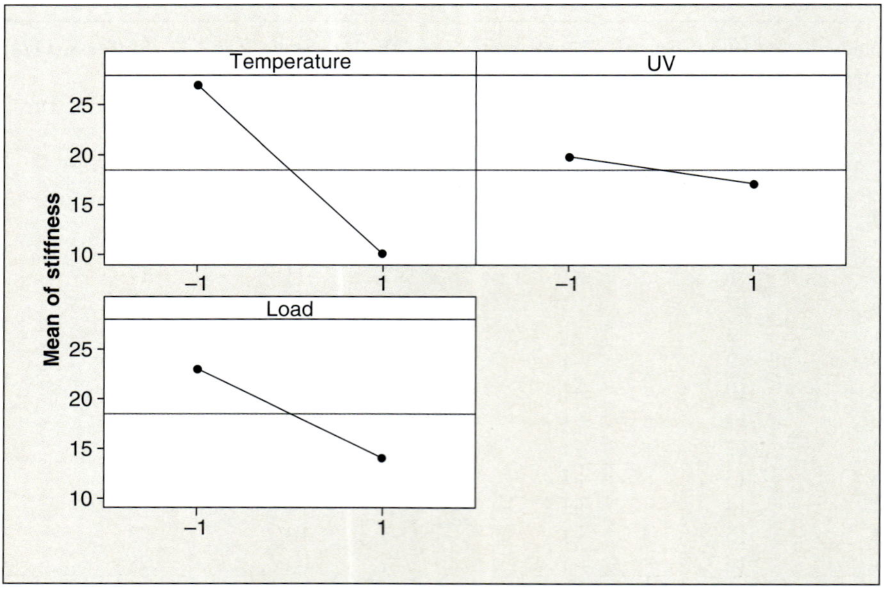

Figure 6.56. Main effects plot (data means) for stiffness.

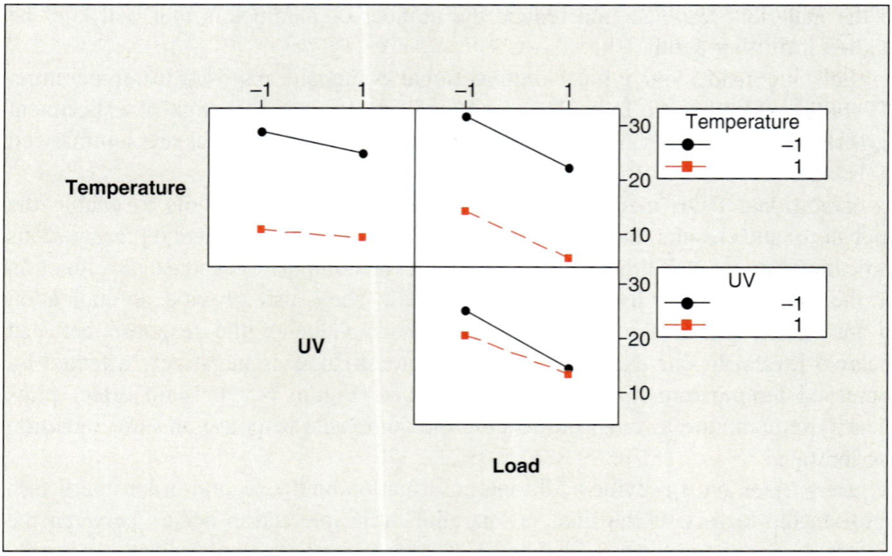

Figure 6.57. Interaction plot (data means) for stiffness.

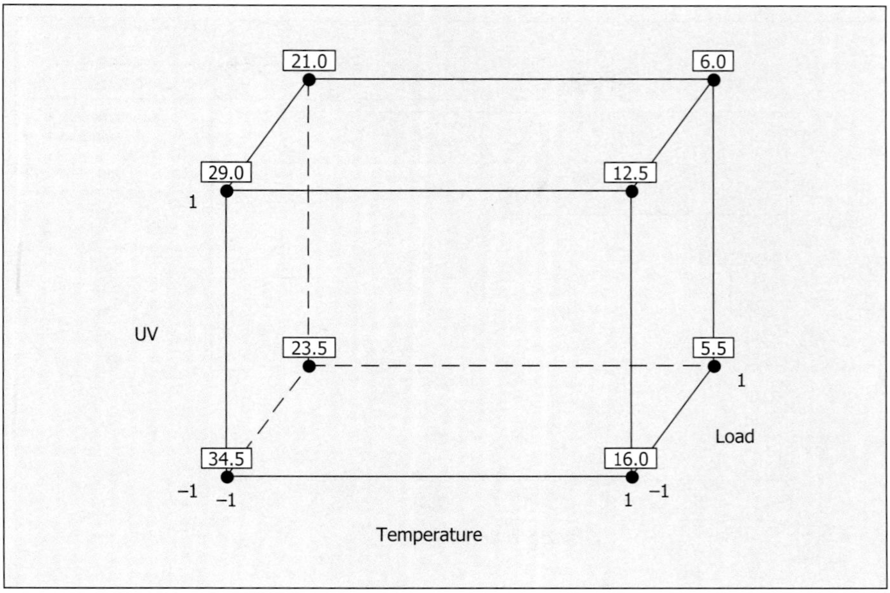

Figure 6.58. Cube plot (data means) for stiffness.

Practically, we can conclude that, in the frame of the current example, there is no need to further pursue combined load and temperature testing: individual load and temperature experiments will suffice.

Cube plots are another type of graphical representation of the results, which translate the same results in a three-dimensional form (Figure 6.58). This tool is generally used to help optimize the factor settings for a given specification.

Pareto charts of the standardized effects help us further identify the factors that are statistically most relevant to the stiffness changes observed. The vertical line represents the limit of statistical relevance. From Figure 6.59, it can be concluded with high confidence that combined UV/load exposure, load, UV and temperature have increasing statistically significant effects on the composite response.

The normal probability plot of factor effects is another representation equivalent to the Pareto diagram showing the significance and contribution of the different factors to the response. Here too, the main effects and the load/UV interaction are found most significant (Figure 6.60).

In the current example, our durability assessment efforts should therefore be concentrated on main effects (temperature, load and UV radiation) and on UV/load exposure interactions. Other interactions such as temperature/load or temperature/UV exposure can be neglected.

Once main contributors to the response have been identified, results can be modeled using response surface plots and regression models leading to prediction equations. These equations are useful as curve fit. It is, however, recommended

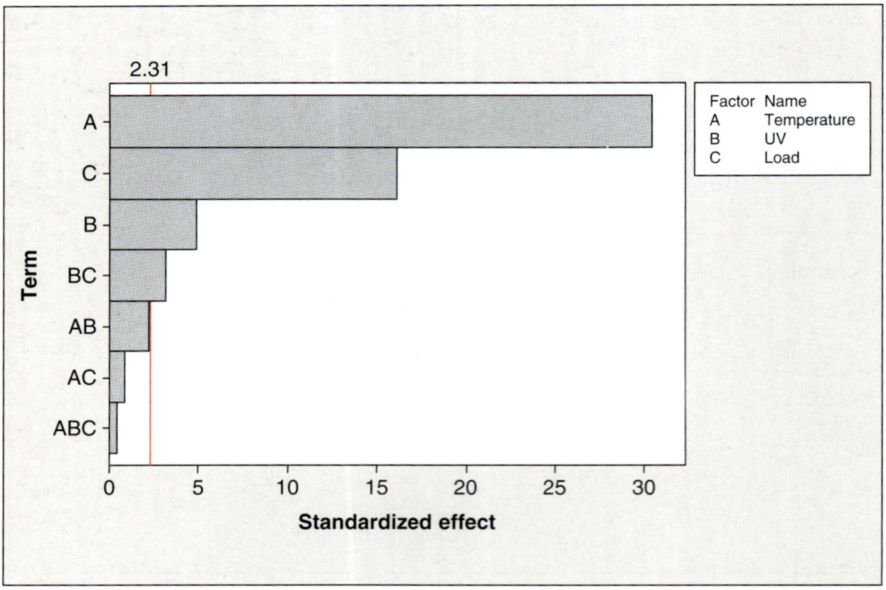

Figure 6.59. Pareto chart of the standardized effects (response is stiffness, $\alpha = 0.05$).

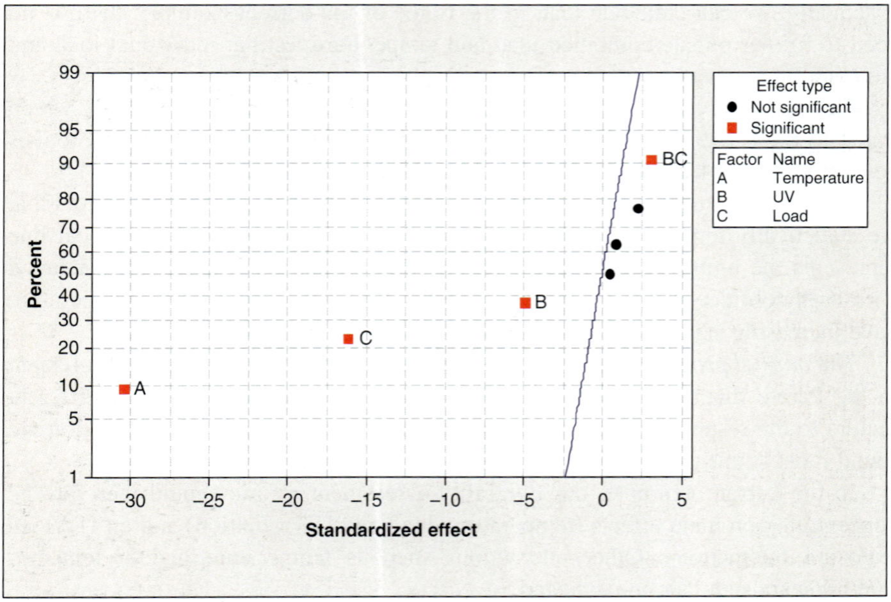

Figure 6.60. Normal probability plot of the standardized effects (response is stiffness, $\alpha = 0.05$).

for this last step to revert as much as possible to physically based models such as those presented in this book, in order to model the materials response to the environmental factors.

6.5 TOOL KIT

Topic	Equation	Assumptions	Importance
Crack growth rate	$\frac{\mathrm{d}a}{\mathrm{d}N} = A(\Delta K)^m$	Brittle materials	Generally not applicable to PMCs
Survival probability in a *dry* fiber bundle	$\sigma = E_\mathrm{f}\varepsilon \exp\left[-\frac{l}{l_0}\left(\frac{\varepsilon}{\varepsilon_0}\right)^m\right] = E_\mathrm{f}\varepsilon R(\varepsilon)$	Fiber failure obeys Weibull distribution	Expresses statistical nature of fiber failure
Miner law	$\sum_{i=1}^{k} \frac{n_i}{N_i} = 1$	For linear materials	Enables block summation
Failure function	$Fa = Fa\left(\frac{\sigma_{ij}}{X_{ij}}\right)$ and $0 \leq Fa \leq 1$	Depends on the choice of the failure criterion	Enables damage tracking
Remaining strength integral	$Fr = 1 - \int_0^{\tau_1}(1 - Fa)\left(\frac{\sigma_{ij}(\tau)}{X_{ij}(\tau)}\right) j\tau^{j-1}\mathrm{d}\tau$ with failure at $Fa = Fr$	Assumed form for residual strength	Enables damage tracking and life prediction
Remaining strength	$Fr = 1 - \sum_{i=1}^{i=N_\mathrm{blocks}} \Delta Fr_i$	Damage equivalence	Enables incremental approaches

REFERENCES

1. Reifsnider, K.L., *Fatigue of Composite Materials*. Elsevier, 1991.
2. Harris, B., *Fatigue in Composite Materials: Science and Technology of the Fatigue Response of Fibre-Reinforced Plastics*. CRC Press, 2003.
3. Harris, B., A historical review of fatigue behavior of fiber-reinforced plastics. In B. Harris, (Ed.), *Fatigue in Composites*. Woodhead Publishing Ltd, Boca Raton, 2003.
4. Taljera, R., *Fatigue of Composite Materials*, Technomic, 1987.
5. Broutman, L.J., *Fracture and Fatigue (Composite materials)*. Academic Press, 1974.
6. Black, S., How are composite bridges performing. *Composites Technology*, December 2003, 16–22.
7. Pultruded enclosure protects bridge. *Reinforced Plastics*, June 2004, 4.
8. Historic bridge gets composite deck. *Reinforced Plastics*, October 2004, 7.
9. Mei, C., H.F. Wolfe and I. Elishakoff, *Vibration and Behavior of Composite Structures/Ad-14*, American Society of Mechanical Engineers Winter Meeting, American Society of Mechanical Engineers Aerospace Division Structures, Elsevier.

10. Jarzynski, J., Vibration measurements of a composite panel, School of Mechanical Engineering, Georgia Institute of Technology, 1994.
11. Jones, D.I., *Handbook of Viscoelastic Vibration Damping*, John Wiley & Sons, 2001.
12. http://www.washingtonpost.com/wp-srv/weather/longterm/historical/data/zurich_switzerland.htm.
13. Callister, W.D., Jr, *Materials Science and Engineering – An introduction*, 4th ed. John Wiley & Sons, USA, 1997.
14. Reifsnider, K.L. and S.W. Case, *Damage Tolerance and Durability of Material Systems*. Wiley & Sons, New York, 2002.
15. Kerr, J.R. and J.F. Haskins, Time-Temperature-Stress Capabilities of Composite Materials for Advanced Supersonic Technology Application. NASA Contractor Report 178272, May 1987.
16. Sperling, L.H., *Introduction to Physical Polymer Science*, 2nd ed. John Wiley & Sons, Inc., USA, 1992.
17. Hinton, M.J., P. Soden and A.S. Kaddour (Eds), 1998, Failure criteria in fibre-reinforced polymer composites: Part C, *Composites Science and Technology*, 65(3–4), 2004.
18. Lifshitz, J.M., *Time Dependent Fracture of Fibrous Composites, in Fracture and Fatigue*. Academic Press, New York, 1974, 266–278.
19. Raghavan, J. and Meshii, Prediction of creep rupture of unidirectional carbon fiber reinforced polymer composite, *Materials Science and Engineering*, A, 1994, 197, 237–249.
20. Christiensen, R.M., Lifetime predictions for polymers and composites under constant load. *Journal of Rheology*, 1981, 25(5), 517–528.
21. Russell, B.E., C.A. Mahieux and K.L. Reifsnider, Stress Rupture of PMC's in End-Loaded Bending. *Journal of Applied Composites*, 1998, 5, 151–159.
22. Mahieux, C.A., B.E. Russell and K.L. Reifsnider, Stress rupture of unidirectional high performance thermoplastic composites in end-loaded bending at elevated temperatures. Part I Experimental Characterization of the failure mode. *Journal of Composite Materials*, 1998, 32(14).
23. Mahieux, C.A. and K.L. Reifsnider, Effects of Out-of-plane Deformation on Stress Rupture of Unidirectional Polymer Matrix Composites. Composites and Functionally Graded Materials, ASME 97, MD-Vol. 80, 1997.
24. Miner, M.A. and S.M. Calif, Cumulative damage in fatigue. *Journal of Applied Mechanics*, A-159-A-164, September 1945.
25. Hashin, Z. and A. Rotem, A cumulative damage theory of fatigue failure. *Materials Science and Engineering*, 1978.
26. Hashin, Z., A reinterpretation of the Palmgren-Miner Rule for fatigue life prediction. *Journal of Applied Mechanics*, 1980.
27. Reifsnider, K.L. and W.W. Stinchomb, A critical element model of the residual strength and life of fatigue loaded coupons. In H.T. Hahn (Ed.), *Composite Materials: Fatigue and Fracture*, ASTM STP 907, ASTM, Philadelphia, 1986, 298–303.
28. Reifsnider, K.L., Use of mechanistic life prediction methods for the design of damage tolerant composite material systems. In M.R. Mitchell and O. Buck (Eds), *Composite Materials: Fatigue and Fracture*, ASTM STP 1157, American Society for Testing and Materials, Philadelphia, 1992, pp. 205–223.
29. Reifsnider, K.L., A micro-kinetic approach to durability analysis: The critical element method, in *Progress in Durability of Composite Systems*, A.H. Cardon, K.L. Reifsnider and H. Fukuda (Eds), Balkema, Rotterdam, 1996, pp. 3–11.

30. Case, S., N. Iyengar and K.L. Reifsnider, Life Prediction Tool for Ceramic Matrix Composites at Elevated Temperatures, Seventh Symposium on Composites: Fatigue and Fracture, ASTM STP 1330, R.B. Businell (Ed.), 1998, 165–178.
31. http://www.enrg.lsu.edu-publications-online-outlook_hydrocarb_res.pdf.
32. Black, S., Composite drilling risers ready for commercialization. *High-Performance Composites*, July 2002, 40–42.
33. Vennett, R.M., J.G. Williams, K.H. P. Lo, Ganguly, Economic benefits of using composites for offshore development and operations. In S.S. Wang, J.G. Williams and K.H. Lo (Eds), *Composite Materials for Offshore Operations-2*, ABS. Houston, 1999, 3–16.
34. Kalman, M. and J. Belcher, Flexible risers with composite armor for deep water oil and gas production. In *Composite Materials for Offshore Operations*, 2nd ed., S.S. Wang, J.G. Williams and K.H. Lo (Eds), ABS. Houston, 1999, p. 161.
35. Mahieux, C.A., K.L. Reifsnider and J.J. Jackson, Property modeling across transition temperatures in PMC's: Part III Bending fatigue. *Applied Composite Materials*, July 2001, 8(4).
36. Jackson, J.J. and K.L. Reifsnider, End displacement bending fatigue life prediction of AS4/PPS composite material at elevated temperature, NSF-SURP program. 1997.
37. Budiansky, B. and N.A. Fleck, Compressive kinking of fiber composites: A topical review. *App. Mech. Rev.*, 1994, 47(6).
38. Antony, J., *Design of Experiments for Engineers and Scientists*. Butterworth-Heinemann, Amsterdam, 2003.
39. Cox, D.R. and N. Reid, *The Theory of the Design of Experiments*. Chapman & Hall, Boca Raton, 2000 (Monographs on statistics and applied probability, 86).

INDEX